INTERNATIONAL SERIES OF MONOGRAPHS ON PHYSICS

SERIES EDITORS

J. BIRMAN	CITY UNIVERSITY OF NEW YORK
S. F. EDWARDS	UNIVERSITY OF CAMBRIDGE
R. FRIEND	UNIVERSITY OF CAMBRIDGE
M. REES	UNIVERSITY OF CAMBRIDGE
D. SHERRINGTON	UNIVERSITY OF OXFORD
G. VENEZIANO	CERN, GENEVA

INTERNATIONAL SERIES OF MONOGRAPHS ON PHYSICS

161. V.Z. Kresin, H. Morawitz, S.A. Wolf: *Superconducting state—mechanisms and properties*
160. C. Barrabès, P.A. Hogan: *Advanced general relativity—gravity waves, spinning particles, and black holes*
159. W. Barford: *Electronic and optical properties of conjugated polymers, Second edition*
158. F. Strocchi: *An introduction to non-perturbative foundations of quantum field theory*
157. K.H. Bennemann, J.B. Ketterson: *Novel superfluids, Volume 2*
156. K.H. Bennemann, J.B. Ketterson: *Novel superfluids, Volume 1*
155. C. Kiefer: *Quantum gravity, Third edition*
154. L. Mestel: *Stellar magnetism, Second edition*
153. R.A. Klemm: *Layered superconductors, Volume 1*
152. E.L. Wolf: *Principles of electron tunneling spectroscopy, Second edition*
151. R. Blinc: *Advanced ferroelectricity*
150. L. Berthier, G. Biroli, J.-P. Bouchaud, W. van Saarloos, L. Cipelletti: *Dynamical heterogeneities in glasses, colloids, and granular media*
149. J. Wesson: *Tokamaks, Fourth edition*
148. H. Asada, T. Futamase, P. Hogan: *Equations of motion in general relativity*
147. A. Yaouanc, P. Dalmas de Réotier: *Muon spin rotation, relaxation, and resonance*
146. B. McCoy: *Advanced statistical mechanics*
145. M. Bordag, G.L. Klimchitskaya, U. Mohideen, V.M. Mostepanenko: *Advances in the Casimir effect*
144. T.R. Field: *Electromagnetic scattering from random media*
143. W. Götze: *Complex dynamics of glass-forming liquids—a mode-coupling theory*
142. V.M. Agranovich: *Excitations in organic solids*
141. W.T. Grandy: *Entropy and the time evolution of macroscopic systems*
140. M. Alcubierre: *Introduction to 3+1 numerical relativity*
139. A.L. Ivanov, S.G. Tikhodeev: *Problems of condensed matter physics—quantum coherence phenomena in electron-hole and coupled matter-light systems*
138. I.M. Vardavas, F.W. Taylor: *Radiation and climate*
137. A.F. Borghesani: *Ions and electrons in liquid helium*
135. V. Fortov, I. Iakubov, A. Khrapak: *Physics of strongly coupled plasma*
134. G. Fredrickson: *The equilibrium theory of inhomogeneous polymers*
133. H. Suhl: *Relaxation processes in micromagnetics*
132. J. Terning: *Modern supersymmetry*
131. M. Mariño: *Chern-Simons theory, matrix models, and topological strings*
130. V. Gantmakher: *Electrons and disorder in solids*
129. W. Barford: *Electronic and optical properties of conjugated polymers*
128. R.E. Raab, O.L. de Lange: *Multipole theory in electromagnetism*
127. A. Larkin, A. Varlamov: *Theory of fluctuations in superconductors*
126. P. Goldbart, N. Goldenfeld, D. Sherrington: *Stealing the gold*
125. S. Atzeni, J. Meyer-ter-Vehn: *The physics of inertial fusion*
123. T. Fujimoto: *Plasma spectroscopy*
122. K. Fujikawa, H. Suzuki: *Path integrals and quantum anomalies*
121. T. Giamarchi: *Quantum physics in one dimension*
120. M. Warner, E. Terentjev: *Liquid crystal elastomers*
119. L. Jacak, P. Sitko, K. Wieczorek, A. Wojs: *Quantum Hall systems*
117. G. Volovik: *The Universe in a helium droplet*
116. L. Pitaevskii, S. Stringari: *Bose-Einstein condensation*
115. G. Dissertori, I.G. Knowles, M. Schmelling: *Quantum chromodynamics*
114. B. DeWitt: *The global approach to quantum field theory*
113. J. Zinn-Justin: *Quantum field theory and critical phenomena, Fourth edition*
112. R.M. Mazo: *Brownian motion—fluctuations, dynamics, and applications*
111. H. Nishimori: *Statistical physics of spin glasses and information processing—an introduction*
110. N.B. Kopnin: *Theory of nonequilibrium superconductivity*
109. A. Aharoni: *Introduction to the theory of ferromagnetism, Second edition*
108. R. Dobbs: *Helium three*
107. R. Wigmans: *Calorimetry*
106. J. Kübler: *Theory of itinerant electron magnetism*
105. Y. Kuramoto, Y. Kitaoka: *Dynamics of heavy electrons*
104. D. Bardin, G. Passarino: *The Standard Model in the making*
103. G.C. Branco, L. Lavoura, J.P. Silva: *CP Violation*
102. T.C. Choy: *Effective medium theory*
101. H. Araki: *Mathematical theory of quantum fields*
100. L.M. Pismen: *Vortices in nonlinear fields*
 99. L. Mestel: *Stellar magnetism*
 98. K.H. Bennemann: *Nonlinear optics in metals*
 94. S. Chikazumi: *Physics of ferromagnetism*
 91. R.A. Bertlmann: *Anomalies in quantum field theory*
 90. P.K. Gosh: *Ion traps*
 87. P.S. Joshi: *Global aspects in gravitation and cosmology*
 86. E.R. Pike, S. Sarkar: *The quantum theory of radiation*
 83. P.G. de Gennes, J. Prost: *The physics of liquid crystals*
 73. M. Doi, S.F. Edwards: *The theory of polymer dynamics*
 69. S. Chandrasekhar: *The mathematical theory of black holes*
 51. C. Møller: *The theory of relativity*
 46. H.E. Stanley: *Introduction to phase transitions and critical phenomena*
 32. A. Abragam: *Principles of nuclear magnetism*
 27. P.A.M. Dirac: *Principles of quantum mechanics*
 23. R.E. Peierls: *Quantum theory of solids*

Superconducting State

Mechanisms and Properties

VLADIMIR Z. KRESIN
Lawrence Berkeley National Laboratory
University of California at Berkeley

HANS MORAWITZ
IBM Almaden Research Center

STUART A. WOLF
University of Virginia, Charlottesville

OXFORD
UNIVERSITY PRESS

Great Clarendon Street, Oxford, OX2 6DP,
United Kingdom

Oxford University Press is a department of the University of Oxford.
It furthers the University's objective of excellence in research, scholarship,
and education by publishing worldwide. Oxford is a registered trade mark of
Oxford University Press in the UK and in certain other countries

© Vladimir Z. Kresin, Hans Morawitz, Stuart A. Wolf 2014

The moral rights of the authors have been asserted

First Edition published in 2014

Impression: 1

All rights reserved. No part of this publication may be reproduced, stored in
a retrieval system, or transmitted, in any form or by any means, without the
prior permission in writing of Oxford University Press, or as expressly permitted
by law, by licence or under terms agreed with the appropriate reprographics
rights organization. Enquiries concerning reproduction outside the scope of the
above should be sent to the Rights Department, Oxford University Press, at the
address above

You must not circulate this work in any other form
and you must impose this same condition on any acquirer

Published in the United States of America by Oxford University Press
198 Madison Avenue, New York, NY 10016, United States of America

British Library Cataloguing in Publication Data
Data available

Library of Congress Control Number: 2013944189

ISBN 978-0-19-965255-6

Printed in Great Britain by
Butler Tanner and Dennis Ltd

Links to third party websites are provided by Oxford in good faith and
for information only. Oxford disclaims any responsibility for the materials
contained in any third party website referenced in this work.

This book is dedicated to the memory of

Boris Geilikman
Leonard Schiff
Bernard Serin

Preface

The first edition of our book *Mechanisms of Conventional and High T_c Superconductivity* was published in 1993. In the years since then, the field of superconductivity has undergone truly intensive developments, motivating us to prepare a second edition. The present monograph offers many new topics and numerous revisions. Several new chapters have been written, and descriptions in earlier chapters have been updated.

It should be emphasized that this book is neither a textbook nor a broad review of the field. We focus here mainly on the subject of mechanisms of superconductivity and related spectroscopies. There are a number of very important topics which are not reflected in this book, such as vortex physics, Josephson tunneling, the physics of heavy fermions, and so on. There are, however, excellent books that cover these topics, such as those by Barone and Paterno (1982), and Tinkham (1996). And, of course, admittedly the selection of material was greatly affected by the scientific interests of the authors.

We assume that the reader is familiar with the fundamentals of solid-state physics and the physics of superconductivity (as contained, for example, in the book by Ashcroft and Mermin, 1976). The theoretical section also assumes a knowledge of statistical mechanics and many-body techniques, at the level of, for example, the books by Abrikosov *et al.* (1975), Lifshitz and Pitaevskii (2002), and Mahan (1993). Note, however, that we always endeavor to supply a qualitative explanation of the main concepts and results. We hope that this will enable the book to be read without submerging in the calculational details, making it useful and interesting for a broader readership.

Chapters 2, 3, 7–9, 12, 14, 15, and Sections 4.1–4.3 were written by VZK, Chapters 6 and 13 were written by SAW, and Chapter 5 was written by HM. Chapters 1, 10, and 11 were written jointly by VZK and SAW, and Section 4.4 was written jointly by VZK and HM.

Acknowledgments

Over the last few years we have benefited from many fruitful discussions. In particular we are indebted to A. Bianconi, A. Bill, A. Bussman-Holder, R. Dynes, J. Friedel, L. Gor'kov, H. Keller, W. Little, K. Mueller, and Y. Ovchinnikov. We are especially grateful to Vitaly V. Kresin for his help and many discussions.

We are very grateful to Oxford University Press for making this book possible. We would especially like to thank commissioning editors Sönke Adlung and Jessica White for all of their help in setting up and carrying out the production of this volume. We are also very grateful to Robert Marriott for his editing of the manuscript, to Shyam Krishnan and Philo Antonie for their help during the proof and final printing stages of the production, and to Christine Boylan for her skills in preparing the index.

We thank Alistair McGregor (Berkeley) and Lillian Gao (Charlottesville) for their careful assistance in the preparation of the manuscript.

This book, as well as our other scientific endeavors, would not have been possible without the forbearance of Lilia Kresin, Iris Wolf, and Terry Morawitz.

Contents

1	**Historical perspective**	1
2	**Electronic states, phonons, and electron-phonon interaction**	5
2.1	Adiabatic approximation: Hamiltonian	5
2.2	Adiabatic approximation and non-adiabaticity: Born–Oppenheimer and "crude" approaches	6
2.3	Electron–phonon coupling	9
2.4	Electron–phonon interaction and renormalization of normal parameters	11
2.5	The "Migdal" theorem	15
2.6	Polaronic states	16
	2.6.1 Concept	16
	2.6.2 Dynamic polaron	17
3	**Phonon mechanism**	20
3.1	Superconductivity as a "giant" non-adiabatic phenomenon	20
3.2	The BCS model	21
3.3	Phonon mechanism: main equations	22
3.4	Critical temperature	26
	3.4.1 Weak coupling	26
	3.4.2 Intermediate coupling ($\lambda \lesssim 1.5$)	28
	3.4.3 Coulomb interaction	29
	3.4.4 Very strong coupling	31
	3.4.5 The general case	33
	3.4.6 About an upper limit of T_c	35
3.5	Properties of superconductors with strong coupling	36
3.6	The Van Hove scenario	39
3.7	Bipolarons: BEC versus BCS	39
3.8	Superconducting semiconductors	40
3.9	Polaronic effect and its impact on T_c	42
	3.9.1 Double-well structure	42
	3.9.2 Superconducting state	44
4	**Electronic mechanisms**	47
4.1	The Little model	47
4.2	"Sandwich" excitonic mechanism	50
4.3	Three-dimensional systems: electronic mechanism	50
4.4	Plasmons	52
	4.4.1 Plasmons in layered systems: dispersion law and "electronic sound"	53

	4.4.2	Plasmons in layered conductors: pairing	57
	4.4.3	The 3D case: "demons"	58

5 Magnetic mechanism 59
5.1 Introduction 59
5.1.1 Localized versus itinerant aspects of the cuprates 60
5.2 Fermi liquid-based theories 62
5.2.1 The spin-bag model of Schrieffer, Wen, and Zhang (1989) 62
5.2.2 The t-J model (Emery, 1987; Zhang and Rice, 1988) 66
5.2.3 Two-dimensional Hubbard model studies by Monte Carlo techniques 70
5.2.4 Spiral phase of a doped quantum antiferromagnet (Shraiman and Siggia, 1988–89) 77
5.2.5 Slave bosons 82
5.3 Non-Fermi-liquid models 85
5.3.1 The resonant valence bond (RVB) model and its evolution 85
5.3.2 Anyon models and fractional statistics 86
5.4 Conclusions 87

6 Experimental methods: Spectroscopic 88
6.1 Tunneling spectroscopy 88
6.1.1 Experimental method 88
6.1.2 Energy gap and transition temperature 90
6.1.3 Inversion of the gap equation and $\alpha^2 F(\Omega)$ 91
6.1.4 Electron-phonon coupling parameter λ 94
6.2 Scanning tunneling microscopy and spectroscopy 96
6.3 Infrared spectroscopy 97
6.4 Ultrasonic attenuation 99
6.5 Angle-resolved photoemission 100
6.6 Muon spin resonance (μSR) 100
6.6.1 μSR studies of superconductivity 102

7 Multigap superconductivity 103
7.1 Multigap superconductivity: general picture 103
7.2 Critical temperature 104
7.3 Energy spectrum 105
7.4 Properties of two-gap superconductors 108
7.4.1 Penetration depth; surface resistance 108
7.4.2 Strong magnetic field: Ginzburg–Landau equations for a multigap superconductor 110
7.4.3 Heat capacity 111
7.4.4 Experimental data 111
7.5 Induced two-band superconductivity 112
7.6 Symmetry of the order parameter and multiband superconductor 113

8	**Induced superconductivity: proximity effect**	**114**
	8.1 Proximity "sandwich"	114
	8.2 Critical temperature	115
	8.3 Proximity effect versus the two-gap model	119
	8.4 Pair-breaking: gapless superconductivity	119
9	**Isotope effect**	**122**
	9.1 General remarks	122
	9.2 Coulomb pseudopotential	122
	9.3 Multi-component lattice	123
	9.4 Anharmonicity	123
	9.5 Isotope effect in proximity systems	124
	9.6 Magnetic impurities and isotope effect	125
	9.7 Polaronic effect and isotope substitution	126
	9.8 Penetration depth: isotopic dependence	128
10	**Cuprate superconductors**	**131**
	10.1 History	131
	10.2 Structure of the cuprates	132
	10.3 Preparation of bulk and film cuprates	133
	10.4 Properties of the cuprates	134
	10.4.1 Phase diagram	134
	10.4.2 Critical field H_{c2}	135
	10.4.3 Two-gap spectrum	136
	10.4.4 Symmetry of the order parameter	136
	10.5 Isotope effect	138
	10.5.1 Polaronic state	138
	10.5.2 Isotopic dependence of the penetration depth	140
	10.6 Mechanism of high T_c	140
	10.7 Proposed experiment	145
11	**Inhomogeneous superconductivity and the "pseudogap" state of novel superconductors**	**147**
	11.1 "Pseudogap" state: main properties	148
	11.1.1 Anomalous diamagnetism above T_c	148
	11.1.2 Energy gap	150
	11.1.3 Isotope effect	152
	11.1.4 "Giant" Josephson effect	152
	11.1.5 Transport properties	153
	11.2 Inhomogeneous state	154
	11.2.1 Qualitative picture	154
	11.2.2 The origin of inhomogeneity	155
	11.2.3 Percolative transition	156
	11.2.4 Inhomogeneity: experimental data	156
	11.3 Energy scales	157
	11.3.1 Highest-energy scale (T^*)	158

	11.3.2	Diamagnetic transition (T_c^*)	158
	11.3.3	Resistive transition (T_c)	159
11.4	Theory		159
	11.4.1	General equations	160
	11.4.2	Diamagnetism	160
	11.4.3	Transport properties; "giant" Josephson effect	162
	11.4.4	Isotope effect	166
11.5	Other systems	167
	11.5.1	Borocarbides	167
	11.5.2	Granular superconductors; Pb+Ag system	167
11.6	Ordering of dopants and potential for room-temperature superconductivity	168
11.7	Remarks	171

12 Manganites — 172
- 12.1 Introduction — 172
- 12.2 Electronic structure and doping — 173
  - 12.2.1 Structure — 173
  - 12.2.2 Magnetic order — 176
  - 12.2.3 Double-exchange mechanism — 176
  - 12.2.4 Colossal magnetoresistance (CMR) — 177
- 12.3 Percolation phenomena — 178
  - 12.3.1 Low doping: transition to the ferromagnetic state at low temperatures — 178
  - 12.3.2 Percolation threshold — 179
  - 12.3.3 Increase in temperature and percolative transition — 180
  - 12.3.4 Experimental data — 181
  - 12.3.5 Large doping — 182
- 12.4 Main interactions: Hamiltonian — 183
- 12.5 Ferromagnetic metallic state — 184
  - 12.5.1 Two-band spectrum — 184
  - 12.5.2 Heat capacity — 186
  - 12.5.3 Isotope substitution — 187
  - 12.5.4 Optical properties — 189
- 12.6 Insulating phase — 190
  - 12.6.1 Parent compound — 190
  - 12.6.2 Low doping: polarons — 191
- 12.7 Metallic A-phase: S–N–S Josephson effect — 193
  - 12.7.1 Magnetic structure — 193
  - 12.7.2 Josephson contact with the A-phase barrier — 193
- 12.8 Discussion: manganites versus cuprates — 195

13 Novel superconducting systems — 197
- 13.1 Fe-based pnictide and chalcogenide superconductors — 197
- 13.2 Magnesium diboride: MgB_2 — 199
- 13.3 A-15 structure superconductors — 201

	13.4	Granular superconductors	202
	13.5	Sr_2RuO_4: a very novel superconductor	203
	13.6	Ruthenium cuprates	204
	13.7	Intercalated nitrides: self-supported superconductivity	205

14 Organic superconductivity   206
   14.1   History   206
   14.2   Organic superconductors: structure, properties   207
   14.3   Intercalated materials   210
   14.4   Fullerides   212
   14.5   Small-scale organic superconductivity   213
   14.6   Pair correlation in aromatic molecules   214

15 Pairing in nanoclusters: nano-based superconducting tunneling networks   218
   15.1   Clusters: shell structure   218
   15.2   Pair correlation   220
       15.2.1   Qualitative picture   220
       15.2.2   Main equations: critical temperature   222
       15.2.3   Energy spectrum; fluctuations   225
   15.3   How to observe the phenomenon?   226
   15.4   Cluster-based tunneling network: macroscopic superconductivity   227
   15.5   Cluster crystals   228

Appendices   229
   Appendix A: Diabatic representation   229
   Appendix B: Dynamic Jahn–Teller effect   231

References   233

Index   255

1
Historical perspective

The story of superconductivity started at the beginning of the twentieth century. The first important preliminary step was to obtain liquid helium at 4.2 K. This event took place on July 10, 1908 in the Kamerlingh Onnes Laboratory, in Leiden. To follow this important result, Kamerlingh Onnes decided to study the temperature dependence of the resistance at such low temperatures, and in 1911 he and his student Gilles Kolst discovered that at temperatures close to 4 K the electrical resistance of mercury vanished (Fig. 1.1). According to Kamerlingh Onnes: "The resistance disappeared... Thus the mercury at 4.2 K has entered a new state which, owing to its particular electric properties, can be called a state of superconductivity" (Kamerlingh Onnes, 1911).

Recently, scientists celebrated the centenary of this discovery. There has been significant progress during this century, so that at present we are dealing with a rather complex and strongly developed field.

The event in 1911 was the start of the race to discover new and higher transition temperature superconductors as well as to develop a theory to explain their properties. There were many important discoveries during these years, the first of which was the Meissner effect (Meissner and Ochsenfeld, 1933) whereby magnetic flux is excluded from the inside of a superconducting body. Since 1911 many new superconducting materials have been discovered, and the list now totals more than 6,000. During the

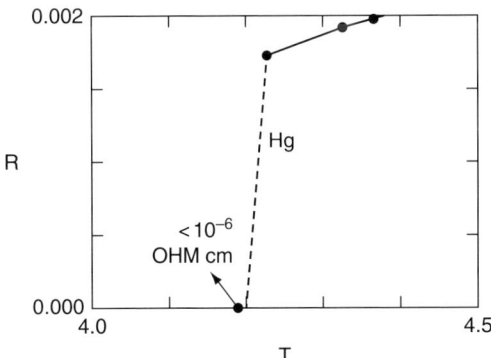

Fig. 1.1 Discovery of superconductivity (H. Kamerlingh Onnes, 1911): resistance of mercury versus temperature.

2 Historical perspective

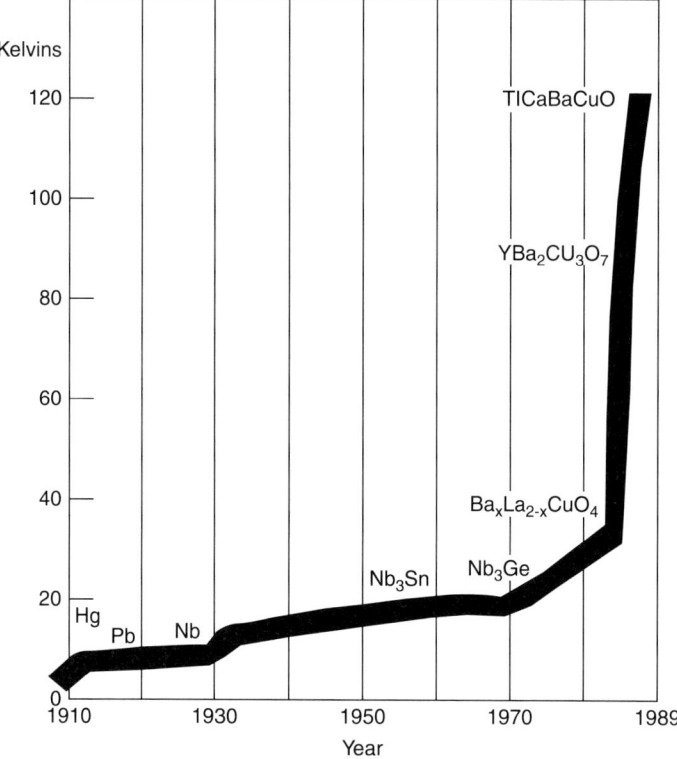

Fig. 1.2 Increase in maximum value of T_c with time.

first fifty years the maximum transition temperature had risen slowly (about 3 K per decade), reaching about 18 K in the late 1950s in both the B1 structure NbCN and the A15 structure Nb_3Sn (see Fig. 1.2).

When discussing theoretical studies one should mention several important developments preceding the microscopic theory. For example, the two-fluid model (Gorter and Casimir, 1934) provided a useful phenomenological picture, and the London equations (1935) provided a description of the Meissner effect.

An important theoretical advance came with the development of the Ginzburg–Landau theory (1950), which contains a phenomenological description of the superconducting state. It was the solution to these equations that provided the description of the magnetic vortices and the two fundamentally different types of superconductors distinguished by their very different behavior in a magnetic field. These two kinds of superconductor are Type I, which exhibits only the Meissner state, and Type II (Abrikosov, 1957), which exhibits a mixed state in which flux is not fully excluded but is incorporated into the superconductor in quantized flux entities called vortices.

The main challenge was to explain the phenomenon of superconductivity. This problem was investigated by the most prominent physicists of the twentieth century. Albert Einstein, for example, considered peculiar molecular currents (1922). It is interesting that according to his picture the sample containing a contact of two independent superconducting pieces should not carry a supercurrent. He convinced Meissner and Holm to perform the experiment (1932, see Huebener and Luebbig, 2008) but, to his disappointment, such a current did flow from one piece to the other. It is interesting that, in reality, Meissner and Holm observed the Josephson effect, which was rediscovered and appreciated twenty years later, after the creation of the microscopic theory.

Many other prominent scientists also tried to resolve the puzzle—among them, Landau, Bloch, and Heisenberg (see Sommerfeld and Bethe, 1933).

In 1950 an important advance came with the major discovery of the isotope shift in T_c (Maxwell, 1950, Reynolds et al., 1950, Fröhlich, 1950). This discovery was one of the important clues to the microscopic mechanism, since it showed that lattices, and hence phonons, were strongly involved in superconductivity.

In 1957 a microscopic description of the superconducting state based on the phonon-mediated pairing of opposite-momentum, opposite-spin electrons on the Fermi surface was developed and published in a landmark paper by Bardeen, Cooper, and Schrieffer (BCS, 1957). This paper included many predictions about the behavior of this unique paired-electron state that were almost immediately verified, and provided the rapid acceptance of this theory.

In the early 1960s, superconducting junctions were shown to exhibit some additional very unusual properties. Both single-electron excitations (quasiparticles) and pairs could tunnel across a thin insulating barrier separating two superconducting electrodes. Quasiparticle tunneling was discovered and understood by Giaever (1960), whereas Josephson (1962) predicted the tunneling of pairs and the very unusual behavior of such junctions. In addition, tunnel junctions provided the ultimate test of the generalized strong-coupling theory, as well as providing a basis for determining the underlying mechanism (see Chapter 5).

Soon afterward, W. Little (1964) described a model of high-T_c organic polymers. This paper marked the beginning of the search for high-T_c superconductivity, and introduced the non-phonon (electronic) mechanism as a key ingredient for such a search. The paper also predicted the phenomenon of organic superconductivity. The first organic superconductor (TMTSF-PF$_6$; $T_c = 1.5$ K) was discovered by Jerome et al. (1980), and organic superconductors have now been synthesized with transition temperatures close to 40 K (see Chapter 14).

The discovery of a new family, high-T_c oxides (cuprates), came in 1986—75 years after the initial discovery of superconductivity!—when Bednorz and Mueller (1986) discovered that La(Ba,Sr)CuO compounds were found to be superconducting and have a maximum T_c of $\sim$40 K. The cuprates themselves have proven very rich in new superconductors, and there are presently more than fifty such compounds. Wu et al. (1987) reported on a YBaCuO compound with a transition temperature of about 90 K, well above the temperature of liquid nitrogen. The highest transition temperature cuprate is a HgBaCaCuO compound with a transition temperature $T_c \approx 133$ K

(Schilling *et al.* 1993), and under pressure the value of T_c was raised to 150 K (Gao et al., 1994).

During the last twenty years, many new superconducting systems have been discovered—among them, fullerenes, MgB_2 compounds, pnictides, and so on. We will discuss the properties of these new systems in Chapter 13. As a whole, we are witnessing a unique time in the history of superconductivity, as never before have so many novel superconducting systems been studied at the same time.

Thus superconductivity is a very rich field, and one can expect many new surprises in the future.

2
Electronic states, phonons, and electron-phonon interaction

2.1 Adiabatic approximation: Hamiltonian

Adiabatic approximation is the foundation of the theory of solids and molecular systems. It allows us to describe the electronic states and phonons, classify the energy spectra, and introduce the electron–phonon interaction, which plays a key role in transport properties and also in superconductivity.

The starting point of the theory of metals is the approximation introduced by Born and Oppenheimer (1927) (see, for example, reviews by Born and Huang (1954), Ziman (1960), and Grimvall (1981)). This approximation is based on the fact that the masses of the ions and the conduction electrons are very different. The electrons follow adiabatically the motion of the ions. However, this adiabatic behavior is not perfect. The residual non-adiabaticity is a measure of the "friction" between the electrons and the lattice, representing what we refer to as the electron–phonon interaction.

The full Hamiltonian of the electron–ion system is

$$\widehat{H} = \widehat{T}_r + \widehat{T}_R + V(r, R) \qquad (2.1)$$

Here, (r, R) are the sets of electronic and ionic coordinates, respectively, $\widehat{T}$ is the kinetic energy operator, and $V(r, R) = V_1(r) + V_2(R) + V_3(r, R)$ is the total potential energy (V_1 and V_2 correspond to the electron–electron and ion–ion interactions; V_3 describes the electron–ion interaction).

Our goal is to describe the interaction between the conduction electrons and the crystal lattice in a metal. At first glance it may seem that it should be possible to treat this interaction in a manner analogous to that used in the theory of radiation, where the full Hamiltonian contains free electron and photon fields, plus a small interaction term. However, the present problem turns out to be quite delicate, the reason being that the very procedure of introducing the phonon field requires great care.

To proceed, we need to introduce the phonon subsystem. Furthermore, a rigorous description of the electron–phonon interaction requires that the Hamiltonian be written in the form

$$\widehat{H} = \widehat{H}_0 + \widehat{H}' \qquad (2.2)$$

where

$$\widehat{H}_0 = \widehat{H}_0^{el}(r) + \widehat{H}_0^{ph}(R) \qquad (2.3)$$

and
$$\hat{H}' \equiv U(r, R) \tag{2.4}$$

Here $\hat{H}_0^{el}(r)$ and $\hat{H}_0^{ph}(R)$ describe the free electron and phonon fields, so that
$$\hat{H}_0^{el}(r) = \hat{T}_r + U^{el}(r) \tag{2.5}$$

$$\hat{H}_0^{ph}(R) = \hat{T}_R + U^{ph}(R) \tag{2.6}$$

U^{el} and U^{ph} are the effective potential energies.

A rigorous theory has to be based on the representation (2.2), with the operator (2.4) corresponding to the electron–phonon interaction. The effects of this interaction can be calculated by perturbation theory (if $\hat{H}'$ is small), or with the help of the diagrammatic technique. However, it turns out that the problem of presenting the Hamiltonian (2.1) in the form (2.2), and, therefore, correctly introducing $\hat{H}'$ is quite non–trivial.

2.2 Adiabatic approximation and non-adiabaticity: Born–Oppenheimer and "crude" approaches

The total wavefunction of the system $\Psi(r, R)$ is the solution of the Schrödinger equation:
$$\hat{H}\Psi(r, R) = E\Psi(r, R) \tag{2.7}$$

The total Hamiltonian is defined by eqn. (2.1). Let us neglect, in the zeroth-order approximation, the term corresponding to the kinetic energy of a heavy ionic system. Then we can introduce a complete orthogonal set of electronic wavefunctions, which are the solutions of
$$\hat{H}_r \psi_n(r, R) = \varepsilon_n(R)\psi_n(r, R) \tag{2.8}$$

where $\hat{H}_r = \hat{T}_r + V(r, R)$.

The functions $\psi_n(r, R)$ and their eigenvalues $\varepsilon_n(R)$ (electronic terms) depend parametrically on the nuclear (ionic) positions.

The full wavefunctions $\Psi_n(r, R)$ associated with the nth electronic state can be sought in the form of an expansion in the complete orthogonal set of the solutions of the electron Schrödinger equation (2.8). That is:
$$\Psi_n(r, R) = \sum_m \phi_{nm}(R)\psi_m(r, R) \tag{2.9}$$

Substituting the expansion (2.9) into eqn. (2.7), multiplying by $\psi_\alpha(r, R)$ and integrating over r yields the following equation:
$$[\hat{T}_R + \varepsilon_\alpha(R)]\phi_{n\alpha}(R) = E\phi_{n\alpha}(R) \tag{2.10}$$

We neglect the sum $\sum_m C_{\alpha m}\Phi_{nm}(R)$ where $C_{\alpha m}$ contains derivatives of ψ_n with respect to the nuclear coordinates.

According to eqn. (2.10), $\Phi_{n\alpha}(\boldsymbol{R})$ can be treated as a nuclear wavefunction. As for the electronic term $\varepsilon_\alpha(\boldsymbol{R})$, it appears as the potential energy for the motion of the heavy particles.

If we also neglect off-diagonal contributions to the expansion (2.9), we arrive at the Born–Oppenheimer expression for the total wavefunction:

$$\Psi_{nv}(\boldsymbol{r}, \boldsymbol{R}) = \psi_n(\boldsymbol{r}, \boldsymbol{R})\phi_{nv}(\boldsymbol{R}) \tag{2.11}$$

where v denotes the set of quantum numbers describing nuclear motion. Equations (2.8), (2.10), and (2.11) define the Born–Oppenheimer adiabatic approximation. It is obvious from the derivation that expression (2.11) is not an exact solution of the total Schrödinger equation.

Let us evaluate the non-adiabaticity neglected in the course of deriving the approximation (2.11). The non-adiabaticity operator can be found in the following way (see, for example, Kresin and Lester, 1985). The zeroth-order operator, which has the function (2.11) as its eigenfunction, has the form

$$\widehat{H}_0 = \widehat{T}_r + V(\boldsymbol{r}, \boldsymbol{R}) + \widehat{H}_R^0 \tag{2.12}$$

where the operator $\widehat{H}_R^0$ is defined by the relation

$$\widehat{H}_R^0 \psi_n(\boldsymbol{r}, \boldsymbol{R})\phi_{nv}(\boldsymbol{R}) = \psi_n(\boldsymbol{r}, \boldsymbol{R})\widehat{T}_R\,\phi_{nv}(\boldsymbol{R}) \tag{2.13}$$

Indeed, we can see by direct calculation that

$$\begin{aligned}\widehat{H}_0\Psi_{\mathrm{BO}}(\boldsymbol{r}, \boldsymbol{R}) &= \left[\widehat{T}_r + V(\boldsymbol{r}, \boldsymbol{R})\right]\psi_n(\boldsymbol{r}, \boldsymbol{R})\phi_{nv} + \widehat{H}_R^0\psi_n(\boldsymbol{r}, \boldsymbol{R})\phi_{nv}(\boldsymbol{R})\\&= \varepsilon_n(\boldsymbol{R})\psi_n(\boldsymbol{r}, \boldsymbol{R})\phi_{nv}(\boldsymbol{R}) + \psi_n(\boldsymbol{r}, \boldsymbol{R})\widehat{T}_R\phi_n(\boldsymbol{R})\\&= \psi_n(\boldsymbol{r}, \boldsymbol{R})\left[\widehat{T}_R + \varepsilon_n(\boldsymbol{R})\right]\phi_{nv}(\boldsymbol{R})\\&= E_{nv}\Psi_{\mathrm{BO}}(\boldsymbol{r}, \boldsymbol{R})\end{aligned}$$

Based on eqns. (2.2), (2.12), and (2.13), we can now determine the non-adiabaticity operator $\widehat{H}'$:

$$\widehat{H}' = \widehat{T}_R - \widehat{H}_R^0 \tag{2.14}$$

Hence

$$\widehat{H}'\Psi_{\mathrm{BO}}(\boldsymbol{r}, \boldsymbol{R}) = \widehat{T}_R\psi_n(\boldsymbol{r}, \boldsymbol{R})\phi_{nv}(\boldsymbol{R}) - \psi_n(\boldsymbol{r}, \boldsymbol{R})\widehat{T}_R\phi_{nv}(\boldsymbol{R}) \tag{2.15}$$

or, in matrix representation,

$$<mv'\left|\widehat{H}'\right|nv> = \int \psi_m^*(\boldsymbol{r}, \boldsymbol{R})\phi_{mv'}^*(\boldsymbol{R})\widehat{T}_R\psi_n(\boldsymbol{r}, \boldsymbol{R})\phi_{nv}(\boldsymbol{R})d\boldsymbol{R}d\boldsymbol{r}\\-\delta_{mn}\int \phi_{mv'}^*(\boldsymbol{R})\widehat{T}_R\phi_{nv}(\boldsymbol{R})d\boldsymbol{R} \tag{2.15a}$$

Let us estimate the accuracy of the Born–Oppenheimer approximation. The characteristic scale of the nuclear wavefunction $\phi_{nv}(\boldsymbol{R})$ is of the order of the vibrational amplitude a, whereas that of the electronic wavefunction is the lattice period L. As a result, we obtain:

$$\hat{H}' \Psi_{BO}(r, R) \sim \frac{1}{M} \frac{\partial \psi}{\partial R} \frac{\partial \phi}{\partial R}$$

$$\sim \frac{1}{M} \frac{\psi}{L} \frac{\phi}{a} \sim \frac{1}{Ma^2} \frac{a}{L} \Psi_{BO} \sim \frac{a}{L} \Omega \Psi_{BO}$$

That is,

$$\hat{H}' \propto \kappa = \frac{a}{L} \ll 1 \tag{2.16}$$

Hence non-adiabaticity is proportional to the small parameter k = a/L (we have used the relation $a = [1/M\Omega]^{1/2}$; Ω is the vibrational frequency).

Moreover, since $\Omega = (K_n/M)^{1/2}$, where K_n is the force constant due to the Coulomb internuclear interaction, we can write $a^2 = (K_n/M)^{-1/2}$. Similarly, $L^2 \cong (K_e/m)^{-1/2}$. Since $K_e = K_n$ (the Coulomb interaction does not depend on the mass factor), we arrive at $k \cong (m/M)^{1/4}$. Here and below, we put $\hbar = k_B = 1$.

It is useful also to present the following classical estimate (Migdal, 2000). Equation (2.15) can be written in the form:

$$\hat{H}' \Psi_{BO}(r, R) = [\hat{T}_R, \psi_n]\phi_{nv}(R) \tag{2.15b}$$

If we assume that the motion of heavy particles is classical, then $i[\hat{T}_R, \psi_n] \to (\partial \psi_n/\partial R)v$, where v is the velocity of the heavy particles. Thus the qualitative aspect of the adiabatic approximation is indeed connected with the small velocity of the ions relative to the much greater velocity of the electrons. Averaging the total potential energy $V(r,R)$ over the fast electronic motion, we obtain the effective potential $\varepsilon(R)$ for the nuclear motion.

The adiabatic method described above is a rigorous approach, which provides a foundation for the theory of solids. Nevertheless, from the practical point of view this approach has a definite shortcoming. Indeed, one can see from eqns. (2.12) and (2.13) that the zeroth-order Hamiltonian does not have the desired form (2.3).

Another conventional form of adiabatic theory (the so-called "crude" or "clamped" approximation) is based on the fact that the ions are located near their equilibrium positions. In this version of the adiabatic theory the expansion of the total wavefunction uses the electronic wavefunctions $\psi_n(r, R) \cong \psi_n(r, R_0)$, where R_0 corresponds to some equilibrium configuration. That is:

$$\psi_n(r, R) = \sum_m \tilde{\phi}_{nm}(R) \psi_m(r, R_0) \tag{2.17}$$

The total wavefunction in the zeroth-order approximation is $\Psi_n(r, R) = \psi_n(r, R_0) \tilde{\phi}_{nv}(R)$. Strictly speaking, the nuclear wavefunctions ϕ_{nv} (see eqn. (2.10)) and $\tilde{\phi}_{nv}$ are different. The non-adiabaticity operator in the "crude" approximation can be written in a relatively simple form:

$$H_1 = \sum_{i,\alpha} \frac{\partial V}{\partial R_{i,\alpha}|_0} \Delta R_{i\alpha} \tag{2.17a}$$

The "crude" approach has the same weakness as the Born–Oppenheimer approximation: the zeroth-order Hamiltonian does not have the form (2.3). As a result, neither approach is adaptable to the usual perturbation theory treatment. It is difficult to evaluate higher order corrections or to consider the case of strong non-adiabaticity.

Now we turn to a different approach to adiabatic theory, which will allow us to overcome these problems.

2.3 Electron–phonon coupling

Let us follow the approach developed by Geilikman (1971, 1975). We introduce the functions

$$\Psi^0(\boldsymbol{r},\boldsymbol{R}) = \psi_n(\boldsymbol{r},\boldsymbol{R}_0)\,\phi_{n\nu}(\boldsymbol{R}) \tag{2.18}$$

which form a complete basis set. The functions $\psi_n(\boldsymbol{r},\boldsymbol{R}_0)$ and $\phi_{n\nu}(\boldsymbol{R})$ are defined by the equation

$$\left[\widehat{T}_r + V(\boldsymbol{r},\boldsymbol{R})\right]\psi_n(\boldsymbol{r},\boldsymbol{R}) = \varepsilon_n(\boldsymbol{R})\psi_n(\boldsymbol{r},\boldsymbol{R})_{|\boldsymbol{R}=\boldsymbol{R}_0} \tag{2.19}$$

and by eqn. (2.10). Here n is the set of quantum numbers for the electronic states, ν is the vibrational quantum number, $\boldsymbol{R}_0$ is the equilibrium configuration. Strictly speaking, $\boldsymbol{R}_0$ depends on n, but for levels close to the ground state this dependence may be neglected.

The functions (2.18) form the basis set in the present method. It is interesting that the total function (2.18) is a combination of the Born–Oppenheimer and the "crude" functions. Indeed, the electron wavefunction is that of the "crude" approach, whereas $\phi_{n\nu}(\boldsymbol{R})$ is a solution of the nuclear Schrödinger equation of the Born–Oppenheimer version.

Equation (2.10) together with the wavefunctions $\phi_{n\nu}(\boldsymbol{R})$ serves to define phonons in the adiabatic theory.

We can write down the Hamiltonian in the zeroth-order approximation (the functions (2.18) are its eigenfunctions; see eqn. (2.21a)). Namely:

$$\widehat{H}_0 = \widehat{H}_{0r} + \widehat{H}_{0R} \tag{2.20}$$

where

$$\widehat{H}_{0r} = \widehat{T}_r + V(\boldsymbol{r},\boldsymbol{R}_0) \tag{2.20a}$$

$$\widehat{H}_{0R} = \widehat{T}_R + \hat{A}(\boldsymbol{R}) \tag{2.20b}$$

The operators $\widehat{T}_r$ and $\widehat{T}_R$ are defined according to eqn. (2.1). The operator $\hat{A}$ is defined by the following matrix form:

$$\hat{A}^{n'v'}_{nv} = \delta_{nn'}\int \phi_{n'v'}(\boldsymbol{R})[\varepsilon_n(\boldsymbol{R}) - \varepsilon_n(\boldsymbol{R}_0)]\phi_{nv}(\boldsymbol{R})d\boldsymbol{R} \tag{2.21}$$

It is easy to show that the operator $\widehat{H}_0$ is diagonal in the space of the functions (2.18). Indeed,

$$\widehat{H}_{0;nv}^{n'v'} = \int \psi_{n'}^*(\mathbf{r}, \mathbf{R}_0) \phi_{n'v'}(\mathbf{R}) \left[\widehat{T}_r + V(\mathbf{r}, \mathbf{R}_0) + \widehat{T}_R + \widehat{A}(\mathbf{R})\right]$$
$$\times \psi_n(\mathbf{r}, \mathbf{R}_0) \phi_{nv}(\mathbf{R}) d\mathbf{r} d\mathbf{R} \tag{2.21a}$$
$$= \delta_{nn'} \int \phi_{n'v'}(\mathbf{R}) \left[\widehat{T}_R + \varepsilon_n(\mathbf{R})\right] \phi_{nv}(\mathbf{R}) d\mathbf{R}$$
$$= \delta_{nn'} \delta_{vv'} E_n$$

Note the important fact that the operator $\widehat{H}_0$ represents a sum of two terms, one of which depends on $\mathbf{r}$ and the other on $\mathbf{R}$. These terms correspond to the electron and the phonon fields respectively.

The electron–phonon interaction is given by the following non-adiabatic term in the full Hamiltonian:

$$\widehat{H}' = \widehat{H} - \widehat{H}_0 \tag{2.22}$$

where $\widehat{H}$ and $\widehat{H}_0$ are defined by eqns. (2.1) and (2.20). As a result, we find

$$\widehat{H}' = V(\mathbf{r}, \mathbf{R}) - V(\mathbf{r}, \mathbf{R}_0) - \widehat{A}(\mathbf{R}) \tag{2.23}$$

or, in matrix representation ($\widehat{H}_{na} \equiv \widehat{H}'$),

$$H_{na;nv}^{n'v'} = [V(\mathbf{r}, \mathbf{R}) - V(\mathbf{r}, \mathbf{R}_0)]|_{nv}^{n'v'} - \widehat{A}(\mathbf{R})|_{nv}^{n'v'} \tag{2.23a}$$

The last term is defined by eqn. (2.21).

The non-adiabatic Hamiltonian can be written as an expansion about the equilibrium coordinates:

$$H_{na;nv}^{n'v'} = \frac{\partial V}{\partial \mathbf{R}}\bigg|_0 \delta \mathbf{R}|_{nv}^{n'v'} - \frac{\partial \varepsilon_n}{\partial \mathbf{R}}\bigg|_0 \delta \mathbf{R}|_{nv}^{n'v'} + \frac{1}{2}\frac{\partial^2 V}{\partial \mathbf{R}_i \partial \mathbf{R}_k}\bigg|_0 \delta \mathbf{R}_i \delta \mathbf{R}_k|_{nv}^{n'v'}$$
$$- \frac{1}{2}\frac{\partial^2 \varepsilon_n}{\partial \mathbf{R}_i \partial \mathbf{R}_k}\bigg|_0 \delta \mathbf{R}_i \delta \mathbf{R}_k|_{nv}^{n'v'} + \frac{1}{6}\frac{\partial^3 V}{\partial \mathbf{R}_i \partial \mathbf{R}_k \partial \mathbf{R}_l}\bigg|_0 \delta \mathbf{R}_i \delta \mathbf{R}_k \delta \mathbf{R}_l|_{nv}^{n'v'} + \cdots \tag{2.24}$$

The terms $\frac{\partial \varepsilon_n(\mathbf{R})}{\partial \mathbf{R}}\big|_0$ vanish. The first term in eqn. (2.24) corresponds to the "crude" approximation (see eqn. (2.17a)).

Note that the result (2.24) represents a power-series expansion in terms of the parameter (2.16). Indeed, in eqn. (2.24) the deviation from the equilibrium position is $\delta R \cong a$, while the equilibrium distance is $R_0 \cong L$. It is convenient to rewrite eqn. (2.24) in the following way:

$$H_{na}|_{nv}^{n'v'} = H_1|_{nv}^{n'v'} + H_2|_{nv}^{n'v'} + \cdots \tag{2.25}$$

where

$$H_1|_{nv}^{n'v'} = \frac{\partial V}{\partial \mathbf{R}}\bigg|_0 \delta \mathbf{R}|_{nv}^{n'v'} - \frac{\partial \varepsilon_n}{\partial \mathbf{R}}\bigg|_0 \delta \mathbf{R}|_{nv}^{n'v'} \tag{2.25a}$$

$$H_2\big|_{nv}^{n'v'} = \left[\frac{1}{2}\frac{\partial^2 V}{\partial R_i \partial R_{\mathrm{k}}|0}\delta R_i \delta R_{\mathrm{k}} - \frac{1}{2}\frac{\partial^2 \varepsilon_n}{\partial R_i \partial R_{\mathrm{k}}|0}\delta R_i \delta R_{\mathrm{k}}\right]\bigg|_{nv}^{n'v'} \quad (2.25b)$$

Let us estimate the magnitude of the terms $\widehat{H}_1, \widehat{H}_2, \ldots$.

Starting with $\widehat{H}_2$, we have $H_2 \cong \Omega$. Note that each additional factor δR_i brings in the parameter $\kappa = a/L \ll 1$. Hence,

$$H_1 \sim \Omega/\kappa, \quad (2.26)$$

$$H_2 \sim \Omega, \quad (2.26a)$$

$$H_3 \sim \kappa\Omega \quad (2.26b)$$

and so on.

One can see from eqns. (2.20) and (2.25) that the total Hamiltonian can be written in the form (2.2); that is, as a sum of the terms describing free electron and phonon fields and an additional terms corresponding to the electron–phonon interaction.

It is interesting that the first term in (2.25) (see eqn. (2.26)) can make the contribution larger than the phonon energy. This is a peculiar feature of the theory.

One should stress a very important fact that the first non-adiabatic terms ((2.26) and (2.26a)) do not contain the smallness $\propto \kappa$. Moreover, the first term is proportional to κ^{-1}. This is the reason why one must be careful in analyzing the effects of the electron–phonon coupling.

In addition, because we are dealing with operators, the corresponding contribution depends on the type of the transitions (diagonal versus non-diagonal). For example, one finds that the electron–phonon interaction makes only a very small contribution to the total energy of the system. At the same time, some other effects, such as mass renormalization, turn out to be very significant. The phenomenon of superconductivity is another example of a gigantic non-adiabatic effect (see Chapter 3).

2.4 Electron–phonon interaction and renormalization of normal parameters

Electron–phonon interaction plays an extremely important role in the physics of metals. For example, it is one of the principal relaxation mechanisms defining electrical conductivity. In this type of situation we are dealing with electron scattering by real thermal phonons. But even in the absence of thermal phonons—for example, at T = 0 K—the polarizability of the lattice (changes in the regime of zero-point vibrations of the ions) often plays a significant role. Superconductivity represents probably the most spectacular manifestation of this type of interaction. However, it also affects the normal state properties. For example, the effective mass m^* measured in certain experiments (for example, cyclotron resonance or the de Haas–van Alphen effect) is different from the so-called band value m^b corresponding to a frozen lattice.

Qualitatively speaking, as a result of lattice polarization electrons find themselves "dressed" in ionic coats. This mass renormalization effect turns out to be quite significant for a number of metals. Namely, it is found that $m^*(0) = m^b(1 + \lambda)$, where λ is

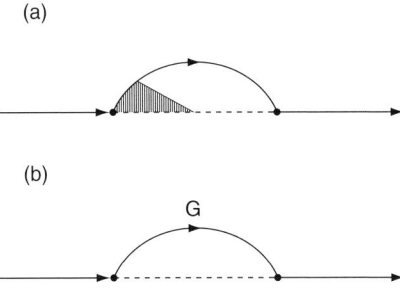

Fig. 2.1 Self-energy parts.

the coupling constant describing the electron–phonon interaction (m* (0) is the value of the effective mass at T = 0 K). For example, in the lead $\lambda \cong 1.4$, so that renormalization leads to the electron becoming about 2.5 times heavier than what would be expected for a frozen lattice. A similar relationship holds for the Sommerfeld constant describing the electronic heat capacity: $\gamma(0) = \gamma^b (1+\lambda)$.

In quantum language, renormalization effects are described by the diagram in Fig. 2.1a; we are dealing with emission and subsequent absorption of a virtual phonon. Analytically, the equation shown in Fig. 2.1b can be written in the form:

$$\sum\nolimits_1 (\bm{k},\omega_n) = \pi T \sum_{\omega_{n'}} \int d\bm{k}' \zeta^2_{q,\kappa'} D(\bm{q},\omega_n - \omega_{n'}) G(\bm{k}',\omega_{n'}) \tag{2.27}$$

Here $\sum_1$ is the self-energy part describing the scattering, D and G are the phonon propagator and one-particle Green's function, $\bm{q} = \bm{\kappa} - \bm{\kappa}'$, $\omega_n = (2n+1)\pi T$, and $\zeta_{q,\kappa'}$ is the matrix element. We employ the thermodynamic Green's functions formalism (see, for example, Abrikosov et al., 1975). A usual equation containing time dependence can be obtained by an analytical continuation to the upper plane. Making transformation to the integration over the phonon momentum $\bm{q}$, and then to phonon frequency, we obtain:

$$\sum\nolimits_1 (\bm{k},\omega_n) = \pi T \sum_{\omega_{n'}} \int d\Omega \frac{g(\Omega)}{\Omega} D(\Omega,\omega_n - \omega_{n'}) G(\bm{k}',\omega_{n'}); \ k \cong k' \cong k_F \tag{2.28}$$

Here we introduced an important function:

$$g(\Omega) = \alpha^2(\Omega) F(\Omega) \tag{2.29}$$

($\alpha^2(\Omega)$ describes the electron–phonon interaction, and $F(\Omega)$ is the phonon density of states). It is defined by the expression (see, for example, Scalapino, 1969; McMillan and Rowell, 1969)

$$\alpha^2(\Omega)F(\Omega) = \frac{\int \frac{dS_{k'}}{|\mathbf{v}_{k'}|} \int \frac{dS_{k'}}{|\mathbf{v}_k|} \frac{1}{(2\pi)^3} \sum_\lambda |\zeta_{k',k,\lambda}|^2 \delta[\Omega - \omega_{\lambda,k'-k}]}{\int \frac{dS_k}{|\mathbf{v}_k|}} \quad (2.30)$$

For many superconductors we can introduce an effective average phonon frequency:

$$\tilde{\Omega} = <\Omega^2>^{1/2} \quad (2.31)$$

$$<\Omega^n> = (2/\lambda)\int g(\Omega)\Omega^{n-1}d\Omega \quad (2.31a)$$

Then eqn. (2.28) can be written in the form

$$\sum_1 = \pi T \lambda \sum_{n' \geq 0} D\left(\tilde{\Omega}, \omega_n - \omega_{n'}\right) G\left(\omega_{n'}\right) \quad (2.32)$$

Here

$$\lambda = 2\int \frac{d\Omega}{\Omega} g(\Omega) \quad (2.33)$$

is the electron–phonon-coupling constant which is rigorously defined by eqn. (2.33).

It should be pointed out that some phenomena (such as high-frequency effects, the Pauli spin susceptibility, and so on) do not involve mass renormalization. A detailed review can be found in the book by Grimvall (1981).

Let us return to the renormalization of the Sommerfeld constant (Grimvall, 1969, 1981, Kresin and Zaitsev, 1978). The entropy of the electron system can be written as (Eliashberg, 1963):

$$S_c = \frac{\nu_0}{T^2}\int_0^\infty \frac{d\varepsilon\, \varepsilon}{\cosh^2(\varepsilon/2T)}[\varepsilon - f(\varepsilon)]. \quad (2.34)$$

Here ν_0 is the unrenormalized density of states at the Fermi level, and $f(\varepsilon)$ is the odd part of the self-energy function, which describes the electron–phonon interaction (see Fig. 2.1).

If we neglect the function $f(\varepsilon)$ on the right-hand side of eqn. (2.34), we recover the usual expressions for the entropy and, consequently, for the heat capacity of a free-electron gas. Renormalization effects are contained in the function $f(\varepsilon)$.

A calculation of the function $f(\varepsilon)$ leads to the following result for the electronic heat capacity:

$$C_e(T) = \gamma(T)T \quad (2.35)$$

where

$$\gamma(T) = \gamma^0\left[1 + 2\int_0^\infty \frac{d\Omega}{\Omega}g(\Omega)Z\left(T/\Omega\right)\right] \quad (2.35a)$$

In this expression, $g(\Omega)$ is defined by eqn. (2.29), and $Z(x)$ is a universal function (Fig. 2.2).

For $T = 0$ K we find

$$\gamma(0) = \gamma^b(1 + \lambda) \quad (2.35b)$$

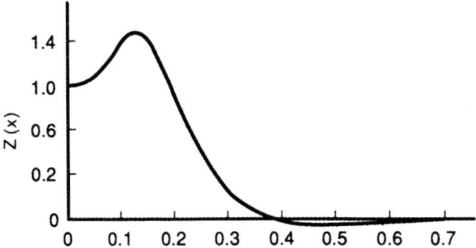

Fig. 2.2 Universal function $Z(x)$.

where λ is the electron–phonon coupling constant (eqn. 2.33).

As can be seen from eq. (2.35), the electron–phonon interaction leads to a deviation from the linear temperature dependence of the electronic heat capacity. This effect is due to the fact that the state of the lattice changes with temperature as thermal phonons appear.

The result (2.35a) can be written in a form that contains only experimentally measured quantities:

$$\gamma(T) = \gamma(0)\left[1 + \rho\left(\frac{k(T)}{k(0)} - 1\right)\right]$$

$$k(T) = 2\int d\Omega g(\Omega)\Omega^{-1} Z\left(T/\Omega\right)$$

(2.36)

where $\gamma(0) = m^* p_F/3$, $m^* = m^b(1+\lambda)$ is the renormalized effective mass, $\rho = \lambda(1+\lambda)^{-1}$ is the renormalized coupling constant, and λ is defined by eqn. (2.33). For $T \to 0$ K,

$$[\gamma(T)/\gamma(0) - 1] \propto T^3 \ln(\Omega_D/T) \tag{2.36a}$$

This dependence was derived by Eliashberg (1963). It is interesting that at sufficiently low temperatures the additional contribution to the electronic heat capacity given, according to (2.36a), by $\Delta C \propto T^3 \ln(\frac{\Omega_D}{T})$, exceeds the lattice heat capacity ($\propto T^3$).

As the temperature increases, $\gamma(T)$ goes through a maximum, and upon further rise in temperature the function $Z(T/\Omega)$ decreases and approaches zero. It becomes small for $x \cong 0.3$, whereby $\gamma(T)$ acquires the unrenormalized value γ^b. Qualitatively, this is explained by the fact that intense thermal motion wash out the ionic "coat," so that the electrons find themselves "undressed" at high temperatures.

One can calculate also the temperature dependence of the effective mass, which is described by the relation:

$$\frac{m^*(T)}{m^*(0)} = 1 + 2\int \frac{d\Omega}{\Omega} g(\Omega) G\left(T/\Omega\right) \tag{2.37}$$

where $G(x) = (2\pi x)^2 \sum_{n=0}^{\infty} \frac{(2n+1)}{\{1+[(2n+1)\pi x]^2\}^2}$ is the so-called Grimvall function (Grimvall, 1969; see also Allen and Cohen, (1970)). The calculated dependence $m^*(T)$ is in good agreement with the cyclotron resonance measurements by Sabo for Zn (1969) and by Krasnopolin and Khaikin for Pb (1973).

So far we have been discussing corrections to the electronic parameters induced by electron–phonon interaction. As we have seen, non-adiabaticity leads to significant changes in the effective mass, electronic heat capacity, the Fermi velocity $v_F = p_F/m^*$, and so on. Let us now look at the effect of this interaction on the phonon spectrum. The results here turn out to be completely different (Geilikman, 1971; Brovman and Kagan, 1967). The correction to the phonon frequency $\Delta\Omega$ can be evaluated as the variation of the non-adiabatic contribution to the total energy. A calculation of this correction leads to $\Delta\Omega \propto \kappa^2$ (κ is defined by eqn. (2.16)); thus the shift in the phonon frequency is relatively small.

This is an important result, which has a direct bearing on the problem of lattice instability. It also has an interesting history. The fact of the matter is that if one directly uses the Frohlich model in which the full Hamiltonian is made up of an electronic term, a phonon term with an acoustic dispersion law ($\omega = u_i q, q \to 0$), and an interaction term given by H_1 (see eqn. (2.17a)), then we obtain the result that $\Omega = \Omega_0 (1 - 2\lambda)^{1/2}$.

In other words, we find a significant shift in the phonon frequency. Furthermore, for $\lambda > \lambda_{\max} = 0.5$ the frequency is imaginary; that is, the lattice becomes unstable. Thus in the early years following the BCS work it was assumed that λ cannot exceed $\lambda_{\max}$, which then leads to an upper limit on the superconducting critical temperature T_c, $T_{c,\max} \cong \Omega_D \exp(-1/\lambda_{\max}) \cong 0.1\Omega_D$ achievable by the phonon mechanism. This result is in contradiction with subsequent experimental observations; at present we know many superconductors with $\lambda > 0.5$ (see, for example, E. Wolf, 2012).

The problem with the above analysis is that the Frohlich Hamiltonian is not a valid tool. If we start with the rigorous adiabatic theory, the interaction is given not by H_1 but by the full expression H_{na}; see eqn. (2.25). The use of this expression leads to the aforementioned weak renormalization of the phonon frequency $\sim \kappa^2$. A rigorous systematic application of the adiabatic theory leads to only weak corrections due to non-adiabaticity.

2.5 The "Migdal" theorem

In the previous section we took into account the lowest-order terms in the electron–phonon interaction. This can be seen directly in Fig. 2.3. Indeed, the electron–phonon scattering is described by the factor containing just the matrix element of the electron–phonon interaction (the so-called "single vertex", Γ_0), and the higher-order terms, such as Γ', are neglected. According to Migdal (1958) this approximation is justified, because higher-order corrections contain an additional small adiabatic parameter $\widetilde{\Omega}/E_F$.

Indeed, consider the term Γ', which can be analytically written in the form

$$\Gamma'(p,q) = \zeta^3 \int d\mathbf{k} d\omega D(p-k) G(k) G(k+q)$$

where $k = \{\mathbf{k}, \omega\}$.

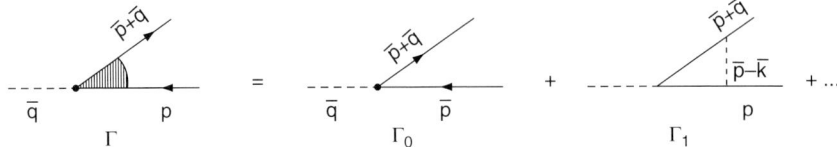

Fig. 2.3 Electron–phonon vertex.

The presence of the D-function provides the cut-off in the integration at $\omega \cong \tilde{\Omega}$; indeed, at $\omega \gg \tilde{\Omega}$ this function decreases rapidly $\left(\sim \tilde{\Omega}^{-2}\right)$. On the other hand, the characteristic scale for the electronic Green's function is $\sim E_F^{-1}$. As a result, the quantity Γ' can be estimated as $\Gamma' \propto \lambda^2 \left(\frac{\tilde{\Omega}}{E_F}\right) \Gamma_0$; $\lambda^2 = \frac{\zeta \nu_0}{\tilde{\Omega}}$, and, indeed, contains the small adiabatic parameter $\tilde{\Omega}/E_F$. A more detailed derivation and some special cases (for example, when $u \gg v_F$, u is the sound velocity) are described by Scalapino (1969), Grimvall (1981).

The "Migdal" theorem was one of the first applications of the quantum field theory to condensed-matter physics. The analysis was based on the pioneering work by Galitskii and Migdal (1958). This theorem is essential for the analysis of the superconducting state (see Chapter 3).

2.6 Polaronic states

2.6.1 Concept

A strong electron–lattice interaction could lead to the formation of polarons. The concept of polarons was introduced and studied by Landau (1933; see Landau, 1966) and by Landau and Pekar (1948). A polaron can be created if an electron is added to the crystal with a small carrier concentration (see, for example, Ashcroft and Mermin, 1976). Because of strong local electron–ion interactions, the electron appears to be trapped and can be viewed as being dressed in a "heavy" ionic "coat". In reality, we are dealing with a strong (non-linear) manifestation of the electron–lattice interaction.

The concept of polarons is an essential ingredient of the physics of high-T_c superconductivity. In fact, the concept of a Jahn–Teller polaronic state (see, for example Hock et al., 1983) was a main motivation for Bednorz and Mueller to search for superconductivity in oxides, and this led to their breakthrough discovery (1986; see Chapter 10).

The formation of polarons is a strong non-adiabatic phenomenon. As we know, the usual adiabatic approximation (see Sections 2.1–2.3) allows us to separate electronic and ionic motions. Indeed, this approximation is based on the fact that ionic motion is much slower than the motion of electrons; it allows us, as a first step, to neglect the kinetic energy of ions and to study the electronic structure for a "frozen" lattice. The electronic energy (electronic terms) appears to be function of the ionic positions

($\varepsilon_{el} = \varepsilon_n(\boldsymbol{R})$). Next, we can study the ionic dynamics; it turns out that the electronic terms $\varepsilon_n(\boldsymbol{R})$ form the potential for the ionic motion. The total wavefunction Ψ can be written as a product: $\Psi = \psi_{el}\phi_{ionic}$ (see eqns. (2.11) and (2.18)). However, such a separation of electronic and ionic terms is invalid for polaronic states: the ions actively participate in their formation.

In order to study the case of small carrier concentration it is convenient to use the so-called Holstein Hamiltonian (1959)

$$\widehat{H} = J \sum_{i,\delta} a^+_{j+\delta} a_j + \sum_q \Omega_q b^+_q b_q + \sum_{j,q} M_q a^+_j a_j e^{iqR}(b_q + b^+_{-q}) \qquad (2.38)$$

introduced to describe small polarons in semiconductors. The first term describes the kinetic energy (hopping) of the particles; the second term is the phonon energy, and the last term corresponds to the interaction. By direct comparison of eqns. (2.1), (2.8), (2.10) and eqn. (2.38) we can see that the polaronic state described by the Hamiltonian (2.38) corresponds to the energy scales opposite to these in the usual adiabatic theory; that is, to the case when the electronic energy is small relative to the vibrational scale. The Hamiltonian (2.38) was derived by Holstein explicitly for such a case.

2.6.2 Dynamic polaron

Consider what can be dubbed the "dynamic polaron" state. In this case ions do not reside, as usual, in a definite local minimum of the potential, but are instead presented with two closely spaced minima ("double-well" structure, Fig. 2.4). Such a situation indeed occurs in practice, and is especially interesting for layered structures (such as the cuprates; see Chapter 10).

Note that "double well" structure is a result of the crossing of electronic terms. The ionic configuration at this crossing corresponds to a degeneracy of the electronic states, which is a key ingredient of the Jahn–Teller effect (1937; see, for example, Landau and Lifschitz, 1977; Bersuker, 2006).

In order to describe such a case, it is convenient to use a so-called diabatic representation (see, for example, O'Malley, 1967, 1961; Kresin and Lester, 1984; see also Appendix A). In this representation we are dealing directly with the crossing

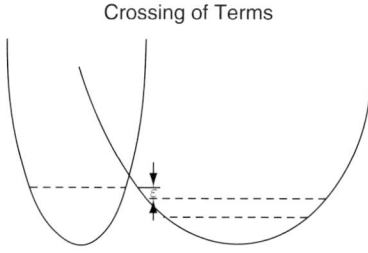

Fig. 2.4 Electronic terms (diabatic representation). (Figure reproduced from Kresin and Wolf, 1994.)

of electronic terms. The operator $\widehat{H}_{el} = \widehat{T}_r + V(r, R)$, $\widehat{T}_r$ is a kinetic energy operator, $V(r, R)$ is a total potential energy, r and R are the electronic and nuclear coordinates, correspondingly) has non-diagonal terms (unlike the usual adiabatic picture when $\widehat{H}_{el}$ is diagonal). The charge transfer in this picture is accompanied by the transition to another electronic term. Such a process is analogous to the Landau–Zener effect (see Landau and Lifshitz, 1977).

The total wavefunction can be written in the form

$$\Psi(r, R, t) = a(t)\Psi_1(r, R) + b(t)\Psi_2(r, R) \tag{2.39}$$

Here, $\Psi_i(r, R) = \psi_i(r, R)\Phi_i(R)$, $i = \{1, 2\}$, $\psi_i(r, R)$, $\Phi_i(R)$ are the electronic and vibrational wavefunctions that correspond to two different electronic terms (see Fig. 2.4).

Let us stress, at first, that it is impossible to separate the electronic and vibrational motion; one can see it directly from eqn. (2.39). Indeed, contrary to the usual adiabatic picture (see eqns. (2.11), (2.18)), the wavefunction (2.39) is not a product of the electronic and nuclear wavefunctions.

Assume that at $t = 0$ we have $a(0) = 1$ and $b(0) = 0$. The $b(t)$ is equal to (Landau and Lifshitz, 1977)

$$|b(t)|^2 = \tilde{b}^2 \left\{ 1 - \cos 2\left[\left(\frac{\varepsilon^2}{4}\right) + \left(\widehat{H}_{el|12}\right)^2\right]^{1/2} t \right\} \tag{2.40}$$

where

$$\tilde{b}^2 = \frac{L_0^2 F_{12}^2}{2\left[(\varepsilon^2/4) + L_0^2 F_{12}^2\right]} \tag{2.40a}$$

Here ε is the energy difference (see Fig. 2.4), and $\widehat{H}_{(el|12)}$ is the matrix element, describing the transition between the terms (in the diabatic representation). One can show (see Appendix A) that $\widehat{H}_{(el|12)}$ can be written in the term:

$$\widehat{H}_{(el|12)} \equiv L_0 F_{12} \tag{2.41}$$

where

$$L_0 = \int dr\, \psi_2^*(r, R)\widehat{H}_{el}\psi_1(r, R)|_{R_0} \tag{2.42}$$

is the electronic constant (R_0 corresponds to the crossing configuration), and

$$F_{12} = \int dR\, \Phi_2^*(R)\Phi_1(R) \tag{2.43}$$

is the Franck–Condon factor describing the overlap of the vibrational wavefunctions.

The probability $[b(t)]^2$ oscillates according to eqn. (2.40). Its average value, which corresponds to equilibrium-doped state, is equal to $\tilde{b}^2$ (see eqn. (2.40a)).

If terms are asymmetric ($\varepsilon \neq 0$) the quantity $\tilde{b}^2$ depends on the Franck–Condon factor. The presence of the Franck–Condon factor is essential, and leads to a peculiar isotope effect in the high T_c cuprates and manganites (see Chapters 10 and 12).

The described dynamic polaronic effect (see eqns. (2.39) and (2.40)) corresponds to the dynamic Jahn–Teller effect (see, for example, Bersuker, 2006; Salem, 1966; see also Appendix B). It is essential for studying the superconducting state in layered materials, and especially, in the high-T_c cuprates (see Sections 3.8 and 3.10).

3
Phonon mechanism

The superconducting state arises from an attractive interaction between electrons. At first glance it seems strange, because we are accustomed to seeing electrons repel each other. However, the electrons are located inside a crystal, and the presence of such a medium can change the sign of the interaction. Qualitatively, it could be understood as a negative sign of the dielectric constant. The discovery of the isotope effect which preceded the creation of the microscopic theory, and tunneling spectroscopy (see Chapter 6), show directly that the pairing can be provided by electron–lattice interaction. Its strength must be sufficient to overcome the Coulomb repulsion forces. The latter are actually somewhat weakened by a peculiar logarithmic factor (see Section 3.4.3). Still, the inter-electron attraction mediated by phonon exchange must be sufficiently strong: its energy must be of the order of ε_F, since the Coulomb repulsion corresponds to this energy scale.

There are two fundamental questions, which should be addressed before introducing the main equations of the microscopic theory and evaluating the major parameters. The first question is formulated above and is concerned with the scale of the electron–phonon interaction relative to the Coulomb repulsion.

For superconductors the interplay between the electron–lattice and Coulomb interactions leads to the resultant electron–electron attraction and formation of the bound electronic pairs. The pairing occurs for any small electron–electron attraction, and it should also be explained. Indeed, according to quantum mechanics, the formation of a bound state requires the attractive potential to exceed some threshold. The absence of such a threshold in metals forms the essence of the Cooper theorem (Cooper, 1956; see also, for example, Ashcroft and Mermin, 1976). Qualitatively, this is due to the fact that the pairing occurs for the electrons moving on the Fermi surface. The presence of the Fermi surface leads to the two-dimensional picture. As for the two-dimensional scenario, we know (see, for example, Landau and Lifshitz, 1977) that in this case any weak attraction leads to the formation of a bound state.

3.1 Superconductivity as a "giant" non-adiabatic phenomenon

Here we focus on the scale of the electron–phonon interaction. Indeed, one may wonder whether the interaction is capable of producing such a large effect ($\sim \varepsilon_F$). It turns out that the answer is "yes", and the reason has to do with the presence of strong non-adiabaticity.

Consider electrons near the Fermi level and their transitions between energy states in the region $\sim \tilde{\Omega}(\tilde{\Omega} \cong \Omega_D)$ near the Fermi level, so that $\Delta\varepsilon < \tilde{\Omega}$. Transitions of this kind are non-adiabatic, since the change in the electronic energy is smaller than the characteristic phonon energy. As we shall see, these non-adiabatic transitions are responsible for the observed large effect.

Let us estimate the strength of the phonon-mediated electron–electron interaction. The matrix elements of interest, M_{fi}, can be written in the form:

$$M_{fi} = \frac{\left|\hat{H}_1\right|_{fk}\left|\hat{H}_1\right|_{ki}}{\varepsilon_{p_1} - \varepsilon_{p_3} - \Omega_q} + \frac{\left|\hat{H}_1\right|_{fl}\left|\hat{H}_1\right|_{li}}{\varepsilon_{p_2} - \varepsilon_{p_4} - \Omega_q} \qquad (3.1)$$

Here $i \equiv \{p_1, p_2\}$ and $f \equiv \{p_3, p_4\}$ the initial and final states, and k,l are the virtual (intermediate) states, $\boldsymbol{p}_i$ are the corresponding momenta. The following conservation rules hold: $\boldsymbol{p}_3 - \boldsymbol{p}_1 = \boldsymbol{p}_2 - \boldsymbol{p}_4 = \boldsymbol{q}, \varepsilon_{p_1} + \varepsilon_{p_2} = \varepsilon_{p_3} + \varepsilon_{p_4}$, and $\boldsymbol{q}$ is the phonon momentum. The electron–phonon interaction is described by the operator $\hat{H}_1$ (see eqns. (2.25) and (2.26)); that is, by the first term in the total Hamiltonian $\hat{H}_{na}$.

Note that expression (3.1) contains terms non-diagonal in the operator $\hat{H}_1$. These non-diagonal terms are finite. Since we are considering virtual phonon exchange processes, the conservation of energy for the elementary act of electron–phonon scattering does not need to be held. The main contribution to the interaction comes from the non-adiabatic region. Unlike the case of small molecules, the existence of such a region is guaranteed by the continuous nature of the electron and phonon spectra.

Let us estimate the matrix element (3.1) in this region. Using eqn. (2.26), we obtain

$$M_{fi} \approx \frac{(\tilde{\Omega}/k)^2}{\Omega} = \frac{\tilde{\Omega}}{k^2} = \frac{\tilde{\Omega}}{(\tilde{\Omega}/\varepsilon_F)} \approx \varepsilon_F \qquad (3.2)$$

Hence the phonon-exchange interaction is very strong ($\sim \varepsilon_F$). This is due to the large contribution of $\hat{H}_1$, eqn. (3.2), which comes from the non-adiabatic domain $\Delta\varepsilon_{el} \lesssim \Omega_D$. Therefore, indeed, we are dealing with a giant non-adiabatic effect.

The semi-qualitative arguments presented above explain how a strong attraction can arise and overcome the Coulomb repulsion. A detailed quantitative analysis must be based on the methods of many-body theory, since effects of this magnitude cannot be treated rigorously by perturbation theory.

3.2 The BCS model

The Bardeen–Cooper–Schrieffer (BCS) theory (Bardeen et al., 1957) was developed in the so-called weak coupling approximation. Indeed, it is based on the Hamiltonian:

$$H_{int} = \lambda \sum_{k,k'} a^+_{k,\frac{1}{2}} a^+_{-k,\frac{1}{2}} a_{k',-\frac{1}{2}} a_{-k',\frac{1}{2}} \qquad (3.3)$$

which describes the effective electron–electron attraction, where λ is the coupling constant. The Hamiltonian (3.3) does not contain the field of bosons (phonons form such a field for conventional superconductors), which provide the attraction. Such an exclusion of the field could be justified for weak interaction only ($\lambda \ll 1$).

Note also that the BCS theory contains many universal relations such as, for example,

$$2\varepsilon(0) = 3.52\, T_c \tag{3.4}$$

$$\frac{\Delta C}{C_{n_{IT_c}}} = 1.43 \tag{3.4a}$$

Here, $\varepsilon(0)$ is the energy gap at $T = 0$ K, $\Delta C = C_s - C_n$ is a jump in heat capacity at T_c ($C_n = \gamma T_c$). This universality is not surprising, because the starting Hamiltonian (3.3) contains only a single parameter, the coupling constant λ. Instead of λ one can select another parameter which can be measured experimentally, namely, the value of the critical temperature, T_c. As a result, all quantities can be expressed in terms of T_c, and this leads to the mentioned universality.

3.3 Phonon mechanism: main equations

As mentioned previously, the BCS model describes the Fermi gas with weak attraction. The attraction is provided by exchange by bosons between the electrons (for example, by phonons). Strictly speaking, the nature of this attraction is not essential in the BCS model and enters the theory only as the corresponding energy scale; it determines the cut-off. Nevertheless, specific mechanisms of superconductivity are defined by the type of the bosons. This chapter and Chapters 4 and 5 are devoted to descriptions of various mechanisms.

The rest of this chapter is concerned with the phonon mechanism. The purpose of the analysis is two-fold. First there will be demonstrated that for weak electron–phonon interaction we obtain the same result as in the BCS theory. Second—and this is very important—the analysis will allow us to go beyond the weak coupling approximation. Indeed, there are many superconducting materials where the electron–phonon interaction responsible for pairing is not weak, and their properties should be described properly.

As noted previously, the BCS theory contains many universal relations, such as eqns. (3.4) and (3.4a). All properties could be described by a single parameter: namely, by the value of the critical temperature, T_c. For the strong coupling case the situation is different. The universality is lost, because, in addition to T_c, the properties depend also on the phonon spectrum, and more specifically on the value of the characteristic phonon frequency.

As a first step, let us formulate the main equations describing the phonon mechanism. We employ the thermodynamic Green's function formalism (see, for example, Abrikosov et al., 1975; Lifshitz and Pitaevsky, 2002), because our special focus will be on the evaluation of the critical temperature. The pairing is described by the diagram for the so-called pairing self-energy part (Fig. 3.1.); F is the pairing Green's function introduced by Gor'kov (1958), and Γ is the total vertex describing the pairing interaction. For the phonon mechanism $\Gamma \cong D$, D is the phonon propagator, with the form:

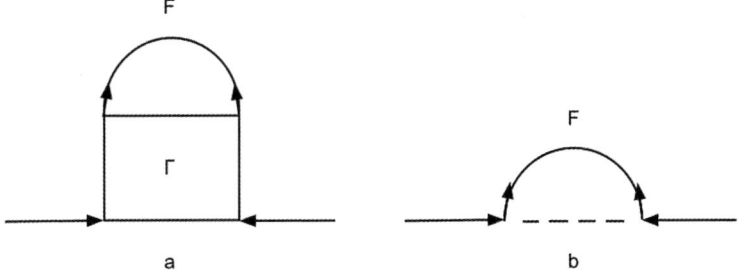

Fig. 3.1 a) Pairing self-energy part; b) phonon mechanism, $\Gamma = D$.

$$D(\Omega, \omega_n - \omega_{n'}) = \frac{\Omega^2}{\Omega^2 + (\omega_n - \omega_{n'})^2} \quad (3.5)$$

where Ω is the phonon frequency, and $\omega_n = (2n+1)\pi T$.

One should also consider the effect described in Section 2.4. An electron moving through the crystal lattice polarizes the latter. Of course, this polarization also acts back on the electron and affects its motion. In quantum language, polarization can be described as emission and subsequent absorption of virtual phonons. The self energy part Σ_1, depicted in Fig. 2.1, describes this process.

Green's function G and the pairing function F are given by the diagrammatic equations (Fig. 3.2).

The pairing is described by the following analytical equation (Eliashberg, 1960, 1961)

$$\Delta(\omega_n, T) Z = T \sum_{\omega_{n'}} \int d\mathbf{k}' |\tilde{\zeta}|^2 D(\omega_n - \omega_{n'}; \Omega) \frac{\Delta(\omega_n, T)}{\omega_{n'}^2 + \xi_{k'}^2 + \Delta^2(\omega_n, T)} \quad (3.6)$$

Here the D-function is defined by eqn. (3.5), $\tilde{\zeta} = \tilde{\zeta}_{k,k'}$ is the matrix element (see eqn. (2.27)), $\xi_k'^2$ is the electron energy referred to the Fermi level, and $\Delta(\omega_n, T)$ is the order

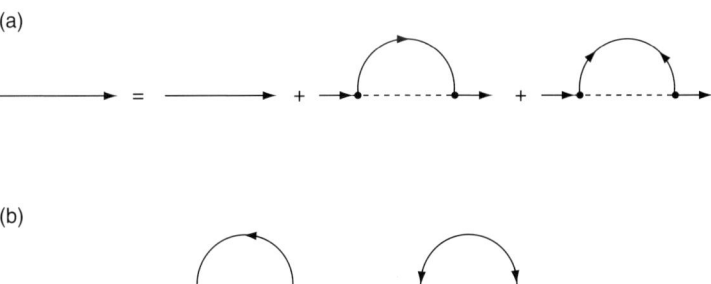

Fig. 3.2 Diagrammatic equations for a) Green's function G, and b) pairing function F^+.

parameter. One can make a transformation to the integration over phonon momentum $q = k' - k$ (see eqns. (2.27) and (2.28)), and then to the phonon frequency, and we obtain

$$\Delta(\omega_n) Z = \pi T \sum_{n'} \int d\Omega \frac{g(\Omega)}{\Omega} D(\omega_n - \omega_{n'}, \Omega) \frac{\Delta(\omega_{n'})}{|\omega_{n'}|}, \quad g(\Omega) = \alpha^2(\Omega) F(\Omega) \quad (3.7)$$

Here, $\Delta(\omega_n)$ is the so-called thermodynamic pairing order parameter; the function $g(\Omega)$ is defined by eqns. (2.29) and (2.30).

The expression for Z has the form:

$$Z = 1 + \pi T \sum_{\omega_{n'}} \int d\Omega \frac{g(\Omega)}{\Omega} D(\omega_n - \omega_{n'}, \Omega) \frac{\omega_{n'}}{|\omega_{n'}|} \quad (3.8)$$

Equations (3.7) and (3.8) are written for $T = T_c$. For $T < T_c$ the substitution $|\omega_n| \to \left[\omega_n^2 + \Delta^2(\omega_n)\right]^{1/2}$ has to be made in the denominator. As noted previously, $\omega_n = (2n+1)\pi T$.

Equation (3.7) contains a very important function, $g(\Omega) = \alpha^2(\Omega) F(\Omega)$; see eqn. (2.29). Here, $\alpha^2(\Omega)$ describes the electron–phonon interaction, and $F(\Omega)$ is the phonon density of states. One can also introduce the quantity $\lambda = \int d\Omega g(\Omega) \Omega^{-1}$ (see eqn. 2.33), which is called the coupling constant. Note that the coupling constant does not directly appear in eqns. (3.7) and (3.8). It arises explicitly, for example, if we consider a model with a single phonon frequency; for example, a delta-function like peak with $g(\Omega) = \lambda \Omega \delta(\Omega - \Omega_1)$.

We can also consider the case of two peaks. In this case the right-hand side of eqn. (3.7) will consist of two terms, each containing the coupling constant of a single peak:

$$\alpha^2(\Omega) F(\Omega) = \frac{\lambda_1 \Omega_1}{2} \delta(\Omega - \Omega_1) + \frac{\lambda_2 \Omega_2}{2} \delta(\Omega - \Omega_2) \quad (3.9)$$

In this case, $\lambda = \lambda_1 + \lambda_2$.

Bergmann and Rainer (1973) studied the impact of different frequency parts of g(Ω) on T_c, and proved that all regions, including the low-frequency part ($\Omega < T_c$), make a positive contribution to T_c.

In real life one deals with complicated phonon spectra, and the superconducting parameters correlate directly with the function $\alpha^2(\Omega) F(\Omega)$. Strictly speaking, it is not possible to rewrite eqns. (3.7) and (3.8) in a form which would explicitly contain the coupling constant. Nevertheless, the coupling-constant concept remains very useful. This is due to the differences in the behavior of the phonon Green's function (3.5) and of the function $g(\Omega) = \alpha^2(\Omega) F(\Omega)$. Green's function D is a relatively smooth function of frequency. The behavior is usually quite different for the function $g(\Omega)$, and especially for the phonon density-of-states factor $F(\Omega)$. The latter typically contains a number of rather sharp peaks whose origin is as follows.

The phonon spectrum of a solid consists of several branches corresponding to longitudinal and transverse acoustic and optical phonons. At small wavevectors q, the acoustic branches have simple linear dispersion laws; for example, $\Omega_{\text{tr}} = u_{\text{tr}} q$. For large q, on the other hand, there are deviations from the linear behavior giving rise to

Phonon mechanism: main equations

high density of states regions ($q^2 dq/d\Omega$ is large). In these regions the frequency varies only weakly with the momentum, with the result that the density of states becomes very large, as we have stated. These regions, where the function $g(\Omega) = a^2(\Omega)F(\Omega)$ contains peaks, make the strongest contribution to electron pairing.

Let us now consider the factor $g(\Omega)D(\Omega)/\Omega$. Breaking up the integration over frequencies into separate intervals containing different peaks, and making use of the smoothness of the phonon Green's function, we can write the right-hand side of eqn. (3.9) as a sum of terms of the form $\lambda_i D(\Omega_i)$, where the coupling constants are defined by expressions of the type (2.33), each referring to a different peak.

For many materials one can introduce an effective average phonon frequency, thereby casting eqn. (3.9) into a form corresponding to a single characteristic frequency and a single coupling constant defined by (2.33). As was shown by Louie and Cohen (1977), the function $K_{n-n'} = 2 \int d\Omega g(\Omega)\Omega \times [\Omega^2 + (n-n')^2(2\pi T)^2]^{-1}$ can be replaced by $K^0_{n-n'} \approx \lambda D(\tilde{\Omega}, \omega_n - \omega_{n'})$, where the D-function and $\tilde{\Omega}$ are defined by eqns. (3.5) and (2.31) respectively. With high accuracy: $K_n = K^0_n(1-r)$ with $r \ll 1$, $(r \propto \delta = \frac{\tilde{\Omega}}{\langle\Omega\rangle} - 1)$. As a result, eqn. (3.7) can be written in the form:

$$\Delta(\omega_n)Z = 1 + \pi T\lambda \sum_{\omega'_n \geq 0} D\left(\omega_n - \omega_{n'}; \tilde{\Omega}\right) \frac{\Delta |\omega_{n'}|}{|\omega_{n'}|} \tag{3.10}$$

The phonon propagator is defined by eqn. (3.5). Equation (3.8) for the renormalization function Z can be written in a similar way:

$$Z = 1 + \pi T\lambda \sum_{\omega_{n'} \geq 0} D(\omega_n - \omega_{n'}; \tilde{\Omega})\frac{\omega_{n'}}{|\omega_{n'}|} \tag{3.11}$$

Equations (3.7) and (3.8), and correspondingly, (3.10) and (3.11), are the main equations of the strong coupling theory. They are valid for any value of the coupling constant and, therefore, for any relation between T_c and $\tilde{\Omega}$. Note, however, that in agreement with the Migdal theorem (Section 2.5), we neglected the higher-order corrections of the vortex; such approximation is justified if $\tilde{\Omega} \ll E_F$. This is the only condition of the applicability of the theory.

The limit of weak coupling deserves special attention, both for its own sake and in order to establish contact with the usual formalism of the BCS theory. This limit corresponds to $\lambda \ll 1$. As can be shown self-consistently from the results of weak-coupling theory, this means that $\pi T_c \ll \tilde{\Omega}$, where $\tilde{\Omega}$ corresponds to the important short-wavelength part of the phonon spectrum. In the weak-coupling approximation ($\lambda \ll 1$) we may set $Z \cong 1$ (see eqn. (3.11)). Furthermore, we can neglect the dependence of the phonon Green's function on ω_n; this approximation is accurate to $\sim(T_c/\tilde{\Omega})^2$. Then $\Delta(\omega_n)$ also becomes independent of ω_n, and we arrive at the equation

$$1 = \lambda T \sum_{n \geq 0} \frac{D(\omega_n, \tilde{\Omega})}{|\omega_n|}\bigg|_{T_c} \tag{3.12}$$

The function D guarantees the convergence of the summation at the upper limit by effectively cutting off this summation at $\tilde{\Omega}$. To calculate T_c we can either directly employ eqn. (3.12) (see Section 3.4.1), or we can approximate $D \cong 1$ and terminate the summation at $\tilde{\Omega}$. This results in the well-known equation of the BCS theory:

$$T_c \cong \tilde{\Omega} \exp(-1/\lambda) \qquad (3.13)$$

The weak-coupling theory is pre-exponentially accurate: the approximation $D \cong 1$ affects the pre-exponential factor. Hence the introduction of the coupling constant λ is valid to the same accuracy. We will discuss the weak coupling case in more detail in the next section.

3.4 Critical temperature

Equations (3.10) and (3.11) are valid for any strength of the electron–phonon interaction. Based on these equations, let us evaluate the main parameter of the theory: the critical temperature.

The value of T_c is determined by several quantities: $T_c = T_c(\lambda, \mu^*, \tilde{\Omega})$, where λ describes the strength of the electron–phonon coupling, $\tilde{\Omega}$ denotes the characteristic phonon frequency and sets the energy scale (in a rough approximation, $\tilde{\Omega} \cong \tilde{\Omega}_D$, where $\tilde{\Omega}_D$ is the Debye temperature), and μ^* describes the Coulomb repulsion (see Section 3.4.3). Curiously, it will turn out that the specific forms of the analytic expressions for T_c are different for different coupling strengths.

3.4.1 Weak coupling

We consider first the case of weak electron–phonon interaction. This case corresponds to the original BCS model.

In this section we employ eqn. (3.6) to consider the weak coupling case. This approach allows us to avoid forcing a cut-off at the characteristic phonon energy (it will take place automatically); and in addition, the pre-exponential accuracy of the model becomes apparent immediately.

As has been mentioned, eqn. (3.6) serves as our starting point (Kresin, 1972). The expression for the constant $\tilde{\zeta}^2$ describing the electron–phonon interaction can be written in the following form (Geilikman, 1971):

$$\tilde{\zeta}^2 = \zeta \frac{u^2 q^2}{\Omega_j^2(q)} p_j(q) \qquad (3.14)$$

Here u is the speed of sound, ζ is the Fröhlich parameter (Fröhlich (1950), Geilikman (1975)), q is the phonon momentum, and $p_j(q) \sim 1$. It is useful to separate out the dependence of the electron–phonon matrix element on the phonon momentum and frequency; see, for example, Ziman (1960). Note the important fact that $\tilde{\zeta}^2$ depends on the phonon frequency; this dependence vanishes only for a purely acoustic dispersion law. For simplicity we are not explicitly including the Coulomb interaction; it can be taken into account in the usual manner. In eqn. (3.6) we are summing over all phonon branches j. Henceforth, we will omit the index j.

From eqn. (3.6), after going over to integration over the phonon momentum and ξ, we obtain:

$$\Delta(\omega_n, T) = \frac{\zeta T}{2p_F^2} \int_0^{k_1} q dq \frac{u^2 q^2}{\Omega^2(q)} \rho(q) \sum_{n'} \int d\xi \frac{\Omega^2(q)}{\Omega^2(q) + (\omega_n - \omega_{n'})^2} \qquad (3.15)$$
$$\times \frac{\Delta(\omega_{n'}, T)}{\omega_{n'}^2 + \xi^2 + \Delta^2(\omega_{n'}, T)}, \quad k_1 = \min\{2p_F, q_{max}\}$$

In the weak-coupling approximation, $\zeta \ll 1$, and as a result one can neglect the direct dependence of $\Delta(\omega_n, T)$ on ω_n and take $\Delta(\omega_n, T) = \varepsilon(T)$, where $\varepsilon(T)$ is the energy gap (the correction turns out to be of the order of $\tilde{\lambda} T_c^2/\Omega_D^2$. Since $\varepsilon(T=T_c) = 0$, the critical temperature is determined by the equation

$$1 = \frac{\zeta T}{2p_F^2} \int_0^{k_1} q dq \frac{u^2 q^2}{\Omega^2(q)} \rho(q) \sum_{n'} \int d\xi \frac{\Omega^2(q)}{\Omega^2(q) + \omega_{n'}^2} \frac{1}{\omega_{n'}^2 + \xi^2}\Big|_{T=T_c} \qquad (3.16)$$

The BCS model is recovered if in eqn. (3.16), in addition to assuming acoustic dispersion, we replace the phonon Green's function by unity in the interval $0 < \omega_n < \Omega_D$;. In the approach based on eqn. (3.15), the presence of the D-function automatically removes the logarithmic divergence.

Now summing over $\omega_{n'}$ and integrating over ξ, we find

$$1 = \frac{\zeta}{2p_F^2} \int_0^{k_1} q dq \frac{u^2 q^2}{\Omega^2(q)} \rho(q) \ln\frac{2\Omega(q)\gamma}{\pi T_c} \qquad (3.16a)$$

Introducing an auxiliary constant $\tilde{\Omega}$ such that

$$\ln\frac{2\Omega(q)\gamma}{\pi T_c} = \ln\frac{2\Omega(q)\gamma}{\pi \tilde{\Omega}} - \ln\frac{\tilde{\Omega}}{T_c} \qquad (3.16b)$$

we arrive, with the help of eqn. (3.16a), at the following expression for the critical temperature:

$$T_c = \tilde{\Omega} \exp(-\frac{a+1}{\lambda}), \qquad (3.17)$$

$$a = \frac{\zeta}{2p_F^2} \int_0^{k_1} q dq \frac{u^2 q^2}{\Omega^2(q)} \rho(q) \ln\frac{2\gamma \Omega(q)}{\pi \tilde{\Omega}} \qquad (3.17a)$$

$$\lambda = \frac{\zeta}{2p_F^2} \int_0^{k_1} q dq \frac{u^2 q^2}{\Omega^2(q)} \rho(q). \qquad (3.17b)$$

Note that T_c does not depend on the choice of $\tilde{\Omega}$, which can therefore be chosen arbitrarily. We choose $\tilde{\Omega}$ so as to minimize a (for instance, for acoustic dispersion $\tilde{\Omega} \cong \Omega_D$, while in the case of a sharp peak at $\Omega = \Omega_1$ in the phonon density of states, $\tilde{\Omega} = 2\gamma\Omega_1/\pi$). This leads us to the following result:

$$T_c = \tilde{\Omega} \exp(-1/\lambda) \qquad (3.18)$$

where λ is given by eqn. (3.17b).

Equation (3.18) is the result for T_c in the weak-coupling limit ($\lambda \ll 1$). It is clear that this result is pre-exponentially accurate. Indeed, the presence of the factor a in eqn. (3.17) is equivalent to a change in the pre-exponential factor.

Generally speaking, the critical temperature is very sensitive to the character of the phonon spectrum; the dominant contribution comes from the frequency dependence of the denominator in the exponent.

Note that it is only if the dispersion is purely acoustic, $\omega = uq$ (in this case, $\gamma = 1$), that the exponent does not depend on the phonon frequency. Indeed, in this case (with $2p_F < q_D$), eqns. (3.17) and (3.18) give

$$T_c = \tilde{\Omega} \exp\left(-1/\zeta\right) \qquad (3.19)$$

(We used the fact that $\zeta = \beta p_F$.) Thus the usual BCS expression is recovered only in the weak-coupling approximation, and furthermore, only under the assumption of an acoustic dispersion law.

On the other hand, if the dispersion law $\Omega(q)$ is not acoustic over the entire range of q, as in the case real materials, then, according to eqn. (3.17b), T_c is very sensitive to the form of the function $\Omega(q)$. Here this sensitivity is more pronounced than in the case of strong coupling, because in the weak-coupling regime, renormalization effects play only a small role.

If the phonon density of states is described by a delta-function-like peak at a frequency Ω_1, the effective coupling constant is $\lambda \propto \Omega_1^{-2}$.

3.4.2 Intermediate coupling ($\lambda \lesssim 1.5$)

In the preceding section we discussed the case of weak coupling ($\lambda \ll 1$). If the superconductor is characterized by a larger coupling constant ($\lambda \cong 1$), the approximations employed there are no longer valid. In particular, we can no longer neglect the dependence of Δ on ω_n. An analysis of this case was carried out by McMillan (1968a), who made use of eqn. (3.10) and computed the critical temperature for a phonon spectrum of the type found in Nb. This phonon spectrum (or, more precisely, the phonon density of states F(Ω)) is known from neutron spectroscopy data, and has a structure typical of many metals. This numerically calculated T_c can be fitted analytically by the following formula:

$$T_c = \frac{\Theta_D}{1.45} \exp\left(-\frac{1.04}{\rho - \frac{\mu^*}{1+\lambda} - 0.62\mu^*\rho}\right). \qquad (3.20)$$

Here, μ^* is the Coulomb pseudopotential (see the next section), and $\rho = \lambda(\lambda + 1)^{-1}$.

Dynes (1972) modified the McMillan equation, replacing the Debye temperature by a characteristic frequency Ω_{ch}. As a characteristic frequency Ω_{ch} one can select the value $\Omega_{ch} = <\Omega>$ (see, for example, E. Wolf, 2012); the average is defined by

$$<f(\Omega)> \equiv \frac{2}{\lambda} \int_0^\infty \alpha^2 F(\Omega) f(\Omega)^{-1} d\Omega \qquad (3.21)$$

The equation for T_c then reads:
$$T_c = \frac{<\Omega>}{1.2} \exp\left(-\frac{1.04(1+\lambda)}{\lambda - \mu^*(1+0.62\lambda)}\right). \tag{3.22}$$

We have already mentioned that the McMillan equation was obtained as a fit to the numerical solution for a Nb-like phonon spectrum; since this spectrum is rather typical, this equation has been applied with great success to many superconductors (see, for example, Allen and Dynes, 1975; E. Wolf, 2012).

The paper by McMillan (1968a) also contains a useful expression for the electron–phonon coupling constant:
$$\lambda = \frac{\nu_F <I^2>}{M <\Omega^2>} \tag{3.23}$$

where $<I>$ is the average matrix element of the electron–phonon interaction, ν_F is the density of electronic states, and M is the ion mass. It is important that the dependence of T_c on the phonon frequency is not determined solely by the pre-exponential factor: in principle, a strong dependence is also contained in the exponent itself. Note that these two dependences have opposite effects on T_c. This fact has led McMillan to propose a softening mechanism for increasing T_c. The previously accepted notion, based on the original BCS theory, was that a rise in the phonon frequency serves only to increase the pre-exponential factor and, consequently, increases T_c. The dependence (3.23) shows that a rise in the phonon frequency may, generally speaking, lead to the opposite result as well.

The softening mechanism—that is, the decrease in the effective phonon frequency—can lead to an increase in the value of the critical temperature. This has been observed for a number of compounds. For example, the solid solution $Al_{1-x}Si_x$ obtained by substituting Si for Al (Sluchanko et al., 1995) is characterized by an increase in T_c (1.2 K → 11 K). Such a drastic increase is caused by phonon softening (Chevrier et al., 1988).

As mentioned previously, the McMillan equation is valid for $\lambda \leq 1.5$. In connection with this fact, it is interesting to note that from the expression $\lambda = a<\Omega>^{-2}$ (here $a = <I^2>\nu_F M^{-1}$; see eqns. (3.22) and (3.23)) it appears that there exists an upper limit on the attainable value of T_c; this was pointed out in the original paper by McMillan. Indeed, neglecting μ^* for simplicity and calculating $\partial T_c/\partial <\Omega>$, we find $T_c^{max} \cong \Omega_m \exp(-3/2)$, where $\Omega_m \cong (a/2)^{1/2}$. This maximum value corresponds to $\lambda = 2$.

This conclusion, however, has to be taken with a great deal of caution. The fact of the matter is that the value $\lambda = 2$ is outside the range where the McMillan equation is applicable; this is even more so for higher values of λ. We will return to this point later when we discuss the expression for T_c, which holds for large values of λ.

3.4.3 Coulomb interaction

In addition to the attraction mediated by phonon exchange, we must keep in mind the presence of the Coulomb repulsion. It turns out that an important aspect of this problem is the phenomenon of logarithmic weakening of the repulsion. This factor is associated with the difference in the energy scales of the attraction and repulsion effects. The attraction is important in an energy interval $\sim \tilde{\Omega} \cong \Omega_D$. The repulsion, on the other hand, is characterized by the electronic energy scale $\sim \varepsilon_F$.

Let us consider this problem in more detail (Tyablikov and Tolmachev (1958), Bogoluybov et al. (1959), Abrikosov and Khalatnikov (1959), Morel and Anderson (1962), McMillan (1968a), and Geilikman et al. (1975)). The equation for the order parameter $\Delta(\omega_n)$ in the presence of Coulomb repulsion can be written as follows(at $T = T_c$):

$$\Delta(\omega_n) = \frac{1}{1 - f^n/\omega_n} \pi T \sum_{\omega_{n'}} \left[\lambda D\left(\omega_n - \omega_{n'}, \tilde{\Omega}\right) - \frac{1}{2} V_c \Theta\left(\varepsilon_0 - |\omega_{n'}|\right) \right] \frac{\Delta(\omega_{n'})}{|\omega_{n'}|} \quad (3.24)$$

The second term in square brackets is the one describing the Coulomb force $(\varepsilon_0 \cong \varepsilon_F)$. Recall that the phonon Green's function D provides a cut-off at energies $\tilde{\Omega}$. We can seek the solution in the following form:

$$\Delta(\omega_n) = \Delta_0 \frac{\tilde{\Omega}^2}{\tilde{\Omega}^2 + \omega_n^2} + \Delta_\infty \frac{\omega_n^2}{\tilde{\Omega}^2 + \omega_n^2} \quad (3.24a)$$

The second term reflects the presence of the repulsion. It results in the order parameter continuing outside the region $\omega_n \simeq \tilde{\Omega}$.

The quantities Δ_0 and Δ_∞ are defined by

$$\Delta_0 = \rho \pi T \sum_{\omega_{n'}} \frac{\tilde{\Omega}^2}{\tilde{\Omega}^2 + \omega_{n'}^2} \frac{\Delta(\omega_{n'})}{|\omega_{n'}|} - \frac{V_c}{1+\lambda} \pi T \sum_{\omega_{n'}}^{\varepsilon_0} \frac{\Delta(\omega_{n'})}{|\omega_{n'}|}, \quad \Delta_\infty = -V_c \pi T \sum_{\omega_{n'}} \frac{\Delta(\omega_{n'})}{|\omega_{n'}|}. \quad (3.25)$$

Substituting eqn. (3.24a) and eliminating the constants Δ_0 and Δ_∞, we arrive at the following result for T_c:

$$T_c = 1.14 \tilde{\Omega} \exp\left(-\frac{1 + 0.5\rho - 0.35\rho^2 + 0.8\rho\left[\mu^*/(1+\lambda)\right] + 0.4\mu^{*2}}{\rho - [\mu^*/1+\lambda] - 0.5\rho\mu^* - 1.5\mu^*\rho^2}\right). \quad (3.26)$$

Here,

$$\mu^* = \frac{V_c}{1 + V_c \ln(\varepsilon_0/\tilde{\Omega})} \quad (3.27)$$

is the so-called Coulomb pseudopotential. It contains the large logarithmic factor $\sim \ln(\varepsilon_0/\tilde{\Omega})$, which reduces the contribution of the Coulomb repulsion. Usually, $\mu^* \cong 0.1 - 0.15$, though its value might be different.

For the weak coupling case ($\lambda \ll 1$) we obtain

$$T_c = \tilde{\Omega} e^{-\frac{1}{\lambda - \mu^*}} \quad (3.28)$$

For the intermediate case, eqn. (3.26) leads to the expression which is close to eqn. (3.22).

The presence of the large logarithmic factor and corresponding effective weakening of the Coulomb repulsion is very beneficial for the appearance of superconductivity.

It reflects the difference in the energy scales for the attraction (characteristic phonon frequency) and the repulsion (electronic energy, $\sim \varepsilon_F$).

Note that if the superconductor contains, in addition to the phonon subsystem, some high-energy electronic excitations which provide an additional attraction, this leads to an effective decrease in μ^*. We will discuss this question in more detail in Chapter 4.

3.4.4 Very strong coupling

We now consider the case of very strong coupling ($\lambda \gg 1$). We shall see that this regime corresponds to the criterion $2\pi T_c \gg \tilde{\Omega}$.

This case was first analyzed by Allen and Dynes (1975) by means of numerical calculations (for $\mu^* = 0$). An analytical treatment, based on a direct solution of eqn. (3.10), was developed by Kresin et al. (1984).

It is convenient to use the dimensionless form of the eqns. (3.10) and (3.11). It turns out that for $\lambda \gg 1$ the function $\Delta(n)$ decays rapidly. In this case it is very useful to employ the matrix method (Owen and Scalapino, 1971). The equation, which determines T_c, can be written in the following form:

$$\phi_n = \sum_{m \geq 0} \tilde{K}_{nm} \phi_m, \tag{3.29}$$

$$\tilde{K}_{nm} = \frac{\alpha}{(2n+1)^{1/2}(2m+1)^{1/2}} F_{n,m,\nu}, \tag{3.29a}$$

where

$$F_{n,m,\nu} = \frac{1}{\nu^2 + (n-m)^2} + \frac{1}{\nu^2 + (n+m+1)^2} - \delta_{nm} \sum_{i=0}^{2n} \frac{1}{\nu^2 + (n-1)^2} \tag{3.29b}$$

where $\phi_n = \Delta_n/(2n+1)^{1/2}$, $\nu = \tilde{\Omega}/2\pi T_c$, $\tilde{\Omega} = <\Omega^2>^{1/2}$, $\alpha = \lambda \nu^2$, and the coupling constant λ is defined, as usual, by eqn. (2.33).

It should be pointed out that the presence of the renormalization function Z (corresponding to the last term on the right-hand side of eqn. (3.29b)) plays an important role: it results in a cancellation of the diagonal term $n = m$.

Equation (3.29) is valid for any λ. The case of very strong coupling corresponds to $\nu \ll 1$, and therefore the quantity ν can be neglected. In the zeroth-order approximation, only $\phi_0 \neq 0$. Then eqn. (3.29) gives $\tilde{K}_{00} = 1$, where, according to eqns. (3.29a) and (3.29b), $\tilde{K}_{00} = \alpha$. We then obtain

$$T_{c0} = (2\pi)^{-1} \lambda^{1/2} \tilde{\Omega} \tag{3.30}$$

In the next approximation we keep more terms. Specifically, we solve the equation $\det \tilde{M} = 0$, where the matrix $\tilde{M}$ is defined by $\tilde{M} = \tilde{K} - \tilde{1}$. Then $\tilde{M}_{00} = \alpha - 1$, $\tilde{M}_{01} = \tilde{M}_{10} = 0.72\alpha$, and $\tilde{M}_{11} = -0.63\alpha - 1$. After a simple calculation we obtain $\alpha = 0.785$, and based on eqn. (3.29) we arrive finally at the following expression:

$$T_c = 0.18 \lambda^{1/2} \tilde{\Omega} \tag{3.31}$$

The next iteration changes the coefficient in eqn. (3.31) by only a negligible amount.

It is remarkable that even the zeroth-order approximation describes T_c with good accuracy ($\cong 13\%$).

Note that the dependence of T_c on λ and $\tilde{\Omega}$ (eqn. (3.31)) can be seen directly from eqn. (3.10). Indeed, if $\pi T_c \gg \tilde{\Omega}$, then one can neglect $\tilde{\Omega}^2$ in the denominator of the phonon Green's function (eqn. (3.5)), and then see directly the scaling behavior $T_c \propto \lambda^{1/2}\tilde{\Omega}$.

One can see from (3.31) that the dependence T_c on λ is in this case drastically different from that for the weak coupling limit (eqn. 3.13). Equation (3.31) is valid for $\lambda \gtrsim 5$.

It is interesting to consider the effect of the Coulomb interaction on T_c in the limit of strong coupling. The equation can be reduced to the form (see eqn. (3.29))

$$\phi_n = \sum_m \tilde{K}^c_{nm} \phi_m \tag{3.32}$$

$\tilde{K}^c_{nm}$ is described by eqn. (3.29a), but contains $F^c_{m,n,v} = F_{m,n,v} - 2\mu^*$.

Solving eqn. (3.32) by analogy with the case $\mu^* = 0$, we obtain

$$T_c = 0.18 \lambda_{\text{eff}}^{1/2} \tilde{\Omega} \tag{3.33}$$

$$\lambda_{\text{eff}} \simeq \lambda(1 + 2.6\mu^*)^{-1} \tag{3.33a}$$

Since $\mu^* \ll 1$, we can restrict ourselves to the linear term.

We see from eqn. (3.33) that the Coulomb term decreases the effective constant ($\lambda_{\text{eff}} < \lambda$), but this decrease differs in a striking way from that in the weak coupling approximation. In the latter case the effective constant has the well-known form: $\lambda_{\text{eff}} = \lambda - \mu^*$. In the strong coupling limit the decrease is not given by a difference, but rather is described by the ratio (3.33). This gives a stronger dependence on μ^*. If the effect of the Coulomb interaction were described by the difference $\lambda - \mu^*$, as in the case of weak coupling, this effect would be negligibly small in the limit of $\lambda \gg 1$. We see that an increase in the electron–phonon coupling is accompanied by a transition to the dependence (3.33), which is more drastic in the $\lambda \gg 1$ limit than a simple subtraction. As a result, the Coulomb term still makes a noticeable contribution despite the fact that $\lambda \gg 1$.

This change in the dependence of T_c on μ^* is connected with the features of the kernel of eqn. (3.32). In the weak coupling approximation the temperature dependence of the phonon Green's function is negligibly small; and this smallness is due to the presence of the small parameter $(T_c/\tilde{\Omega}_D)^2$. In this case the kernel contains the difference $\lambda - \mu^*$.

In the limit $\lambda \gg 1$, on the other hand, the phonon Green's function depends strongly upon T, and this dependence results in a different relation between λ and μ^*.

Therefore, in the limit of very strong coupling T_c is described by expressions (3.31) and (3.33), which are entirely different from the exponential dependences of the weak- coupling case (see eqns. (3.13) and (3.18)).

3.4.5 The general case

In previous sections we described several special cases applying to various strengths of the electron–phonon coupling. We have seen the interesting fact that the analytical expressions for T_c are very different for different coupling strengths. The weak and intermediate coupling cases are described by the exponential dependences (3.13) and (3.22), whereas the superstrong coupling case is characterized by the dependence $T_c \propto \lambda^{1/2}$, eqn. (3.31).

Clearly, it is attractive to try to obtain an expression valid for an arbitrary value of the coupling strength. Furthermore, from the practical point of view, it is important to be able to analyze the case of $1.5 \lesssim \lambda \lesssim 5$, which is outside the limits of applicability of both the McMillan equation ($\lambda \lesssim 1.5$) equation and the very strong coupling limit, eqn. (3.31), ($\lambda \gtrsim 5$). Such a universal equation has been derived by Kresin (1987a). In the following we describe the derivation and discuss its implications.

We will make use of the matrix representation (3.29). The most interesting case corresponds to $\nu^2 = \left(\frac{\tilde{\Omega}}{2\pi T_c}\right)^2 \lesssim 1$, with $\tilde{\Omega} = <\Omega^2>^{1/2}$. The region of $\nu^2 \gg 1$ can be treated analytically and is described by expression (3.33).

The problem reduces to that of solving the matrix equation: $\det\left|\tilde{K} - \tilde{I}\right| = 0$. In the region $\nu^2 \lesssim 1$ the convergence is relatively rapid, and to a 1% accuracy it is sufficient to stop at $m = 5$. The calculation can be performed for a range of values of v and the corresponding values of α determined. As a result, we obtain a dependence of $T_c/\tilde{\Omega}$ upon λ, as shown in Fig. 3.3.

With high accuracy, the dependence $T_c(\lambda)$ can be described by the following analytical expression:

$$T_c = \frac{0.25\tilde{\Omega}}{(e^{2/\lambda} - 1)^{1/2}} \tag{3.34}$$

The dependence (3.34) is the solution of the general eqn. (3.10). Indeed, it was used as a trial function, and it was then demonstrated that it satisfied eqn. (3.10) with a high degree of precision.

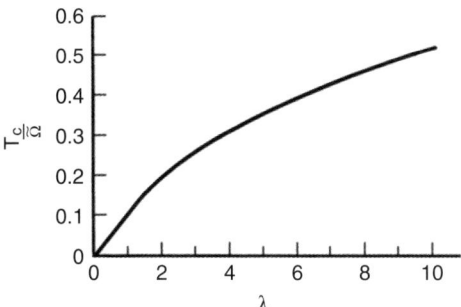

Fig. 3.3 The dependence $T_c(\lambda)$.

34 Phonon mechanism

Let us now consider the effect of the Coulomb interaction on T_c. This question can be analyzed by analogy with the derivation of eqn. (3.34). We need to solve the equation det det $|\tilde{K}_c - \tilde{I}| = 0$; see eqn. (3.32). Usually, $\mu^* \cong 0.1$; that is, $\mu^* \ll 1$. The calculation could be carried out for $\mu^* \lesssim 0.2$. The dependence obtained can be described by the following expression:

$$T_c = 0.25\tilde{\Omega}(e^{2/\lambda_{\text{eff}}} - 1)^{-1/2}, \tag{3.35}$$

where

$$\lambda_{\text{eff}} = (\lambda - \mu^*)[1 + 2\mu^* + \lambda\mu^* t(\lambda)]^{-1}. \tag{3.35a}$$

The universal function $t(\lambda)$ is presented in Fig. 3.4. This function decreases exponentially with increasing λ, and can be written in the form proposed by Tewari and Gumber (1990): $t(\lambda) = 1.5 \exp(-0.28\lambda)$. Thus T_c is described by the analytical expression (3.35), which has the same form as eqn. (3.34), with the replacement $\lambda \to \lambda_{\text{eff}}$.

Equation (3.35) explicitly describes T_c for an arbitrary value of the coupling constant λ. One can see from Table 3.1 that there is a good agreement between eqn. (3.35) and experimental data.

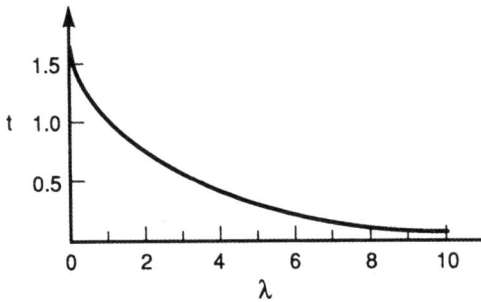

Fig. 3.4 Universal function $t(\lambda)$.

Table 3.1. The ratio $\dfrac{T_c}{\tilde{\Omega}}$

Superconductor	$\left[T_c/\tilde{\Omega}\right]_{\text{th}}$	$\left[T_c/\tilde{\Omega}\right]_{\text{exp}}$
In	0.04	0.04
$Pb_{0.4}Tl_{0.6}$	0.07	0.07
In_2Bi	0.08	0.09
Hg	0.09	0.11
Pn	0.10	0.105
$Pb_{0.9}Bi_{0.1}$	0.125	0.13
$Pb_{0.45}Bi_{0.55}$	0.16	0.15

Let us analyze various limiting cases. If $\lambda \gg 1$, the exponential factor can be expanded in powers of λ^{-1} and we find $T_c \cong 0.18\lambda^{1/2}\tilde{\Omega}$, in accordance with eqn. (3.31). In the opposite limit of weak and intermediate coupling we recover an expression essentially identical to eqn. (3.22).

Note also that based on eqns. (3.34) and (3.23) we find that $\partial T_c/\partial \tilde{\Omega} > 0$ for any λ. Hence phonon softening always results in an increase in T_c. Note that this increase slows down at large λ, and T_c saturates at the value $T_c = 0.18\eta'^{1/2}$; $\eta = <I^2> \nu_F/M$ (see eqn. (3.31)).

3.4.6 About an upper limit of T_c

The question about an upper limit of T_c for the phonon mechanism has some interesting history. Based on the Frohlich Hamiltonian which is a sum of the electronic term, the phonon term with experimentally measured phonon frequency, and the electron–phonon interaction, one can obtain the expression: $\Omega = \Omega_0(1-2\lambda)^{1/2}$ (Migdal, 1960). Based on this expression, one can conclude that the value of the coupling constant λ cannot exceed $\lambda_{\max} = 0.5$, and this implies that the value $T_c \approx 0.1\Omega(\tilde{\Omega} \approx \Omega_D)$ is the upper limit of T_c. Such a point of view, indeed, was almost generally accepted after the appearance of the BCS theory. However, very soon it became clear that something is wrong with this criterion, since there were many superconductors discovered with $\lambda > 0.5$ (for example, Sn, Pb, and Hg). The problem was clarified later by Brovman and Kagan (1967), and, especially, by Geilikman (1971, 1975).

The point is that the expression $\Omega = \Omega_0(1-2\lambda)^{1/2}$ was derived from the Frohlich model. However, the fact is that the electron–ion interaction is also crucial for the formation of a realistic phonon spectrum. In other words, in this model we are double counting. The analysis of the electron–phonon interaction and, correspondingly, the lattice stability has to be carried out with considerable care, and should be based directly on the adiabatic approximation. The conclusion is that the electron–phonon interaction does not lead to the dependence when the phonon frequency becomes equal to zero at some value of the coupling constant. Rigorous analysis shows that this interaction leads only to formation of the observed acoustic dispersion law.

Another faulty restriction on T_c based on the McMillan equation (3.20) was proposed later. Indeed, this equation leads to an upper limit of T_c. As noted previously (Section 3.4.2), the maximum value of T_c corresponds to $\lambda = 2$. However, the McMillan equation is valid only for $\lambda < 1.5$. Therefore, the value $\lambda = 2$ is outside the range of its applicability.

The treatment based on eqns. (3.34) and (3.35) leads to a very different conclusion. As noted in the previous section (see Table 3.1), experimentally large values of λ have been determined. For example, $\lambda \approx 2.1$ for $Pb_{0.65}Bi$, and $\lambda \approx 2.6$ for $Am-Pb_{0.45}Bi_{0.55}$ (Allen and Dynes (1975); see also the review by E. Wolf (2012)). If the material is characterized by relatively large values of both λ and $\tilde{\Omega}$, it might have a very high value of T_c.

3.5 Properties of superconductors with strong coupling

As has been stressed previously, the weak coupling approximation (BCS model) is remarkable for the universality of its results. The best known one relates the energy gap at $T = 0\,\mathrm{K}$ and T_c : $2\varepsilon(0) = 3.52\, T_c$.

The model yields many other universal relations, such as $\varepsilon(T)/Tc = 3.06\,[1 - T/Tc]^{1/2}_{T \to T_c}$, $\Delta C/C_n(T_c) = 1.4$, and so on; see eqns. (3.4) and (3.4a). The origin of this universality is in the fact that all superconducting properties in the BCS model are determined by the single parameter λ, the coupling constant (or, more rigorously, by $\lambda - \mu^*$). Instead of λ, one can equally well select T_c as such a parameter; indeed, it is usually much better to deal with an experimentally measured quantity. As a result, all the other parameters (energy gap, heat capacity, critical field, and so on) are related in a universal way to T_c.

Strong coupling effects destroy this universality. This is caused by the fact that strong coupling theory contains a direct dependence on the structure of the phonon spectrum.

It turns out that an increase in the coupling strength leads to the ratio $\eta = \frac{2\varepsilon(0)}{T_c}$ growing beyond the BCS value of 3.52. Since the ratio η can be measured experimentally, this provides a relatively simple test of the magnitude of the coupling constant. One has to keep in mind, though, that the above argument is valid only if we are dealing with only one gap (and consequently only one coupling constant). This is the case for practically all conventional superconductors. The case of multiband structure requires a separate analysis (see Chapter 7).

Let us examine the relationship between $\varepsilon(0)$ and T_c. We start with the equation (see eqn. (3.7)):

$$\Delta(\omega_n)Z = \pi T \sum_{n'} \int d\Omega \frac{g(\Omega)}{\Omega} D(\omega_n - \omega_{n'}, \Omega) \frac{\Delta(\omega_{n'})}{[\omega_{n'}^2 + \Delta^2(\omega_{n'})]^{1/2}} \quad (3.36)$$

This non-linear equation can be simplified by introducing the coupling constant and characteristic phonon frequency $\tilde{\Omega}$ (see eqn. (3.10)). We will assume that $T_c \ll \tilde{\Omega}$; nevertheless, our goal is to calculate corrections $\sim (T_c/\tilde{\Omega})^2$.

The solution can be found by quasi-linearizing eqn. (3.36). The idea of this method (Zubarev, 1960) is as follows. The equation for $\Delta(\omega)$ at $T = 0\,\mathrm{K}$ can be written in the form:

$$V(\omega) = \frac{\tilde{\Omega}^2}{\tilde{\Omega}^2 + \omega^2} + Z^{-1}\int_{-\infty}^{\infty} d\omega'\, R(\omega, \omega', \Omega) \frac{V(\omega')}{\sqrt{\omega'^2 + \Delta^2(0)V^2(\omega')}} \quad (3.36a)$$

where $V(\omega) = \Delta(\omega)/\Delta(0)$ and the kernel is

$$R(\omega, \omega', \Omega) = \frac{\tilde{\Omega}^2}{\tilde{\Omega}^2 + (\omega - \omega')^2} - \frac{\tilde{\Omega}^2}{\tilde{\Omega}^2 + \omega^2}\frac{\tilde{\Omega}^2}{\tilde{\Omega}^2 + \omega'^2}. \quad (3.36b)$$

It is important that $R(\omega, 0, \Omega) = 0$, and $R(\omega, \omega', \Omega) \propto \tilde{\omega}'^2$ for small ω'. It follows that the main contribution to the integral over ω' comes from the region $\omega' \cong \tilde{\Omega}$, and in the zeroth-order approximation we can neglect the factor $\Delta^2(0)V(\omega')$ in the denominator on the right-hand side of eqn. (3.36a). As a result, we end up with a linear equation for the zeroth-order function $V_0(\omega)$. The solution of the full equation (3.36a) can be sought in the form $V(\omega) = V_0(\omega) + V'(\omega)$, where $V'(\omega) \propto \Delta^2(0)/\tilde{\Omega}^2$. It is also possible to derive a linear integral equation for the function $V'(\omega)$. Thus the quasi-linearization method reduces the solution of the non-linear equation to that of a system of linear equations.

For $T = T_c$, eqn. (3.36) can be transformed with the aid of the Poisson summation formula:

$$\sum_{\omega_n > 0} f(\omega_n) = (1/2\pi T) \int_0^\infty f(z)dz + (1/\pi T) \sum_{s=1}^\infty (-1)^s \int_0^\infty f(z) \cos(sz/T)\, dz.$$

By comparing the equations for $\Delta(\omega)$ at $T = 0\,\mathrm{K}$ and $T = T_c$ we can derive an expression relating the energy gap $\varepsilon(0)$ at $T = 0$ K and the critical temperature T_c (Geilikman and Kresin, 1966; Geilikman et al., 1975):

$$\frac{2\varepsilon(0)}{T_c} = 3.52\left[1 + a\left(\frac{T_c}{\tilde{\Omega}}\right)^2 \ln\frac{\tilde{\Omega}}{T_c}\right] \tag{3.37}$$

where a is a numerical factor ($a \cong 5.3$).

Several comments should be made concerning eqn. (3.37). Clearly, this expression generalizes eqn. (3.4) obtained in the weak-coupling approximation: we recover the latter if the second term in the brackets is neglected. This second term, which reflects the effects of strong coupling, has a peculiar form. In addition to the ordinary quadratic term, it contains the large logarithmic factor $\ln(\tilde{\Omega}/T_c)$.

We see that taking account of the effects of strong coupling destroys the universality of the relation between $\varepsilon(0)$ and T_c. The expression we obtained contains a characteristic phonon frequency, which varies from one superconductor to another.

Furthermore, the relationship (3.37) is no longer that of direct proportionality. This statement can be checked experimentally by studying the dependence of the energy gap and T_c on pressure. Such an experiment (Zavaritski et al., 1971), which utilized a tunnel junction under pressure, has shown that, indeed, in lead the energy gap and the critical temperature do not vary in an identical manner. This is a manifestation of strong coupling effects. The experimental method employed in this work also allowed control of the shift of the characteristic phonon frequency $\tilde{\Omega}$. The results were in very good agreement with eqn. (3.37).

Instead of $\tilde{\Omega}$ we could choose other characteristic frequencies, such as with the averaging carried out with respect to the function $\alpha^2(\Omega)F(\Omega)$: $<\Omega> = \int d\Omega \alpha^2(\Omega)F(\Omega)$, $<\Omega_{\log}>$, and so on (see E. Wolf, 2012; Carbotte, 1990). The comparison between theory and the experimental data depends, of course, on the choice of the characteristic frequency $\tilde{\Omega}$, though the difference is not dramatic. The general qualitative features described above remain valid.

Equation (3.37) shows that one of the effects of strong coupling is to make the ratio $\eta = 2\,\varepsilon(0)/T_c$ exceed the value $\eta_{\mathrm{BCS}} = 3.52$ (the extra term in parentheses is always positive). There is a one-to-one correlation between η and the magnitude of the coupling constant λ. An increase in λ leads to an increase in T_c, and as result, to an increase in the ratio η.

It is important to keep in mind, however, that the result (3.37) has been obtained under the assumption that $T_c \ll \tilde{\Omega}$. As λ, and consequently T_c, increase, we come to a point (at $\lambda \cong 2\text{--}3$) when the above inequality no longer holds. The following question then arises: what happens to the ratio η in the regime of superstrong coupling where the critical temperature is described by eqns. (3.30) and (3.34)?

It can be shown in general (Kresin, 1987c) that the ratio $2\,\varepsilon(0)/T_c$; saturates with increasing λ. The energy gap is defined as the root of the equation $\varepsilon = \Delta(-i\varepsilon)$. We are interested in the behavior of the function for high frequencies, where it is determined from eqn. (3.36):

$$\Delta(\omega) \simeq \frac{\lambda \tilde{\Omega}^2}{\omega^2} \int d\omega' \Delta(\omega') \left|\omega'^2 + \Delta^2(\omega')\right|^{-1/2}. \tag{3.38}$$

As a result, the following expression is obtained for the gap in the regime $\lambda \gg 1$: $\varepsilon(0) = C\lambda^{1/2}$. This dependence is identical with that of T_c (see eqn. (3.30)) in the region of large λ. Consequently, the ratio $\eta = 2\,\varepsilon(0)/T_c$ becomes a universal constant as λ becomes large: $\eta_{|\lambda \gg 1} \cong 13.5$. Note that according to some studies of organics and the cuprates we can observe values close to $\eta_{\max}$ (see Gray et al., 1987).

Hence, the ratio η has universal values in the regions of both weak and strong coupling, and $\eta_{\rm BCS} = \eta_{(|\lambda \ll 1)} = 3.52 \ll \eta_{(|\lambda \gg 1)}$. The transition from weak electron–phonon coupling to strong coupling is accompanied by an increase in η, but the dependence $\eta(\lambda)$ saturates for large λ, even though both T_c and $\varepsilon(0)$ continue to grow.

The temperature dependence of the energy gap is also different from that in superconductors with weak coupling. It has the following form (Geilikman and Kresin, 1968b):

$$\varepsilon(T) = a[1 - (T/T_c)]^{1/2}\big|_{T \to T_c}, \quad a = 3.06\left[1 + 8.8(T_c^2/\tilde{\Omega}^2)\ln(\tilde{\Omega}/T_c)\right]. \tag{3.39}$$

For example, for lead $a \cong 4$. The dependence $\varepsilon(T)$ influences many properties of superconductors, such as heat capacity, thermal conductivity, ultrasound attenuation, and so on. For example, the jump in the heat capacity is related directly to the change in entropy, $S^S - S^n|_{T_c} \propto a^2$. A more detailed evaluation based on the general expression for the thermodynamic potential leads to the result (Kresin and Parchomenko, 1975):

$$\beta = 1.43\left[1 + b\left(\frac{T_c}{\tilde{\Omega}}\right)^2 \left(\ln\frac{\tilde{\Omega}}{T_c} + \frac{1}{2}\right)\right], \quad b \simeq 18. \tag{3.40}$$

For example, for lead $\beta = 2.4$ ($\beta = 1.4$ in the BCS theory) and for gallium $\beta = 2.3$. Ultrasound attenuation as well as the electronic thermal conductivity decrease much more sharply with decreasing temperature below T_c than in weakly coupled superconductors.

The properties of strongly coupled superconductors can be analyzed in detail by tunneling spectroscopy (Scalapino et al., 1966; Scalapino, 1969; E. Wolf, 2012). This important aspect will be discussed in Chapter 6.

3.6 The Van Hove scenario

Properties of low-dimensional systems can be affected strongly by the presence of the Van Hove singularity (VHS; Van Hove, 1953). For the 1D structure this singularity in the electronic density of states has a form $(E - E_s)^{-1/2}$. As for the 2D structure (layered systems) the divergence has a logarithmic dependence.

This issue is important for the description of the A15 compounds A_3B (for example, V_3Si and Nb_3Sn). These materials have the highest values of T_c among conventional superconductors. According to Labbe and Friedel (1966) and Labbe et al. (1967), it can be explained by the fact that A ions form quasi-1D structure, which leads to structural instability, lattice softening, and an increase in T_c (see Chapter 13).

The high-T_c cuprates (Chapter 10) constitute the most important class of layered superconducting systems. According to Barisic et al. (1987) (see the review by Bok and Bouvier, 2012), the presence of the VHS is a key ingredient describing their properties. For the case when the half-width of the peak, $D/2$, is of the same order as the characteristic phonon frequency $\tilde{\Omega}$, but still $D > \tilde{\Omega}$, one can use the following expression for T_c (in a weak coupling approximation):

$$T_c = 1.13 D \exp\left(-\frac{1}{\sqrt{\lambda}}\right)$$

Since $\lambda \ll 1$, the dependence containing $\lambda^{1/2}$ leads to larger values of T_c relative to the BCS expression (3.18). The isotopic dependence of T_c is also different, because the pre-exponential factor is different from that in eqn. (3.18).

3.7 Bipolarons: BEC versus BCS

The bipolaronic scenario ("local" pairs) was proposed as an explanation of superconductivity even before the BCS theory (Schafroth, 1955). A more rigorous concept of a bipolaron, which is a bound state of two polarons, was introduced by Vinetskii (1961) and Eagles (1969). A more detailed model of bipolaronic superconductivity—namely, the picture that bosons (bipolarons) formed on a lattice could form a superconducting system—was proposed by Alexandrov and Ranninger (1981). The quantitative picture of bipolaronic superconductivity is rather elegant, and is very different from the conventional BCS concept. The main difference is the nature of the normal state. As we know, the starting point of the BCS picture is that in the normal state (above T_c or above the critical field) we are dealing with the usual fermions (delocalized electrons) and, correspondingly, with a Fermi surface. According to the bipolaronic picture, the normal state represents a Bose system formed by pairs of polarons: pairing occurs in real space. As a result, the nature of the phase transition at T_c is entirely different. According to the bipolaronic scenario, we are dealing with the Bose–Einstein condensation of bosons, whereas the formation of pairs (Cooper pairs) in usual superconductors occurs at T_c. The Cooper pair is formed by two electrons with opposite momenta, so that the pairs are formed in momentum, not real space.

There is an interesting question of bipolaronic instability. One can expect that for large values of the electron–phonon coupling constant λ the usual Fermi system is unstable with respect to formation of bypolarons. One should note also that an analysis of bypolarons is often based on the Holstein Hamiltonian (2.38). However, as discussed previously (Section 2.6.1) this Hamiltonian is applicable if $E_F \ll \tilde{\Omega}$. Such a situation, indeed, is favorable, because in this case a size of bipolaron could be smaller than the interelectron spacing, and this factor diminishes an overlap of neighboring wavefunctions.

As noted previously, the bipolaronic picture requires a bosonic nature of the carriers. This factor is important, as it contradicts the existence of the Fermi surface. A more general picture was described by Mueller (2007). According to this approach, the compound may contain two components: bipolarons and free fermions. The presence of free fermions explains the presence of the Fermi surface.

As mentioned previously, the bipolaronic superconductivity is caused by the Bose–Einstein condensation (BEC); that is, that picture is similar to that for liquid helium and its superfluidity. In connection with this one can study a general question about the scenario that is intermediate between the Bose–Einstein condensation (BEC) and Cooper pairing (BCS). Such a generalization was considered initially by Leggett (1980) and later by Nozieres and Schmitt-Rink (1985) and Nozieres (1995). The properties of a Fermi gas with an attractive potential have been studied as a function of the coupling strength. BEC and BCS cases correspond to two limits (strong and weak coupling). The analysis shows the importance of one-particle excitations for the Cooper pairing versus collective excitations for the BEC case.

3.8 Superconducting semiconductors

The possibility of observing superconductivity in semiconductors has long attracted a lot of attention, because semiconductors possess a number of interesting properties.

Note that we are discussing degenerate semiconductors, since the analysis based on the Cooper theorem implies the presence of the degenerate electron gas and corresponding Fermi surface. The carrier concentration in semiconductors is lower than in metals, and this is not a favorable feature. Indeed, the coupling constant λ is determined by eqn. (3.23), and the density of states, $\nu_F \sim m^* p_F \sim m^* n^{1/3}$. Thus one can expect that the coupling constant in degenerate semiconductors will be small. Nonetheless, it is also known that semiconductors are usually characterized by large dielectric constants ε. This considerably weakens the Coulomb repulsion, which would otherwise oppose the interelectron attraction. Consequently, the possibility of superconductivity in semiconductors can not be excluded.

The study by Gurevich et al. (1962) focused on polar semiconductors. It was shown that, due to ionic structure, the interaction with optical phonons could be sufficient to overcome the Coulomb repulsion.

It is essential that the Fermi surface of many semiconductors (for example, $SrTiO_3$) consists of a set of disjoint regions ("valleys"). Intervalley carrier transitions, induced by electron–phonon interaction, involve large momentum transfers and are quite effective. The particular role of these transitions was analyzed by Cohen (1964) (see the review by Cohen, 1969).

The first superconducting semiconductor to be discovered was GeTe ($T_c \approx 0.1\,\text{K}$; Hein et al., 1964). This was followed by the discovery of superconductivity in $SrTiO_3$ ($T_c \approx 0.3\,\text{K}$; Schooley et al., 1964). This material is an oxide with a perovskite structure. The transition temperature of these superconductors depends strongly on carrier concentration (see the review by Hulm et al., 1970).

The concentration dependence of T_c was found to be rather peculiar. One can see from Fig. 3.5 that this dependence is characterized by a sharp maximum. The non-monotonic dependence $T_c(n)$ can be explained as follows. According to eqn. (3.15), the electron–phonon coupling constant λ has a form:

$$\lambda = \frac{\zeta}{2p_F^2} \int_0^{k_1} q\,dq \frac{v^2 q^2}{\Omega^2(q)} r(q) \tag{3.15'}$$

It is important that $k_1 = \min\{2p_F, q_D\}$. Note also that $p_F \propto n^{1/3}$ and the Frohlich parameter $\zeta \propto p_F$. If the carrier concentration is low, then $k_1 = 2p_F$. Since $\Omega(q) \propto q$ at small q, we obtain $\lambda \propto p_F \propto n^{\frac{1}{3}}$, the value of the coupling constant, and therefore the value of T_c increases with n. However, the further increase in n leads to a crossover, so that $2p_F > q_D$; then $k_1 = q_D$. Then $\lambda \propto p_F^{-1}$, and T_c decreases with increasing n.

According to Binnig et al., (1980), Nb doping greatly affects properties of $SrTiO_3$. The main feature of Nb-doped $SrTiO_3$ is an appearance of the second energy gap. This was the first observation of the two-gap structure (see Chapter 7). The phenomenon was observed with use of the STM technique (see Section 6.2).

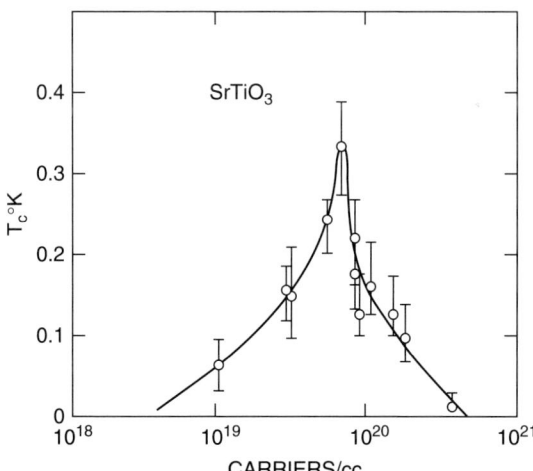

Fig. 3.5 Dependence of T_c on carrier concentration.

SrTiO$_3$ belong to the family of superconducting oxides. The first superconducting oxides were NbO (Miller et al., 1965) and the aforementioned SrTiO$_3$ (Schooley et al., 1964). A very interesting material is BaPb$_{1-x}$Bi$_x$O$_3$ (Sleight et al., 1975); its value of T_c is rather high: $T_c \simeq 12\,\text{K}$ for $x = \simeq 0.25$.

The high-temperature superconductors (Chapter 10) are doped oxides (cuprates), which are characterized by the presence of low-dimensional structures (planes and chains). On the whole, it is interesting to note that oxides occupy a special place in condensed-matter physics. The most interesting ferroelectrics, magnetic systems (manganites, Chapter 12), and the high-temperature superconductors (cuprates, Chapter 10) all contain oxides as their key ingredients.

3.9 Polaronic effect and its impact on T_c

3.9.1 Double-well structure

In Section 2.6 we described the dynamic polaronic state, which is a manifestation of the dynamic Jahn–Teller effect. Here we discuss its impact on the superconducting state, and especially on T_c. More specifically, we address the question of why the presence of polaronic states is beneficial for superconductivity. As is known, the polaronic effect provided the main motivation for the original search for high-T_c in cuprates (Bednorz and Mueller, 1986).

Consider a complex compound where one ionic sub-system is such that an ion is characterized by two close equilibrium positions (double-well potential, see Fig. 2.8). Such strongly anharmonic potential leads to a peculiar non-adiabatic polaronic effect; indeed, in this case the electronic and local lattice degrees of freedom happen to be inseparable.

Especially interesting are the cases of layered systems, and here we focus on such a case. For example, oxygen ions in the cuprates do form such a subsystem; the double-well structure has been observed experimentally (see Chapter 10).

We can show that the presence of such a structure leads to a noticeable increase in the effective strength of electron–lattice coupling relative to the usual case of a single potential minimum for the ionic coordinate (Kresin, 2010). The point is that the pairing interaction is described by the equation which contains the matrix element connecting initial and virtual states. One can show that even in the absence of direct interaction with the double-well structure, the resultant coupling constant will increase. This is due to the increased phase space for virtual transitions, or in other words, due to an increased number of these transitions.

Consider the special case when the electronic term for some ionic subsystem (for example, for oxygen ions in the cuprates) contains two close minima positions. As an example, we focus on the case when the double-well potential corresponds to some direction; for layered systems this direction is perpendicular to the layers (OZ$||$c).As for the dependence of the ionic motion on X,Y, it is described by an usual harmonic dynamics with single minima equilibrium positions. Therefore, $\varepsilon_n(\mathbf{R}) \equiv \varepsilon_n(\boldsymbol{\rho}, Z)$ with the double-well structure in the Z direction, where $\boldsymbol{\rho}$ denotes the in-plane ionic position.

We employ the tight-binding approximation. As a first step we should write down the local wavefunction. In our case, the ion is affected by the double-well potential (the Z direction; Fig. 3.6). At this stage it is very convenient to employ the diabatic representation (see Appendices). As a result of the transformation, the double-well potential is replaced by two crossing harmonic energy terms (Fig. A.1), and each of these terms also contains its vibrational manifold. Therefore, the local state is described by two groups of terms (a and b), so that

$$\Psi_{loc.} = c_a \Psi_a(\mathbf{r}, \mathbf{R}) + c_b \Psi_b(\mathbf{r}, \mathbf{R})$$

where

$$\Psi_{a(b)}(\mathbf{r}, \mathbf{R}) = \psi_{a(b)}\left(\mathbf{r}, \mathbf{R}_0^{a(b)}\right) \chi_{a(b)}\left(z - z_0^{a(b)}\right) \quad (3.41)$$

Note that $\Psi_a = \psi_s \chi_{sv}$; $\Psi_b = \psi_{s'} \chi_{s'v'}$. Here $\{s,s'\}$ and $\{v,v'\}$ are the electronic and vibrational quantum numbers. If both terms are equivalent, then $c_a^2 = c_b^2 = 0.5$. Because of the inter-term tunneling, the energy levels which correspond to isolated terms are split into symmetric and antisymmetric; the scale of the splitting is of order of ε_{ab}, where (see Appendices)

$$\begin{aligned}\varepsilon_{ab} &= \int d\mathbf{R} \chi_a \hat{H}_{r;ab} \chi_b \\ \hat{H}_{r;ab} &= \int d\mathbf{r} \psi_b^* \hat{H}_r \psi_a\end{aligned} \quad (3.42)$$

It is essential that, contrary to the usual adiabatic picture, the operator $\hat{H}_r = \hat{T}_r + V(r, R)$ has non-diagonal terms in the diabatic representation.

We can see directly from eqn. (3.41) that the wavefunction $\Psi_{loc.}$ cannot be written as a product of electronic and ionic wavefunctions. In other words, the electronic and local ionic motions cannot be separated as in the usual adiabatic theory. Such a polaronic effect is a strong non-adiabatic phenomenon.

In the tight-binding approximation, each local term, including those formed by the transformation (Fig. 3.6), is broadened into the energy band. Note at first that the local electronic function (see eqn. (3.41)) $\psi_i(r, R_0^i)$ ($i = a, b$) can be written in the form: $\psi_i(r - R_0^n)$, R_0^n corresponds to the crossing point for nth ion ($i = a, b$). Indeed, because the electronic wavefunction has a scale of the length of the bond that greatly

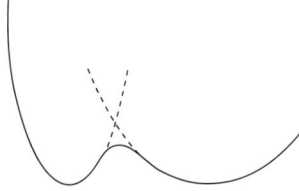

Fig. 3.6 Transformation to the diabatic representation: solid line, initial term (double-well structure); broken lines, crossing terms.

exceeds the amplitude of vibrations and the distance "δ" between the minima, one can neglect the difference between R_0^n and R_0^i.

The total wavefunction also contains the vibrational part. First let us separate the local vibrational mode that corresponds initially to the double-well structure, or, in the diabatic approximation, to two crossing terms. As for the dependence of ionic motion on X and Y, it can be described by a set of normal modes. Then the total wavefunction has the form:

$$\Psi(\mathbf{r},\mathbf{R}) = u_k(\mathbf{r},\mathbf{R})e^{i\mathbf{kr}}\Phi_{vib.}$$
$$u_k(\mathbf{r},\mathbf{R}) = \sum_n \Psi_{loc.}(\mathbf{r},\mathbf{R})e^{i\mathbf{k}(R_0^n - \mathbf{r})} \quad (3.43)$$

$\Psi_{loc.}$ is determined by eqn. (3.41). The local vibrational wavefunctions are centered at z_0^i and at $z_0^i - d$. If the terms are similar, then $\chi^b(z) = \chi^a(z-d)$.

We should stress a key difference between the ionic motions described by the function $\Phi_{vib}(R_x, R_y) = \Pi\varphi(\Omega_m)$ (Ω_m are the frequencies of the usual normal modes) and the function $\chi(R_z)$. There is no in-plane ionic hopping between various ionic sites. However, the oxygen can tunnel between the two minima, and this is reflected in a noticeable overlap of vibrational functions χ^i corresponding to the two crossing terms (Fig. 3.6).

3.9.2 Superconducting state

Let us consider the case when the main contribution to the interaction is coming from the interaction with usual normal (harmonic) modes Q_m (one can easily describe a more general case). Then the interaction Hamiltonian can be written in the usual form (see eqn. (2.17a)):

$$\hat{H}' = (\partial V/\partial Q_m)\delta Q_m \quad (3.44)$$

Summation over m is implied. The equation for the pairing amplitude has the form (see eqn. (3.6)):

$$\Delta(\omega_n) = T\sum_{\omega_{n'}}\int d\xi |\zeta|^2 \nu_F D(\omega_n - \omega_{n'}, \tilde{\Omega}) F^+(\omega_{n'}) \quad (3.45)$$

Here, $D = \tilde{\Omega}^2[\tilde{\Omega}^2 + (\omega_n - \omega_{n'})^2]^{-1}$ is the phonon propagator, $\omega_n = (2n+1)\pi T$, $\tilde{\Omega}$ is the characteristic phonon frequency, $F^+(\omega_n) = \Delta(\omega_n)\left[\omega_n^2 + \xi_{mv}^2 + \Delta^2(\omega_n)\right]^{-1}$ is the Gor'kov pairing function, ξ_{nv} is the excitation energy relative to the Fermi energy, and ν_F is the density of states). It is essential that ξ_{nv} is the energy of virtual transitions, which contribute to pairing. Note that the integrand contains the matrix element ζ, which has the form:

$$\zeta = \int \Psi^{*f}\hat{H}'\Psi^i d\mathbf{r}d\mathbf{R} \quad (3.46)$$

where $\hat{H}'$ is defined by eqn. (3.44), i and f denote the initial and virtual states, and the total function has the form (3.43).

Let us introduce some characteristic frequency $\tilde{\Omega}$ and a single mode $\tilde{Q}$, which provides a main contribution to the pairing (see eqn. (3.44)). Then the product $\Pi \varphi_{\nu_m}(Q_m)$ can be replaced by the function $\varphi_v(\tilde{Q})$. Correspondingly, $H' = (\partial V/\partial \tilde{Q})_0 \delta \tilde{Q}$. In other words, we model the phonon spectrum as consisting of two modes. One of them, harmonic mode $\tilde{Q}$, provides a major contribution to the coupling interaction, and the second mode which is characterized by double-well structure (Fig. 3.6), provides an additional number of the virtual states. This model can be easily generalized. One can see from eqn. (3.46) that the average value of the coupling matrix element is a sum:

$$\zeta_{av.} = \zeta_0 + \zeta_1 \tag{3.47}$$

where

$$\zeta_0 = \int dr d\tilde{Q} \sum_n \psi_s(\mathbf{r} - \mathbf{R}_0^n) \psi_{s'}(\mathbf{r} - \mathbf{R}_0^n) \left(\frac{\partial V}{\partial \tilde{Q}}\right)_0 \delta \tilde{Q}_{av}; \quad \zeta_1 \approx \zeta_0 F^{ab} \tag{3.48a}$$

where

$$F^{ab} = 2 \sum_{s_i, v_i} \int dz \chi_{sv}(z - z_0^a) \chi_{s_1 v_1}(z - z_0^b) \tag{3.48b}$$

As usual (see, for example, Anselm, 1982), the displacement $\delta \tilde{Q}$ can be expanded in the Fourier series.

It is very essential that the Franck–Condon factor $F^{ab} \neq 0$. Indeed, eqn. (3.45) assumes summation over virtual states; that is, integration over ξ' and summation over v'. The vibrational wavefunctions for the same term are orthogonal. However, this is not the case for different crossing terms.

The additional contribution ζ_1 corresponds to virtual transitions, which are accompanied by the change of electronic terms upon ionic tunneling between the two minima. In other words, ζ_1 corresponds to transitions to the states with various v' (for given v). It is important to note that the contribution comes only from the terms with the same symmetry; that is, from the terms with the same sign of the coefficient c_b (see eqn. (3.41)).

Therefore, there is an additional contribution (relative to the usual case) caused by the transitions between the local vibrational levels. Such an increase in phase space for virtual transitions leads to an increase in the value of the coupling matrix element ζ.

Consider in more detail the Franck–Condon factor F^{ab}. Assume for concreteness that the initial state corresponds to $\nu_a = 0$; eqn. (3.45) contains a sum $\sum_{v'} F^{ab}_{v'0}$ which represents the contributions of virtual transitions to the manifold of the coherence states. In the diabatic representation, a single non-harmonic term is replaced by two crossing terms, and each of them can be treated in harmonic approximation. The Franck–Condon factor is equal (see, for example, Landau and Lifshitz, 1977): $F^{ab} = (\beta^v/v!) \exp(-\beta)$, where $\beta = (d/a)^2$ (a is the amplitude of vibrations, and d is the distance between the minima). One can see that $F_{v'0}$ contains a small factor, $\exp(-(d/a)^2)$, which describes the probability of the ionic tunneling between two minima. But this smallness is compensated by summation over all virtual transitions. If, for example, $(d/a) = 2$, then $\zeta_1 \approx 0.7\zeta_0$. If $\zeta_0 \approx 1$, then $\zeta = \zeta_0 + \zeta_1 \approx 1.7$. The value of T_c is determined by the coupling constant $\lambda \propto \zeta^2$, and we obtain $\lambda \approx 3$. Therefore, the polaronic effect leads to a large increase in the effective strength of the electron–lattice coupling. As stressed previously, this enhancement arises due to the increase in phase space for virtual transitions, which promotes the pairing.

4
Electronic mechanisms

Up to this chapter we have focused on the phonon mechanism of superconductivity. However, electron pairing can be created not only by phonon exchange but also by exchange of other excitations. In this chapter we will discuss the electronic mechanism of superconductivity. The most common (but not unique) instance of such a mechanism arises in the presence of two groups of electronic states. Excitations within one of these groups serve as "agents," giving rise to pairing in the other group.

The study of electronic mechanisms of superconductivity began with the work of Little (1964). This work also served to initiate the general discussion of the relationship between dimensionality and the superconducting transition. In the following we shall describe some relevant models.

4.1 The Little model

Little (1964) has considered the case of a one-dimensional organic polymer. Remarkably, at that time no superconducting polymers existed; there also were no organic superconductors. The first superconducting polymer was synthesized in 1975 (Greene et al., Gill et al.). Organic superconductivity was discovered in 1980 (Jerome et al.), and at present represents one of the most actively developing areas of research (see, for example, Ishiguro et al., 1998; Jerome, 2012; see also Chapter 14). Therefore, the paper by Little (1964), in addition to introducing a novel mechanism of pairing, contains also the prediction of superconductivity in organic materials.

In the Little model the conduction electrons in the primary chain (see Fig. 4.1) are paired due to their interaction with electronic excitations in the side branches. The critical temperature is given by a BCS-like expression (3.13), but in the present case the pre-exponential factor corresponds to the electronic, rather than ionic, energy scale, so that

$$T_c = \Delta E_{el} \exp(-\lambda_{el}) \qquad (4.1)$$

The fact that $\Delta E_{el} \gg \tilde{\Omega}$ is responsible for the hope that high critical temperatures may be achieved.

Here it is of worth to make a brief historical digression. The thing is that the dream of high temperature of superconductivity culminating in an ideal case of observing the phenomenon at room temperature was born a long time ago, even before the appearance of the BCS theory. One of the first questions with which this theory was confronted was that of the upper limit of the critical temperature. In other words,

48 *Electronic mechanisms*

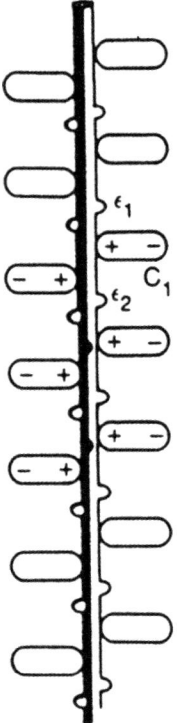

Fig. 4.1 Little's model.

it was essential to understand whether superconductivity is a low-temperature phenomenon or is not subject to such a restriction. The initial answer was, unfortunately, not optimistic. Indeed, as noted previously (Section 3.4.6), in the early 1960s the common understanding was that the value of the electron–phonon coupling cannot exceed $\lambda_{\max} = 0.5$, and therefore, according to expression (3.13), the critical temperature was not expected to exceed some optimum value $T_{c;\max} \approx 0.1\,\tilde{\Omega}$, $\tilde{\Omega} \approx \Omega_D$. As a result, the dream about high-temperature superconductivity was thought as being unrealistic. As stated previously (Section 3.4.6), this conclusion was erroneous, and it became clear that even with phonons one can reach a high T_c. However, in early 1960s this was not thought to be the case. The work by Little (published in 1964) revived the dream of high T_c and initiated a search for this phenomenon.

The idea to search for an electronic mechanism is very fruitful, but its realization is quite non-trivial. Clearly, the pre-exponential factor acts so as to raise T_c, though the coupling constant λ also depends sensitively on ΔE_{el}, and it is essential that this constant should not turn out to be too small.

Furthermore, we are really dealing with the difference $\lambda - \mu^*$. Because of the inequality $\Delta E_{el} \gg \hbar\Omega_D$, the logarithmic weakening of the Coulomb repulsion is not as strong as in the phonon case. This factor also decreases the exponent. Finally, certain fundamental problems concerning the superconducting transition arise in low-dimensional cases.

The problem of the intensity of the pairing interaction has been discussed by Gutfreund and Little (1979). According to quantum-mechanical calculations, in the ordinary model of the type shown in Fig. 4.1 the average spacing between the conducting chain and the charge localized on the side branches is quite large (~ 4.5 A). As a result, the value of λ is quite low. As a consequence, a different approach was developed by Little (1983). Here, pairing takes place in a chain containing a transition metal atom chain (see Fig. 4.2). The conduction electrons move in the $d_{z^2} - p_z$ orbital band. Strong coupling occurs due to the overlap of the d_{xz} and d_{yz} orbitals with those of the ligands. In this situation it is critical that the energy levels of the conduction electron orbitals be close to those of the ligands' electrons. It is also necessary to have very dense packing of the polarizable ligands. All this works to strengthen the coupling and to optimize the delicate balance of factors controlling the electron–electron interaction.

Another important problem involves low dimensionality and fluctuations (Ferrell, 1964; Rice, 1965; Hohenberg, 1967). The fact of the matter is that fluctuations of the order parameter destroy long-range order: the correlation function $K(x_1 - x_2)$ vanishes for $|x_1 - x_2| \to \infty$ at finite temperatures. This difficulty can be overcome by creating "quasi-one-dimensional" systems made up of bundles of superconducting threads (Dzyaloshinski and Katz, 1968). The probability of jumps between the threads is small, but is sufficient to stabilize the system with respect to fluctuations.

In the one-dimensional case the Fermi "surface" is made up of two points at $\pm k_F$. This imposes severe restrictions on the allowed values of transferred momenta. The different possible types of instability in the one-dimensional case form the subject of g-ology (Gutfreund and Little, 1979). One should distinguish various coupling constants λ_{SS}, λ_{TS}, λ_{CDW}, λ_{SDW}; SS and TS correspond to singlet and triplet

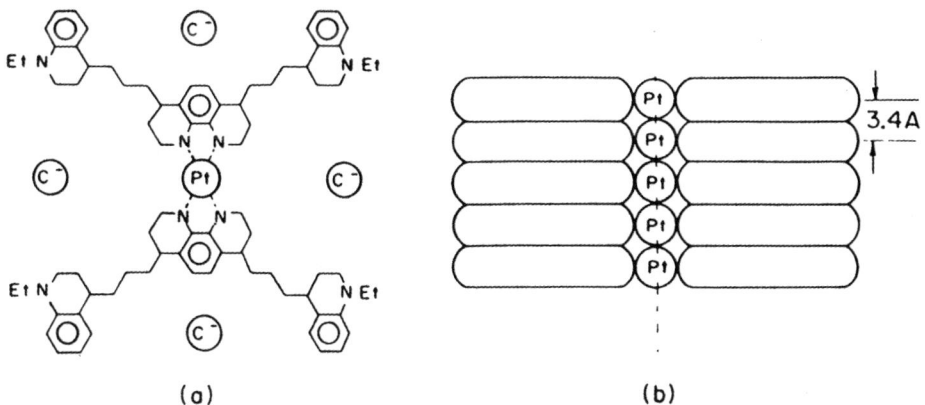

Fig. 4.2 Transition-metal atom chain: (a) top view; (b) side view.

superconducting states, and CDW and SDW denote transitions with the appearance of charge- and spin-density waves, respectively. A rigorous treatment of various instabilities has been carried out by Bychkov *et al.* (1966) and by Dzyaloshinski and Larkin (1972).

4.2 "Sandwich" excitonic mechanism

As described previously, the main feature of the Little model was the presence of two groups of electronic states. Based on this idea, Ginzburg (1965) (see also Ginzburg and Kirzhnits, 1982) has analyzed the case of two-dimensional conducting layers with coupling due to the exchange of excitons in an adjacent dielectric film. Thus the idea is to create a heterogeneous layered structure with alternating metallic and dielectric planes.

A similar geometry was considered by Allender *et al.* (1973), but with semiconducting rather than dielectric layers. Clearly, the effectiveness of such a system is highly dependent on the thickness of the metallic film which must be on the order of the screening length; that is, it may not exceed 10–15 Å. In this model an important parameter is the penetration depth D of an electron tunneling from the metallic film into the semiconducting covering. This depth can be shown to be equal to

$$D = \frac{\pi}{3} \frac{k_F}{2m\Delta} \tag{4.2}$$

Substituting $\Delta \approx 1$ eV and $k_F \approx 1.5 \text{ Å}^{-1}$, we find $D \approx 5$ Å. Thus we are dealing with quite substantial penetration depths.

4.3 Three-dimensional systems: electronic mechanism

A system with two different coexisting electron groups can be realized in three dimensions as well. This case was first studied by B. Geilikman (1965, 1971, 1973). Here we have to consider two special cases: 1) the group providing attraction is localized, and 2) both groups are delocalized. Let us briefly discuss these two cases.

Speaking of the first case, we can consider, for example, a metal with non-metallic impurities. The interaction of the conduction electrons with the electrons in the non-metal atoms leads to an additional effective interaction between the conduction electrons. This effective interaction has the form:

$$\Gamma_{p1,p2;p3,p4} \approx \Gamma'_c + \Gamma''_{p1,p2;p3,p4} \tag{4.3}$$

where

$$\Gamma''_{p1,p2;p3,p4} \approx \sum_{n,\lambda,\lambda'} V_{p_1,n\lambda;p_3,n\lambda'} V_{n\lambda',p_2;n\lambda,p_4} \Pi_{\lambda\lambda'},$$

$$\Pi_{\lambda\lambda'} = \frac{n_\lambda - n_{\lambda'}}{E_\lambda - E_{\lambda'} + \hbar\omega},$$

$$\omega = \xi_3 - \xi_1,$$

and p_i are the momenta of the conduction electrons. The first term in eqn. (4.3) corresponds to the usual screened Coulomb repulsion, and the second term to the attractive interaction. Here, $V_{p1,n\lambda;p3,n\lambda'}$ is the matrix element of the interaction between the delocalized conduction electrons and the localized electrons, λ is the set of quantum numbers of the electron in the non-metal atom at lattice site n, E_λ is the energy of this electron, and $\Pi_{\lambda\lambda'}$ is the polarization operator. Because of the summation over n, $\Gamma \propto N$, N being the concentration of the non-metal atoms.

Superconducting pairing can arise if the attraction outweighs the Coulomb repulsion. Note that a number of factors serve to decrease the value of the effective coupling constant. These include the exchange interaction, the smallness of the ratio a/L (where a is the radius of the orbit of the localized electrons, and L is the lattice parameter), and so on. In addition, the spacing $\Delta E = E_{\text{loc}}$ should be sufficiently small. On the other hand, this smallness leads to a decrease in T_c; see eqn. (4.1). Therefore, there exists an optimum value for ΔE. It appears that ΔE should be of the order of 0.1–$0.5\,\text{eV}$. Values of such a scale are possible, for example, in the case of fine-structure levels.

It should be pointed out that not only electronic but also vibrational states can be localized. For example, a molecule can be placed on the surface of a thin film or introduced into the sample. Then the interaction of the carriers with molecular vibrations can provide additional inter-electron attraction (Kresin, 1974). As a result, the presence of such molecules leads to an increase in T_c. This effect has been observed (Gamble and McConnell, 1968). The inclusion of organic molecules (tetracyanoguinodimethan) raised the value of the critical temperature of Al films up to $T_c = 5.24\,\text{K}$.

A special case is that of a superconducting molecular crystal. Systems of this type have lattices consisting of molecules with their own internal electronic and vibrational levels.

It is interesting to consider the case of pairing provided both by the ordinary phonon mechanism and by an additional interaction with localized electronic states. Making use of the step method (see Section 3.4.3), one finds in the weak-coupling approximation:

$$T_c \cong \tilde{\Omega} e^{-1/g_o}$$

where

$$g_o = g_{ph} + (g_e - g_c) \Big/ \Big[1 - (g_e - g_c)\ln(\Delta E_a/\hbar\tilde{\Omega})\Big] \qquad (4.4)$$

Note that the difference $g_e - g_c$ enters the expression for g_o with a minus sign. This means that the additional attraction is logarithmically enhanced.

The case when both electronic groups are delocalized is realized, for example, in a metal with two overlapping bands of very different widths, such as an s-band and a p-band. Virtual electronic transitions in the narrow band caused by the Coulomb interaction of the a and b electrons can lead to pairing in the wide band. The calculations (Geilikman, 1965, 1973) were performed within the random-phase approximation; they are at least of qualitative interest when applied to real metals. One can consider also multi-valley semiconductors, where the large value of the dielectric constant ε makes

RPA more applicable (Geilikman and Kresin, 1968a). The calculation leads to the following expression for the effective interaction of the "a" electrons:

$$\Gamma_{aa} = (V_{aa} + \Pi_{bb}R)S^{-1}, \quad \Gamma_{bb} = (V_{bb} + \Pi_{aa}R)S^{-1}$$
$$S = 1 + V_{aa}\Pi_a + V_{bb}\Pi_b + \Pi_a\Pi_b R, \quad R = V_{ab}^2 - V_{aa}V_{bb} \quad (4.5)$$

Here Π_{aa} and Π_{bb} are the polarization operators for the "a" and "b" electrons, and V_{ik} are the Coulomb matrix elements. Conditions under which the effective interaction becomes attractive turn out to be rather non-trivial.

Systems with overlapping bands can give rise to low-lying collective branches (acoustic plasmons). The exchange of these collective excitations can also lead to pairing. This mechanism will be discussed in the next section.

4.4 Plasmons

Usual plasmons in 3D metals are described in many textbooks (see, for example, Ashcroft and Mermin, 1976; Mahan, 1993). As we know, plasmons describe collective oscillations of the electrons with respect to the positive background. Recall that the usual electronic quasi-particle excitations correspond to the poles of a one-particle Green's function. Correspondingly, the spectrum of plasma oscillations which represent the collective motion is given by the poles of a two-particle Green's function (Fig. 4.3) or by poles of the vertex Γ.

Note that if we imagine the system of non-interacting electrons then the one-particle spectrum does not disappear, but is reduced to the kinetic energy of a free electron. As for collective excitations, such as plasmons, their existence is caused explicitly by the interaction only. That is why the calculation is directly related to the analysis of the Coulomb vertex Γ.

Usually, the value of the plasma frequency for an isotropic three-dimensional system is quite high ($\sim 5 - 10\,\text{eV}$). However, when dealing with more complicated band structures one encounters additional low-lying plasmon branches. Such low-lying branches could also be observed in low-dimensional systems, even in the simplest one-band case.

In this section we consider a different electronic mechanism of superconductivity: namely, the case when pairing is caused by an exchange by plasmons. We focus mainly on layered conductors. The reason for such a focus is a two-fold. Firstly, many novel superconducting systems contain conducting layers. In addition, the plasmon spectrum for such conductors drastically differs from that for usual three-dimensional metals. We can demonstrate that it contains an acoustic branch which is similar to acoustic phonons. As a consequence, the presence of such a low-lying plasmon branch leads to

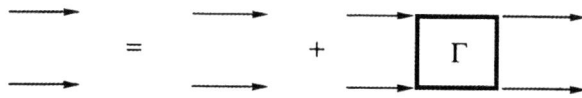

Fig. 4.3 Two-particle Green's function; Γ is a total vertex.

its noticeable contribution to the pairing in many-layered systems (see Chapters 10 and 13). It turns out that for the family of superconducting nitrides the exchange by acoustic plasmons is playing a key role (see Section 13.7).

4.4.1 Plasmons in layered systems: dispersion law and "electronic sound"

As noted previously, the situation in layered conductors is very different from that in bulk three-dimensional materials. In contrast to isotropic metals, these materials have considerable anisotropy in their conductivities along and perpendicular to the layer planes. As a consequence, the screening properties of such a system are quite different from the isotropic case. The energy spectrum has been analyzed by Visscher and Falicov (1971), Fetter (1974), and Kresin and Morawitz (1990).

The term "layered electron gas (LEG)" describes the basic physical concept: namely, an infinite set of two-dimensional layers of carriers (2D electron gas), separated by electronically inactive (insulating) spacer layers.

We have assumed that we are dealing with the in-plane motion of electrons only; the interlayer hopping, in a first approximation, is neglected. At the same time, the Coulomb interaction is, as usual, three-dimensional. In other words, in the LEG model the Coulomb interaction between charge carriers on different layers is included exactly, while the electronic collective motion is calculated for a single layer with two-dimensional plane-wave wavefunctions for the carriers. In addition, the dielectric response of the intervening non-metallic region is included by using the low-frequency limit of its dielectric constant (for example, for the cuprates, $\varepsilon \approx 3-6$) to screen the interaction between conducting layers.

A direct consequence of the layering of the carriers is that the spatial periodicity along the normal to the planes allows the definition of a corresponding wavevector $q_z = \pi/L$, where L is the interlayer spacing.

Unlike the one-particle excitations, the spectrum of plasma oscillations, which represents the collective motion, corresponds to the poles of a two-particle Green's function. Since the existence of plasmons is caused by interaction only, the calculation is directly related to the analysis of the Coulomb vertex $\Gamma \equiv \Gamma_c$ (Fig. 4.4). We employ the random-phase approximation (RPA)—a method that allows us to obtain the complete solution. RPA provides a rigorous result for $r_s \ll 1$ (high-density limit; $r_s = r_0/r_B$; see, for example, Mahan, 1993), r_B is the Bohr radius, and $r_0 \approx 0.6 n^{-1/3}$, n is the carrier concentration). For real metals, $5.5 \gtrsim r_s \gtrsim 2$. Nevertheless, based on

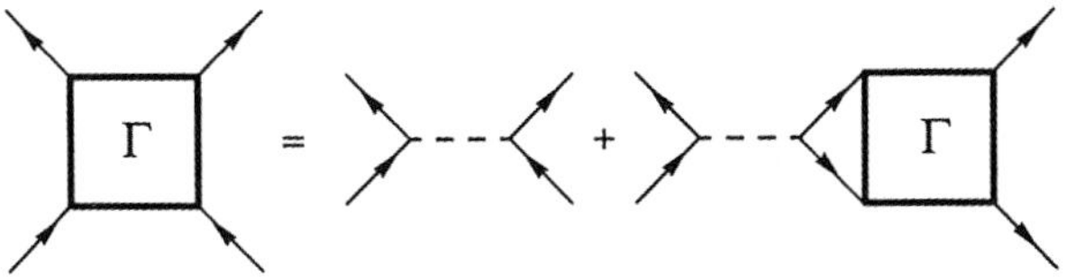

Fig. 4.4 Diagrammatic equation for the total vertex.

RPA, one can obtain a good qualitative description of the screening and the spectrum (plasmons).

The expression for Γ_c has the following form:

$$\Gamma_c(q; |n-m|) = \frac{V_c(q)}{\epsilon(q, |n-m|)} \qquad (4.6)$$

The Coulomb term is written in its most general form as the ratio of the bare Coulomb interaction and the dielectric function. Both the Coulomb interaction and the dielectric function have to be calculated for a layered structure. The Coulomb potential $V_c(q)$ is the Fourier transform of the 3D Coulomb interaction:

$$V_c(r) = e^2/\varepsilon_M |r| \qquad (4.7)$$

(ε_M is the dielectric constant of the spacers), and takes the form:

$$V_c(q) = \frac{\lambda_c}{\nu} \frac{R(\tilde{q}; q_z)}{\tilde{q}} \qquad (4.8)$$

with

$$R(\tilde{q}, q_z) = \frac{\sinh(\tilde{k}_{\tilde{F}}\tilde{q})}{\cosh(\tilde{k}_{\tilde{F}}\tilde{q}) - \cos(q_z L)}(1 - \delta_{q,0}) \qquad (4.8a)$$

We introduce dimensionless quantities $\tilde{q} = \frac{q_{||}}{2k_F}$, $\tilde{k}_{\tilde{F}} = 2k_F L$ ($q_{||}$ is the in-plane momentum, L is the interlayer distance, and k_F is the in-plane Fermi wavevector), as well as $\nu = \frac{m_b}{2\pi\hbar^2}$, the 2D electronic density of states (m_b is the band mass). Note that $R(\tilde{q}; q_z)$ contains the factor $(1 - \delta_{q,0})$, reflecting the presence of the neutralizing positive ion counter-charges. Eqn. (4.8) contains the dimensionless Coulomb interaction constant defined by

$$\lambda_c = \frac{e^2}{2\varepsilon_m v_F}$$

v_F is the Fermi velocity. Note that $\lambda_c \sim r_s$, where $r_s = \sqrt{2}/k_F r_B$ ($r_B = \varepsilon_M/mc^2$ is the Bohr radius) is the well-known dimensionless electron density radius defined here for a layered electron gas.

The dielectric function $\varepsilon(q, \omega_n)$ which describes the electronic screening of the Coulomb interaction and also enters eqn. (4.6) can be written in general form as

$$\epsilon(q, \omega_n - \omega_m) = 1 - V_c(q)\Pi(q, \omega_n - \omega_m) \qquad (4.9)$$

The plasmons are defined by the poles of Γ ($\Gamma \equiv \Gamma_c(q, \omega)$ is the analytical continuation of the function $\Gamma_c(q, \omega_n)$) or, equivalently, by the zeros of the dielectric function.

Let us introduce the dimensionless parameter $\alpha = \omega/q_{||}v_F$. One has to distinguish two cases: $\alpha < 1$ and $\alpha > 1$. Plasmons appear in the region of $\alpha > 1$. For simplicity, consider the case $\alpha \gg 1$. Then the polarization operator is given by

$$\Pi(q_{||}, \omega) = \nu(v_F q_{||}/\omega)^2 \qquad (4.10)$$

where ν is the density of states.

Let us demonstrate that the layering indeed leads to the dispersion relation which is entirely different from that for 3D bulk metals (then $\omega_{pl} \cong \omega_0 + aq^2$; usually, the value of ω_0 is high; $\omega_0 \cong 5 - 10\,\text{eV}$); we use here the semi-quantitative treatment. The expression for Coulomb potential can be written in the form

$$V(q) = e^2 / \left(q^2_{\|} + q_z^{\,2}\right) \qquad (4.11)$$

Let us remember that we are considering the 2D in-plane electronic motion, but 3D Coulomb interaction. Then one can see from eqns. (4.9) and (4.11) that the dependence $\omega\left(q_{\|}\right)$ is determined by the value of q_z. For example, for $q_z = 0$ we obtain the usual 3D value of the plasmon frequency, so that $\omega = \omega_0$ for $q_{\|} = 0$. However, for $q_z = \pi/L$ we obtain $\omega \propto q_{\|}$; that is, the acoustic dispersion law.

Rigorous evaluation leads to the result:

$$\omega = q_{\|}v_F\left[1 + (\nu V_c)^2(0.25 + \nu V_c)^{-1}\right]^{1/2} \qquad (4.12)$$

where $V_c \equiv V_c(\tilde{q}, q_z)$ is the Coulomb interaction defined in eqns. (4.8) and (4.11). If $\nu V_c \gg 1$ we obtain the dispersion law, $\omega = \hbar q_{\|}v_F\sqrt{1 + \nu V_c}$, which corresponds to the hydrodynamic approximation for small $q_{\|}$; see Fetter (1974). For $\omega \gg \hbar q_{\|}v_F$, eqn. (4.12) reduces to the expression $\omega \cong \hbar q_{\|}v_F\sqrt{\nu V_c}$, which at $q_z = 0$ leads to the usual "optical" plasmon with $\Omega^2_{pl} = \omega^2\left(q_{\|} = 0, q_z = 0\right) = \frac{4e^2\varepsilon_F}{\epsilon_M L}$. For $q_z = \pi/L$, on the other hand, we obtain the acoustic dispersion law.

One can see directly from eqns. (4.7), (4.7a), and (4.11) that the plasmon spectrum of the layered conductor is described not by single branch $\omega = \omega_0 + aq^2$, but the plasmon band with branches corresponding to $0 < q_z < \pi/L$ (Fig. 4.5a). The plasmon band $\omega = \omega(q_{\|}, q_z)$ is confined between the upper branch with $q_z = 0$ (in-phase motion of the charge carriers) and the lower branch at $q_z = \frac{\pi}{L}$ (out-of-phase motion of carriers). It is also possible to include a small hopping term t in the c-axis direction. One now find a small gap at $q_{\|} = 0$ of order $4t$.

One can show that the density of states for this plasmon band peaks near its boundaries; that is, near $q_z = \pi/L$ and $q_z = 0$ (Morawitz et al., 1993). Then, qualitatively, the plasmon spectrum can be visualized as a set of two branches (Fig. 4.5b). The upper branch ($q_z = 0$) is similar to the usual plasmon branch for an isotropic 3D metal; it corresponds to the in-phase motion of neighboring layers.

The most important feature of the plasmon spectrum of a layered metal is an appearance of the lower branch ($q_z = \pi/L$), which has an acoustic nature. This branch corresponds to the out-of-phase motion of neighboring layers; that is, they oscillate relative to each other. This mode can be called an "electronic sound".

There are several novel consequences arising from the splitting of the collective excitation spectrum into layer plasmon branches. These arise specifically because of the existence of a low-frequency electronic response. The first of these is the possibility of pairing the carriers by exchange of acoustic plasmons.

A second consequence of the low-frequency plasmon spectrum is the possibility of a temperature-dependent term in the electron loss function. Electron-energy-loss spectroscopy is a conventional technique for studying plasmon excitations in solids. For

56 Electronic mechanisms

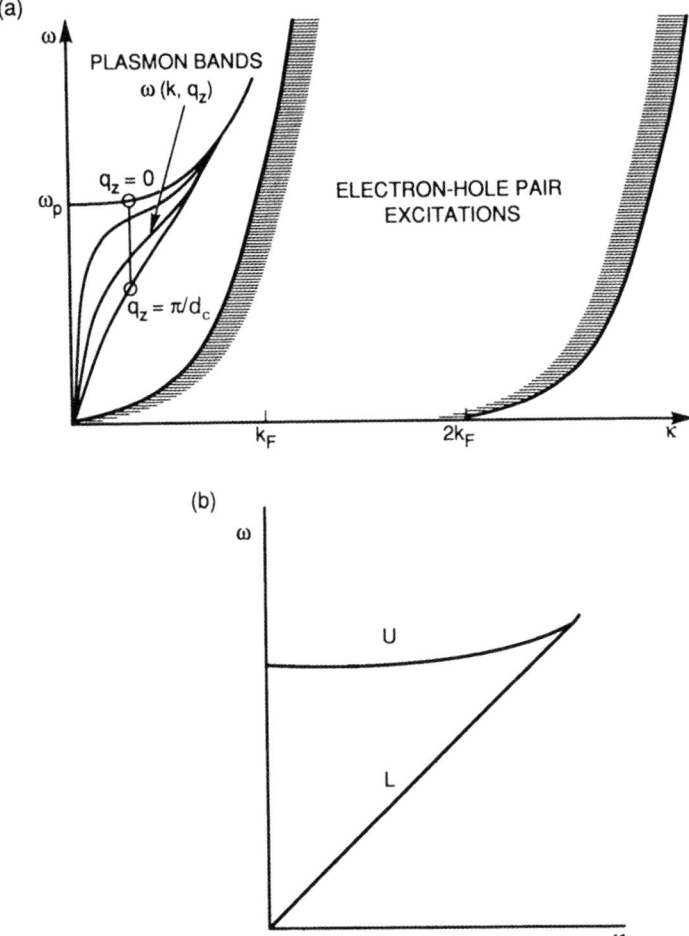

Fig. 4.5 Plasmons in a layered conductor: (a) plasmon band; (b) U and L branches.

the ordinary 3D metals, $\omega_{pl} \gg T$. As a result, the amount of total losses is independent of temperature. Because of the thermal population of the acoustic plasmon branches, the picture is entirely different for layered conductors. The passage of a charged particle through a layered conductor is accompanied by the energy loss:

$$\frac{\partial E}{\partial t} = (2\pi)^{-3} \int (\varepsilon_p - \varepsilon_{p-q}) W_q d^3 q \tag{4.13}$$

(q is a momentum transfer, and W_q is a total probability), which is temperature dependent (Kresin and Morawitz, 1991). Such a dependence ($\sim T^3$) could be observed, and is a direct manifestation of a peculiar plasmon spectrum in layered conductors.

Therefore, the plasmon spectrum in layered conductors such as cuprates represents the plasmon band (Fig. 4.5a). The density-of-states peaks near the upper ($q_z = 0$) and lower ($q_z = \pi/L$) boundaries. Qualitatively, one visualizes the plasmon band as a set of two branches (Fig. 4.5b), while the upper (U) branch is similar to that in usual metals. A very important feature of layered metals is the appearance of the lower (L) branch, which has an acoustic dispersion law and can be called "electronic" sound.

4.4.2 Plasmons in layered conductors: pairing

The plasmon mechanism implies that pairing occurs via the exchange of plasmons; in other words, plasmons play a role similar to that of phonons. In principle, the pairing can be provided by contributions of both channels, that is, by phonons and plasmons.

Let us make the following general remark. In recent years we have witnessed the discovery of many new superconducting materials: high-temperature cuprates, fullerides, borocarbides, ruthenates, MgB_2, metal-intercalated halide nitrides, intercalated Na_xCoO_2, and so on. Systems such as organics, heavy fermions, and nanoparticles have also been studied intensively. Many of these systems belong to the family of layered (quasi-2D) conductors and are characterized by, for example, strongly anisotropic transport properties. An interesting question raised by the observation of superconductivity in all the systems is the following: why is layering a favorable factor for superconductivity? In connection with this interesting question one should stress that layering leads to peculiar dynamic screening of the Coulomb interaction, which is important for the description of the superconducting state in layered conductors. As discussed previously, layered conductors have a plasmon spectrum that differs fundamentally from 3D metals. In addition to a high-energy "optical" collective mode, the spectrum also contains an important low-frequency part (see Fig. 4.5). The screening of the Coulomb interaction is incomplete, and the *dynamic* nature of the Coulomb interaction becomes important. The contribution of the plasmons *in conjunction with* the phonon mechanism may lead to high value of T_c.

As noted previously, we consider a layered system consisting of a stack of conducting sheets along the z-axis separated by dielectric spacers. Because of the large anisotropy of the conductivity, it is a good approximation to neglect transport between the layers. On the other hand, the Coulomb interaction between charge carriers is effective both *within* and *between* the sheets. To ensure charge neutrality we further introduce positive counter-charges spread out homogeneously over the sheets. In order to calculate the superconducting critical temperature T_c, one can use the equations for the superconducting order parameter and the renormalization function Z:

$$\phi_n(\mathbf{k}) = T \sum_{m=-\infty}^{\infty} \int \frac{d\mathbf{k}'}{(2\pi)^3} \Gamma\left(\mathbf{k}, \mathbf{k}'; \omega_n - \omega_m\right) \frac{\phi_m(\mathbf{k}')}{\omega_m^2 + \xi_{\mathbf{k}'}^2}\bigg|_{T_c} \quad (4.14)$$

Here, $\phi_n(\mathbf{k}) = \Delta_n(\mathbf{k}) Z, \omega_n(\mathbf{k}) = \omega_n Z, \Delta_n \equiv \Delta(\omega_n)$; we shall not write out the expression for Z. The interaction kernel can be written as a sum of electron–phonon and Coulomb interactions. The Coulomb term contains the plasmon excitations (see eqns. (4.6) and (4.8)).

A detailed analysis (Bill et al., 2003) based on eqn. (4.14) shows that the impact of dynamic screening is different for various layered systems. For example, for the intercalated metal halide nitrides (see Chapter 13) the plasmon contribution dominates.

4.4.3 The 3D case: "demons"

The value of the plasmon frequency for a one-band 3D metal is very high. As a result, the contribution of such plasmons to the pairing is not essential, because the corresponding coupling constant is negligibly small. However, when dealing with more complicated band structures one encounters additional low-lying plasmon branches. In the following we describe such a case.

Consider a metal with two overlapping energy bands, one of which (a) contains light carriers and the other ("b") is narrow and contains heavy carriers. It turns out that its energy spectrum is characterized by the presence of an acoustic plasmon branch ("demons"). This mode corresponds to the collective motion of the light carriers with respect to the heavy ones. It turns out that this mode is described by an acoustic dispersion law, and this acoustic branch is similar to phonons. Indeed, the system is described by the following equations (see, for example, Geilikman, 1966):

$$\Gamma_{aa} = (V_{aa} + \Pi_{bb} R) S^{-1},$$
$$S = 1 + V_{aa}\Pi_a + V_{bb}\Pi_b + \Pi_a \Pi_b R, \quad R = V_{ab}^2 - V_{aa}V_{bb} \tag{4.15}$$

where V_{aa}, V_{bb}, and V_{ab} are the Coulomb matrix elements. In this case, Γ_{aa} is defined by two parameters: $\alpha_a = \omega/qv_{Fa}$ and $\alpha_b = \omega/qv_{Fb}$, where the indices a and b refer to the two bands. Expression (4.15) is a generalization of the equation describing usual 3D plasmons (see, for example, Abrikosov et al., 1975) for the case of two bands.

Let us assume that $m_b \gg m_a$ and $v_{Fa} \gg v_{Fb}$, and consider the case of $\alpha_a \ll 1$ and $\alpha_b \gg 1$. Then,

$$\Pi_{a;0} = -\frac{m_a k_F}{\pi^2}, \quad \Pi_{b;\alpha_b \gg 1} = \nu\left(\frac{v_{Fb}q}{\omega}\right)^2 \tag{4.16}$$

In this case the light carriers provide screening, while the presence of the heavy ones results in the appearance of an additional plasmon branch. Indeed, from eqns. (4.15) and (4.16) and the condition $S(q,\omega) = 0$ we obtain the dependence $\omega \propto q$, that is, the new plasmon branch has an acoustic character.

These so-called "demons"—acoustic plasmons which are due to the presence of two groups of carriers of different masses—are similar to phonons. A similar picture can also arise in multivalley semiconductors, semimetals, and so on.

The possibility of superconducting pairing due to the exchange of "demons" has been considered in a number of papers (Garland, 1963; Geilikman, 1966; Frohlich, 1968; Ihm et al., 1981); see also the review by Ruvalds (1981).

5
Magnetic mechanism

5.1 Introduction

The basic idea that elementary excitations arising from magnetic degrees of freedom (magnons) may affect superconducting pairing between electrons predates the discovery of the superconducting cuprates by about two decades. It was considered in papers by Berk and Schrieffer (1966) and Doniach and Engelsberg (1966), who showed that incipient ferromagnetic order could suppress phonon-induced s-wave pairing. Pairing in ^{3}He associated with nearly ferromagnetic spin fluctuations in the p-wave channel was proposed by Anderson and Brinkman (1973). For the heavy fermion materials, d-wave pairing due to nearly antiferromagnetic spin-fluctuations was proposed by Scalapino, Loh, and Hirsch (1986–87) and Miyake, Schmitt-Rink, and Varma (1986).

We note at the outset that pairing interactions based on the repulsive ($U > 0$) single-band Hubbard model lead to d-wave pairing (Scalapino *et al.*, 1986). It is naturally not surprising that after the superconducting cuprates were found to be high-temperature superconductors (Bednorz and Mueller, 1986) with parent compounds showing antiferromagnetic order, various extensions of this result were applied to them and have led to a breathtaking advance in understanding the manifestations of magnetic interactions in highly correlated materials. In the following we present a very cursory overview of this huge field, and refer the interested reader to a few of the original publications in this field as well as existing summaries specifically aimed at spin-fluctuation-induced superconductivity.

For conceptual convenience it is useful to classify the very large number of theoretical papers on this subject into two categories by their approach: (1) Fermi liquid-based models, and (2) non-Fermi liquid models. In this survey chapter we concentrate on a few of the papers in category (1)—models based on Fermi liquid theory. This is dictated by the fact that Fermi liquid theory is a well-developed and mature framework with no need for a separate conceptual development or justification.

The topics we have, somewhat arbitrarily, chosen (from the huge number of possibilities) to include in our survey characterize those most accessible to the physics community familiar with the current state of understanding in condensed-matter theory. Nevertheless, we stress that the advances made in these particular areas are considerable and that the results are highly original and novel. All the models use a version of the Hubbard model as a starting Hamiltonian, and then use diagrammatic, perturbative, or numerical techniques to study the physical consequences, such as various instabilities (spin-density wave (SDW), superconductivity, energy gap formation, and so on), one- and two-particle Green's functions, and their relationship to experiment.

The models we review are the spin-bag concept proposed by Schrieffer, Wen, and Zhang (1989), the *t-J* model due to Emery (1987) and Rice and Zhang (1988), and the use of slave bosons due to Kotliar, Lee, and Read (1988, 1986). We also provide an overview of the work on the two-dimensional Hubbard model (1989) by Scalapino *et al.* In addition, we briefly include the idea of spiral distortions of the antiferromagnetic background for low hole doping due to Shraiman and Siggia (1988–89).

Category (2) is included for completeness. It may be argued that the most radical new ideas in condensed-matter theory are found in this area. In particular, the resonant valence bond (RVB) concept proposed and elaborated in a series of papers by Anderson (1987) and many others (Kivelson, *et al.*, 1987), as well as the anyon model due to Kalmeyer and Laughlin and coworkers (1987–88), are very exciting new concepts possibly containing the key to the new materials. However, it seems premature to discuss their current state of development, as it is still in a state of rapid change, making a summary very subject to being out of date by the time of publication of this book.

We therefore confine ourselves to a mostly qualitative discussion, and refer the interested reader to recent work by the main proponents in this field.

5.1.1 Localized versus itinerant aspects of the cuprates

It is a difficult and not well-understood problem how to describe carrier systems in a regime intermediate to the limits of an itinerant model and a fully localized model. The standard band structure approaches, used for the itinerant case, are based on a single-particle picture and differ in how they include correlations between the single-particle states in terms of some mean field approximation. On the other hand, for the localized case such as magnetic insulators, a spin Hamiltonian provides a starting point for the localized electrons and can describe their collective excitations, the spinwaves, and magnetic order, for example, ferromagnetic, antiferromagnetic, or some other more elaborate magnetic structure. The question arises of how to move from one description to the other, or more accurately to deal with situations which have aspects of both.

In this chapter we attempt to present—to the best of our ability—a broad overview of the very large number of papers addressing the role of magnetic effects in the cuprates. The motivation for these approaches arises from the experience and insights gained in dealing with strong electron–electron correlation effects in transition metals and transition metal oxides due to pioneering theoretical work by Mott (1968), Anderson (1963), and Hubbard (1964).

The difficulties are associated with the size of the local electron–electron repulsion, which can be estimated from atomic calculations. Typically—for a Cu 3D electron, for example—such energies are 15–20 eV and thus larger than the one-electron conduction bandwidth. We can visualize the difficulties for the traditional band-picture based on the single-particle approach by using a very simple conceptual model: the Hubbard model (1964) in one dimension. This considers a one-dimensional chain of N atoms and includes a nearest-neighbor hopping term t (kinetic energy gained by delocalizing over two sites) and an on-site Coulomb repulsion U between two carriers of opposite spin. The corresponding Hamiltonian for the carriers has the form:

$$H_{\text{Hubbard}} = -t \sum_{i,\sigma}(c^{+}_{i+1,\sigma} c_{i\sigma} + \text{h.c.}) + U \sum_{i} n_{i\uparrow} n_{i\downarrow} \tag{5.1}$$

where $c_{i,\sigma}$ its conjugate are the annihilation and creation operators for the carriers obeying anticommutation relations, $-|t|$ is the gain in energy by delocalizing the carriers (hopping integral), and U is the Coulomb repulsion between two carriers of opposite spin on the same site. If we choose a particular filling—for example, one carrier on each site, which in the absence of correlation effects ($U \equiv 0$) corresponds to a half-filled band (half-filled since each momentum k up to the Fermi momentum k_F can accommodate two spin states because of the spin degeneracy)—we can visualize the basic problem for the case of $U \gg |t|$. If we place one carrier per site we fill all sites, and assuming carriers on adjacent sites to have opposite spins we obtain an antiferromagnetic ground state. Now, to move a carrier costs energy U to transfer any electron from its site to the next compared to a gain of $-|t|$, so this state is an antiferromagnetic insulator.

A key question, then, is how to describe the system for a carrier concentration around this special, simple case, and to predict whether there is a transition as a function of carrier concentration from the insulating (antiferromagnetic state) to a conducting state for a range of values $0 < \eta < \infty$, where $\eta = |t|/U$.

This question remains open for the three-dimensional one-band Hubbard model, although the one-dimensional Hubbard model is well understood and a great deal of progress has been achieved for the two-dimensional Hubbard model, both by analytic and extensive numerical methods, by Scalapino (Scalettar et al., 1989) and a large group of collaborators.

A central argument presented by the proponents of magnetic models for the cuprates is the experimental fact that the starting materials for two of the superconducting compounds $La_{(2-x)}Sr_xCuO_4$ and $YBa_2Cu_3O_{(7-y)}$ ($y < 0.5$) are the antiferromagnetic insulators La_2CuO_4 and $YBa_2Cu_3O_{(7-y)}$ ($y > 0.5$). The antiferromagnetic phase for the undoped La_2CuO_4 material is well described by the 2D Heisenberg Hamiltonian for spin $s = 1/2$ with a reduced magnetic moment per Cu site of $0.4\mu_B$ due to the transverse quantum fluctuations (Chakravarty et al., 1988). In addition, detailed studies of the conduction-band structure of the same two compounds have established several important features of the actual orbital states involved in the formation of one-electron bands around the Fermi energy. These, by now generally agreed, features are:

1. The conduction band consists of a strongly hybridized mixture of Cu $d_{(x^2-y^2)}$ and O $2p_{x,y}$ orbitals (Matheiss, 1987; Yu et al., 1987; Cohen et al., 1989).
2. The Madelung energy is important in bringing the donor one-electron energy levels (copper 3D orbitals) within 1–2 eV of the acceptor one-electron levels (oxygen p orbitals) (Torrance and Metzger, 1989; Kondo et al., 1988; Feiner and deLeeuw, 1989).

3. Apart from the mixing of the CuO orbitals in the plane, there is also sizable bandwidth (about 0.5 eV) for O–O hopping (McMahan et al., 1990; Hybertsen et al., 1989).
4. The out-of-plane orbitals such as Cu d_{z^2} and O p_z orbitals of the apical oxygen atoms linking the CuO planes in the perpendicular direction may also play a role, and cannot be neglected in a rigorous treatment (Weber, 1988; Zaanen et al., 1991; Clougherty et al., 1989).

The consequences of these four points on the low-energy electronic charge and spin excitations are the following.

Because of point (1) the electronic state formed upon hole doping of the antiferromagnetic 2-1-4 compound—for example, by substituting Sr^{2+} for La^{3+}—is a change in the mixture of (d^9L), where d^9 denotes the Cu^{2+} ion hybridized with an oxygen (L for ligand) hole.

Point (2) is a key requirement for the occurrence of the basic (CuO) carrier planes with their short (covalent) Cu–O bond distance of 1.9 Å and the formation of hybridized states between Cu^{2+} and O^{2-} of mixed valence character. The large value of the Madelung energy (about 45 eV) arises from the ionic host lattice with large-valence ions (La^{3+}, Y^{3+}, and so on), as stressed by Matheiss (1987). It was considered in great detail by Torrance and Metfger (1989), Kondo (1988), and Feiner and deLeeuw (1989) in their study of the charge transfer nature of these materials and the occurrence of a phase boundary of metal-insulator character.

Point (3) allows for a second channel for the transport of holes in the CuO_2 planes in addition to CuO hopping. Because of the corner-linked nature of the (CuO_2) checkerboard, it facilitates derealization in the planes by motion on the oxygen sublattice. We note, however, that as long as antiferromagnetic order is present (small doping), the oxygen and copper spins are paired to the Zhang–Rice singlet (1988), and only a correlated motion of both is possible, as discussed below.

Point (4) is a very important aspect of the Jahn–Teller character of the copper ion in the cuprates, which was a central point in the minds of Bednorz and Mueller (1986) as enhancing electron phonon interactions. Note that it arises in the following manner. Because of the Jahn–Teller character of the Cu^{2+} ion, the orbital symmetry between degenerate $d_{(x^2-y^2)}$ and $d_{(z^2)}$ is removed, and as a consequence, the apical oxygen is considerably further away (2.4 Å) than the planar oxygens. Doping and certain lattice modes such as the octahedral tilt mode rehybridize these two orbitals and bring back some orbital degeneracy (Weber, 1988; Zaanen et al., 1991; Clougherty et al., 1989). The latter may play a role in several key physical phenomena, such as electron–phonon coupling, orbital relaxation of NMR T_1 processes, and others.

5.2 Fermi liquid-based theories

5.2.1 The spin-bag model of Schrieffer, Wen, and Zhang (1989)

One of the most carefully presented models for the origin of the pairing interaction between the carriers based on magnetic effects is the spin-bag model (Schrieffer et al., 1989). Physically, the model is presented as a version of the magnetic polaron, in which one carrier locally perturbs the antiferromagnetic order by its own spin, forming a spin bag. A second carrier within the coherence length of this local distortion (estimated to be in the order of two or three lattice constants) experiences an attractive interaction by sharing the bag.

In this model it is assumed that BCS pairing theory can be used and that the physical quasi-particles are spin-$\frac{1}{2}$ fermions dressed by a local antiferromagnetic distortion, which is treated in mean-field approximation. The single-band 2D Hubbard model is chosen to describe the carriers.

$$H = \sum_{k,\alpha} \varepsilon_k C^+_{k,\alpha} C_{k,\alpha} + \frac{U}{2}\frac{1}{N} \sum_{k,k',q} \sum_{\substack{\alpha,\alpha' \\ \beta,\beta'}} \delta_{\alpha\alpha'}\delta_{\beta\beta'} c^+_{k'\alpha'} c^+_{-k'+q\beta'} c_{-k+q\beta} c_{k\alpha}, \qquad (5.2)$$

$$\varepsilon_k = -2t(\cos k_x a + \cos k_y a).$$

Furthermore, it is assumed that one may start from the itinerant (weak coupling) limit of this model, in which $U \ll t$. The half-filled band case (see Fig. 5.1 for its Fermi surface) is known to undergo a spin-density wave (SDW) distortion, which introduces a SDW gap Δ_{SDW} and an antiferromagnetic ground state. For the true ground state $|\Omega\rangle$ with static spin-density wave included, assuming the SDW mean field to be polarized in the z-direction,

$$\langle \Omega | S^2_Q | \Omega \rangle = \sum_k \langle \Omega | c^+_{k+Q\alpha} \sigma^3_{\alpha\alpha'} c_{k\alpha'} | \Omega \rangle = SN, \qquad (5.3)$$

S is a variational parameter.

The Hartree–Fock form of the starting Hamiltonian (5.2)

$$H_{HF} = \sum_{k,\alpha} \varepsilon_k c^+_{k\alpha} c_{k\alpha} - \frac{US}{2} N \sum_{k,\alpha,\alpha'} c^+_{k+Q\alpha} \sigma^3_{\alpha\alpha'} c_{k\alpha'} \qquad (5.4)$$

is diagonalized by the following transformation to H_{SDW}:

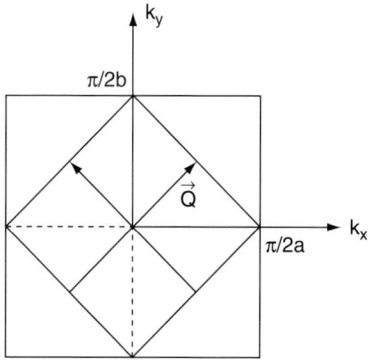

Fig. 5.1 Femi surface at half filling; $Q =$ nesting vector.

$$H_{SDW} = \sum_{k,\alpha'} E_k(\gamma_{k\alpha}^{+c}\gamma_{k\alpha}^c - \gamma_{k\alpha}^{+\nu}\gamma_{k\alpha}^\nu),$$

$$\gamma_{k\alpha}^c = u_k c_{k\alpha} + v_k \sum_\beta (\sigma^3)_{\alpha\beta} c_{k+Q\beta},$$

$$\gamma_{k\alpha}^\nu = v_k c_{k\alpha} - u_k \sum_\beta (\sigma^3)_{\alpha\beta} c_{k+Q\beta};$$

(5.5)

$$\left.\begin{matrix}u_k\\v_k\end{matrix}\right\} = \frac{1}{\sqrt{2}}\left(1 \pm \frac{\varepsilon_k}{E_k}\right)^{1/2}, \quad E_k = (\varepsilon_k^2 + \Delta_{SDW}^2)^{1/2}, \quad \Delta_{SDW} = \frac{U}{2}S.$$

where Δ_{SDW} is the SDW energy gap.

These antiferromagnetic correlations at momentum $Q = (\pi/a, \pi/b, 0)$ are built into a set of quasi-particle states, which have a mean-field character and consist of superposition of momentum states k and $k+Q$ which are either spin-up (even sublattice) or spin-down (odd sublattice). The next step in the spin-bag formalism is the use of the random-phase approximation for the calculation of the various susceptibilities in all available charge and spin channels (Figs. 5.2 and 5.3).

The charge and spin correlation functions (susceptibilities) are defined as

$$\chi^{00}(q,t) = \frac{i}{2N} \langle 0|T[\rho_q(t)\rho_{-q}(0)]|0\rangle,$$

$$\chi^{ij}(q,t) = \frac{i}{2N} \langle 0|T[S_q^i(t)S_{-q}^j(0)]|0\rangle;$$

(5.6)

$\rho_q = \sum_{k,\alpha} c_{k+q\alpha}^+ c_{k\alpha}$ charge density operator,

$S_k^i = \sum_{k,\alpha,\beta} c_{k+q\alpha}^+ \sigma_{\alpha\beta}^i c_{k\beta}$ spin density operator, (5.7)

$\sigma_{\alpha\beta}^i$ pauli matrix.

The next step is the construction of the pair potential for two electrons in the charge and three spin channels. The spin channels correspond to either fluctuations in the spin-density wave amplitude associated with the longitudinal component (σ_z) or the transverse components corresponding to orientational fluctuations ($\sigma_{+,-}$). It is found that for the parameter range considered the longitudinal fluctuations dominate the pair potential.

Using the random-phase approximation (RPA) the correlation functions in the presence of the Coulomb repulsion U may be expressed in terms of the non-interacting

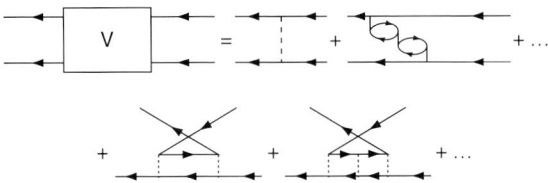

Fig. 5.2 Spin-bag formalism (RPA).

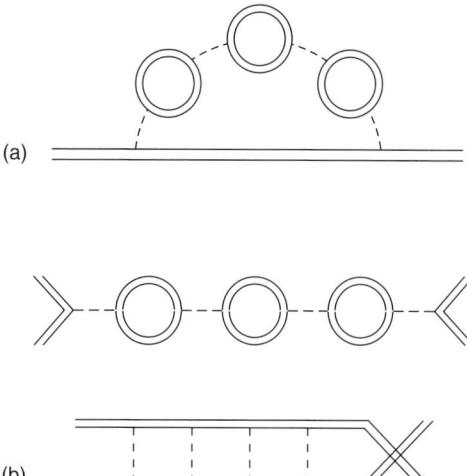

Fig. 5.3 (a) One-loop paramagnon corrections to electron self-energy in the SDW state; (b) Feynman diagrams used in RPA to calculate pairing in the SDW background.

correlation functions $\chi_0^{00}(q,\omega)$,

$$\chi_{RPA}^{00}(q,\omega) = \frac{\chi_0^{00}(q,\omega)}{1+U\chi_0^{00}(q,\omega)} \quad (5.8)$$

in the charge channel, and

$$\chi_{RPA}^{ii}(q,\omega) = \frac{\chi_0^{00}(q,\omega)}{1-U\chi_0^{00}(q,\omega)}\delta^{ij} \quad (5.9)$$

in the spin channel.

The authors then consider the orbital symmetry of the superconducting order parameter Δ for several forms of the carrier Fermi surface. They point out that it is possible to construct an order parameter with no zeros at the Fermi surface, as seems to be implied by experiments sensitive to such features (Krusin-Elbaum et al., 1989). This overcomes ingeniously the apparent difficulty of obtaining a p-wave or d-wave superconducting gap-function in magnetic models and experiments, suggesting s-wave-like symmetry of the gap. The idea is to use those regions of the Fermi surface in which the spin-density wave gap is finite for letting the superconducting gap change sign, leading to the appearance of a nodeless order parameter despite its underlying d-wave structure.

As proposed, the spin-bag model requires certain inequalities to hold for the two different length scales determining the superconducting and antiferromagnetic order. These length scales are set by the superconducting coherence length ξ_{SC} and the antiferromagnetic correlation length ξ_{SDW}. The inequality requires that $\xi_{SDW} < \xi_{SC}$. The cuprates are characterized by an extremely short superconducting coherence length of about 25Å in the CuO$_2$ planes, and an order of magnitude less in the direction perpendicular to the planes. The antiferromagnetic correlation length ξ_{SDW}, deduced from the width of the neutron scattering peaks, is of order 10–20 Å.

Extensions of the spin-bag approach to the interpretation of the photoemission and neutron-scattering results (Olsen et al., 1990; Shirane et al., 1989) were made in Kampft and Schrieffer (1990), and higher-order terms in the interaction terms were investigated in Brenig et al. (1991).

5.2.2 The t-J model (Emery, 1987; Zhang and Rice, 1988)

Although the initial theoretical papers used the Hubbard model to describe the carrier dynamics in the cuprates with the central role being played by the Cu(d^9,d^8), corresponding to Cu^{2+} and Cu^{3+} respectively, it was soon discovered from X-ray absorption and electron loss experiments (Tranquada et al., 1987–88; Neucker et al., 1989; Romberg et al., 1990) that the carriers introduced through doping resided mainly on the oxygens.

This makes necessary the inclusion of the oxygen orbitals in the starting Hamiltonian (1987), and increases the complexity of the parameter space. The work of Zhang and Rice (1988) was intended to resolve the controversy of whether a single-band or multiband Hubbard model was necessary to describe the essential physics. The starting Hamiltonian used is

$$H = \sum_{i,\sigma} \varepsilon_d d^+_{i\sigma} d_{i\sigma} + \sum_{l,\sigma} \varepsilon_p p^+_{l\sigma} p_{l\sigma} + U \sum_{1} d^+_{i\uparrow} d_{i\uparrow} d^+_{i\downarrow} d_{i\downarrow}, \quad (5.10)$$

where $d^+_{i\sigma}$ describes Cu $d(x^2-y^2)$ holes at site i with σ, and $p^+_{l\sigma}$ creates $O(\mathbf{2p}_{x,y})$ holes with spin σ at site l. U is the Coulomb repulsion at the Cu site. The orbital energy of the Cu holes is set at $\varepsilon_p = 0$ to define the energy scale and the case $\varepsilon_p > 0$ studied. The hybridization term H' is

$$H' = \sum_{i,\sigma l \leftarrow \{i\}} (V_i d^+_{i\sigma} p_{i\sigma} + \text{h.c.}), \quad (5.11)$$

where the sum over l is over the four oxygen neighbors of the Cu site at i. The hybridization matrix element V_{il} is proportional to the wavefunction overlap of the Cu and O holes. Taking account of the phases of the wavefunctions, one finds

$$V_{il} = (-1)^{M_{il}} t_0, \quad (5.12)$$

where t_0 is the amplitude of the hybridization, $M_{il} = 2$ if $l = i - 1/(2x)$ or $i - 1/(2y)$, and $M_{il} = 1$ if $l = i + 1/(2x)$ or $i + 1/(2y)$, The Cu–Cu distance is used as the length unit. It is further assumed that the following inequalities hold: $t_0 \ll U$, ε_p, $U-\varepsilon_p$. Undoped La$_2$CuO$_4$ has one hole per Cu. At $t_0 = 0$ (atomic limit) and $(U, \varepsilon_p) < 0$, all Cu sites are singly occupied, and all O sites are empty in the hole representation. If t_0 is finite but small, virtual hopping involving doubly occupied Cu-hole states produces a superexchange antiferromagnetic interaction between neighboring Cu holes. The

Hamiltonian reduces to an $S = 1/2$ Heisenberg model on a square lattice of Cu sites:

$$H_s = J \sum_{ij} (S_i \cdot S_j), \quad J = \frac{4t_0^4}{\varepsilon_p^2 U} + \frac{4t_0^4}{2\varepsilon_p^3}, \tag{5.13}$$

where S_i are spin-$1/2$ operators of the Cu holes, and $<ij>$ are nearest-neighbor pairs.

Doping introduces additional holes into the CuO_2 layers. In the atomic limit $t_0 \to 0$, the additional holes sit at Cu sites if $\varepsilon_p > U$ or at O sites in the opposite case. In the former case, the hybridization may be included by eliminating the O sites to give an effective Hamiltonian for motion on Cu sites only. This is obviously then a single-band Hubbard model. In the second case it is not so clear that one can eliminate the O sites. We consider this case in the following, and show that it also leads to an single-band effective Hamiltonian for the CuO_2 layer.

To explore the second case, consider the energy of an extra hole in La_2CuO_4. To zeroth order in t_0 this energy is ε_p for any O hole state. However, the system may gain energy from Cu–O hybridization, which leads to an antiferromagnetic (AF) superexchange interaction between O and Cu holes. Therefore, the first task is to choose a proper set of the localized O-hole states.

Consider the combinations of four oxygen hole states around a copper ion. They can form either symmetric or antisymmetric states with respect to the central copper ion:

$$P_{i,\sigma}^{(S,A)} = \frac{1}{2} \sum_{l \in \{i\}} (\pm 1)^{M_{il}} P_{l\sigma}, \tag{5.14}$$

where ($\pm$) correspond to the symmetric S or anti-symmetric A state and the phases of the p and d wavefunctions are defined in Fig. 5.4. Both the S and A states may combine with the d-wave hole to form either singlet or triplet spin states. To second-order perturbation theory, the energies of the singlet and triplet states for S are $-8(t_1 + t_2)$ and 0 respectively, where $t_1 = t_0^2/\varepsilon_p$ and $t_2 = t_0^2/(U-\varepsilon_p)$, while A has energy $-4t_1$. In band-structure language, S forms bonding and anti-bonding states,

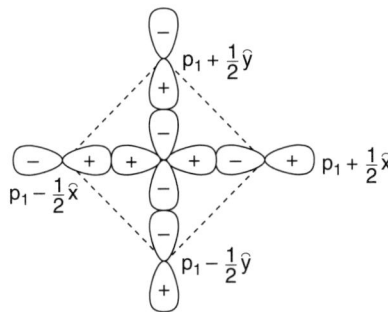

Fig. 5.4 Zhang–Rice hybridization of oxygen hole ($2p^5$) and copper hole (d^9).

while A is non-bonding (the self-energy of the central Cu^{2+} is $-4t_1$ in the absence of the O hole; the A state has the same energy, so it is non-bonding).

The large binding energy in the singlet S state is due to the phase coherence. This energy should be compared with the energy of an O hole at a fixed site l. In the latter case, the binding energy of a singlet combination of an O hole and its neighboring Cu hole is $-2(t_1+t_2)$, $1/4$ of the phase-coherent square S-state. Because the effective hopping energy of the O hole is t_1 or t_2 (depending on the spin configuration), much less than the energy separation of these localized states, we may safely project out the antisymmetric O-hole state, and work in the subspace of the S states. The energy of two O holes residing on the same square—that is, the configuration $P_i^{(S)} P_i^{(S)} d_{i\sigma}$—is $-(6t_1+4t_2)$, which is much higher than the energy of two separated O holes. It follows that the two holes feel a strong repulsion on the same square.

The localized states of the square are not orthogonal, however, because neighboring squares share a common O site. Therefore,

$$\left\langle P_{i\sigma}^{(S)} | P_{i\sigma'}^{(S)+} \right\rangle = \delta_{\sigma\sigma'}(\delta_{ij} - \frac{1}{4}\delta_{\langle ij\rangle,0}). \tag{5.15}$$

In analogy to Anderson's treatment for the isolated spin quasi-particle, one can construct a set of Wannier functions (N_s is the number of squares):

$$\Phi_{i\sigma} = N_s^{-1/2} \sum_k P_{k\sigma} \exp\{i\mathbf{k}\cdot \mathbf{R}_i\}, \quad P_{k\sigma} = N_s^{-1/2}\beta_k \sum_i P_{i\sigma}^{(s)} \exp\{-i\mathbf{k}\cdot \mathbf{R}_i\}, \tag{5.16}$$

and β_k is a normalization factor

$$\beta_k = [1 - \frac{1}{2}(\cos k_x + \cos k_y)]^{-1/2}. \tag{5.17}$$

The functions $\Phi_{i\sigma}$ are orthogonal, and complete in the symmetric hole space. $\phi_{i\sigma}$ combines with the Cu hole at the site i to form a spin singlet $(-)$ or spin triplet $(+)$:

$$\Psi_l^{\pm} = \frac{1}{2}(\phi_{i\uparrow}d_{i\downarrow} \pm \phi_{i\downarrow}d_{i\uparrow}), \tag{5.18}$$

with energies in second-order perturbation theory of

$$E_{\pm} = \sum_w |\langle \Psi_i^{\pm}|H|w\rangle|^2/\Delta E_w, \tag{5.19}$$

where w runs over all possible intermediate states, and ΔE_w is the zeroth-order energy difference between and Ψ and w; that is, $\Delta_w = \varepsilon_p - U$ or ε_p, depending on whether or not the state w contains a doubly occupied Cu hole. For the sake of simplicity the parameters are further simplified by choosing $\varepsilon_p = U - \varepsilon_p$; hence $t_1 = t_2 = t$, and the physics is expected to be essentially the same. One finds for the singlet and triplet energies:

$$E_{\pm} = -8(1 \mp \lambda^2)t, \tag{5.20}$$

$$\lambda = N_s^{-1} \Sigma \beta_k^{-1} \approx 0.96. \tag{5.21}$$

These energies are very close to those of a single square. Since $E_+ - E_- = 16t \gg t$, one may safely ignore transitions between $\{\psi_i^-\}$ and $\{\psi_i^+\}$, the system can be treated within the singlet $\{\psi_i^-\}$ subspace. It is important to stress that it is the phase coherence over the four oxygen sites that produces the large energy separation of the different symmetry states from the spin-singlet state of the Cu hole and the symmetric O hole.

After having obtained the proper Wannier functions with large binding energy, the next problem is the motion of these singlet states due to hopping. Since the O holes are superposed on a background of singly occupied Cu holes, the motion is correlated in the sense that if the state ψ^- moves from site j to i, a Cu hole simultaneously moves from site i to j. This process is represented by $\{\psi_i^-\} d_{j\sigma} \to \{\psi_i^-\} d_{i\sigma}$, with the kinetic energy described by an effective hopping Hamiltonian:

$$H_l = \sum_{i \neq j, \sigma} t_{ij} (\Psi_j^- d_{j\sigma}) + \Psi_j^- d_{j\sigma}, \tag{5.22}$$

with the effective hopping matrix element t_{ij} given within second-order perturbation theory by

$$t_{ij} = \sum_{<w>} \langle \Psi_j^- d_{j\sigma} | H' | w \rangle \langle w | H' | (\Psi_j^- d_{j\sigma})^+ \rangle / \Delta E_w. \tag{5.23}$$

We can evaluate in the original O-site representation $p_{i\sigma}$ and find two different types of two-step hopping processes contributing to t_{ij}. One involves spin exchange between the Cu and O holes, and is expressed by $t_{ij}^{(a)}$. The second is the effective O-hole hopping, and is found for $i \neq j$ as

$$\begin{aligned} t_{ij} &= t_{ij}^{(a)} - \tfrac{1}{2} t \delta_{\langle ij \rangle o}. \\ t_{ij}^{(a)} &= \tfrac{8\lambda t}{N_s} \sum_k \beta_k^{-1} \exp \langle ik \cdot (i-j) \rangle. \end{aligned} \tag{5.24}$$

Evaluating these equations for nearest neighbors i, j, $t_{ij} \approx -1.5\,t$ and all other effective hopping matrix elements are very small; that is, the next-nearest-neighbor $t_{ij} \simeq -0.16\,t$—one order of magnitude smaller.

We see that when a Cu d-hole is created at site i, the singlet state is destroyed at the same site. It follows that the state ψ_i^- is equivalent to the empty state of the d-hole at site i. The effective hopping Hamiltonian, after dropping the empty state operators, is then reduced to the form

$$H_t = \sum_{i \neq j, \sigma} t_{ij} (1 - n_{i,-\sigma}) d_{i\sigma}^+ d_{j\sigma} (1 - n_{j,\sigma}). \tag{5.25}$$

As was discussed at the beginning of this chapter, an effective Heisenberg Hamiltonian $H_J = J \Sigma \mathbf{S}_i \cdot \mathbf{S}_j$ holds for the doped system describing the antiferromagnetic interaction between the d-holes. The singlet state has no magnetic interaction with all other d-holes. In summary, an effective Hamiltonian H_{eff} has been found:

$$H_{\text{eff}} = H_t + H_J, \tag{5.26}$$

where both H_t and H_j refer only to Cu d-holes. This once again is the effective Hamiltonian of the single-band Hubbard model in the large-U limit.

5.2.3 Two-dimensional Hubbard model studies by Monte Carlo techniques

A different approach to the study of the role of magnetic excitations in the layered cuprates has been pioneered by Scalapino and a large number of collaborators (Scalettar et al., 1989; Krusin-Elbaum et al., 1989). It is an attractive alternative to the field-theoretic and phenomenological approaches discussed so far, as it uses numerical methods such as the quantum Monte Carlo simulation of the partition function and various correlation functions for a given model Hamiltonian such as the Hubbard model on a finite square lattice. The results obtained are exact for the size of lattice considered (currently 10×10 is typical maximum size), and can offer considerable insight into the short-range correlations induced in the appropriate correlation functions. Even with the relatively small-size lattices these calculations are numerically very intensive, as the number of configurations sampled varies as $N^3 L$.

It is assumed at the outset that a two-dimensional single-band Hubbard model describes the basic physics of the strongly correlated carriers in both the undoped (antiferromagnetic) starting materials ($La_2CuO_4, YBa_2Cu_,O_6$) and the doped, conducting, and superconducting materials ($La_{(2-x)}Sr_xCuO_4$, $YBa_2Cu_3O_{(7-y)}$, and so on).

The Hubbard model is simulated on a finite $N \times N$ lattice by calculating the partition function Z, the one-particle and two-particle Green's functions, and other quantities of physical interest which can be related to them. A very clear description of this approach was given by Scalapino at the Santa Fe meeting (1957) on correlated systems, with which we follow here.

It is instructive to point out that even a simple-looking model such as the single-band Hubbard model (5.1), which depends only the parameters $U/t, n_i = <n_{i\uparrow} + n_{i\downarrow}>$, the band filling, and the temperature T is amazingly rich in content, and as we will see, is successful in instructing us about some of the physics occurring in strongly correlated systems.

The key quantity to be calculated is the partition function Z, defined as

$$Z = \text{Tr } e^{-\beta(H-\mu N)} = \text{Tr } e^{-\Delta\tau H} e^{-\Delta\tau H} \cdots e^{-\Delta\tau H}, \quad (5.27)$$

where we have divided the interval $(0, \beta)$ into N imaginary time slices $\Delta\tau$ such that $N\Delta\tau = \beta = 1/k_B T$, and T is the temperature. Assuming that $\Delta\tau$ is small, we have approximately,

$$e^{-\Delta\tau H} \approx e^{-\Delta\tau K} e^{-\Delta\tau V} + O(\Delta\tau^2 tU), \quad (5.28)$$

where K is the first and V the second term in eqn. (5.1). Using the ideas introduced by Stratonovich (1957) and Hubbard (1959) (Wang, Evanson, and Schrieffer (1969)), it is possible to define a Hubbard–Stratonovich field $x_i(\tau_i)$ at site i, so that the bilinear term in n_i proportional to U may be written:

$$\exp(-\Delta\tau U n_{i\uparrow} n_{i\downarrow}) = \left(\frac{\Delta\tau}{\pi}\right)^{1/2} \exp[(\Delta\tau U/2)(n_{i\uparrow} + n_{i\downarrow})] \int_{-\infty}^{\infty} dx_{i(\tau_l)} \quad (5.29)$$
$$\times \exp\{-\Delta\tau[x_i^2(\tau_i) + (2U)^{1/2} x_i(\tau_i)(n_{i\uparrow} - n_{i\downarrow})\}.$$

The partition function Z can then be expressed as

$$Z = \sum_{\langle x_i(\tau_i) \rangle} \text{Tr}\, e^{-\Delta\tau h(x(\tau_n))} \ldots e^{-\Delta\tau h(x(\tau l))}, \tag{5.30}$$

where $h(x(\tau_l)) = \sum c_i^+ h_{ij}(x(\tau_l))c_j$ is an effective one-body interaction, which has to be averaged over all Hubbard–Stratonovich fields $h_{ij}(x(\tau_i))$. The remarkable advantage of his transformation is the reduction of the original Hamiltonian containing two-body interactions to an effective one-body Hamiltonian. The latter may, of course, be solved exactly, though we note that we have paid a certain price: namely, the presence of the effective Hubbard–Stratonovich field $h_{ij}(x(\tau_i))$, which has to be averaged over all values.

Carrying out the fermion trace in eqn. (5.30), one finds

$$Z = \sum_S \det M_\uparrow(\langle S \rangle) \det M_\downarrow(\langle S \rangle), \tag{5.31}$$

where $M_\sigma(S) = (1 + e^{\beta\mu}\Pi_1 e^{-\Delta\tau h(1,\sigma)})$. The Hubbard–Stratonovich reformulation of the original problem can be used as a starting point for stochastic sampling procedures. The product of the spin-up and spin-down Fermion determinants acts as a weight function for Monte Carlo importance sampling, just as the Boltzmann weight would do in a classical calculation. Note that it is a key requirement for this approach to be valid that the sign of the weight function $P(\{S\})$ is positive definite. We will return to this aspect later in a discussion of the problem areas of the Monte Carlo approach. The weight function $P(\{S\})$ is defined as

$$P(\{S\}) = \frac{\det M_\uparrow(\langle S \rangle) \det M_\downarrow(\langle S \rangle)}{\sum_{\langle s' \rangle} \det M_\uparrow(\langle S' \rangle) \det M_\downarrow(\langle S' \rangle)} \tag{5.32}$$

and generates configurations in a Monte Carlo algorithm distributed according to this weight. It is then possible to calculate with these configurations thermodynamic one- and two-particle Green's functions and equal-time expectation values.

To demonstrate the power and usefulness of this method, consider the one-particle Green's function

$$G_{ij}(\tau) = -\text{Tr}\, e^{-\beta H} T \langle c_{i\sigma}(\tau) c_{j\sigma}^+(0) \rangle / Z. \tag{5.33}$$

It can be found by summoning the different expressions

$$G_{ij}(\tau; \langle S \rangle) = (1 + e^{\beta\mu} e^{-\Delta\tau h(l-1,0)} \ldots e^{-\Delta\tau h(1,0)})^{-1} \tag{5.34}$$

over configurations $\{S\}$ with weight factor $P(\{S\})$. Then define the Monte Carlo average of G_{ij} below:

$$\overline{G}_{ij} = G_{ij}(\tau_j \langle S \rangle)_{MC} = \frac{1}{M} \sum_s G_{ij}(\tau_j \langle S \rangle), \tag{5.35}$$

where M is the number of Monte Carlo configurations. Figure 5.5 shows schematically how this sum, in principle, includes all Feynman diagrams. The simulation is therefore

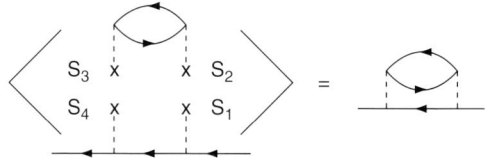

Fig. 5.5 One-particle Green's function: Feynman diagrams.

non-perturbative, and the results are exact except for errors (of order $\Delta\tau^2$) due to the finite step-size $\Delta\tau^2$. It is, of course, essential that all regions of phase space are sampled.

Other quantities of physical interest are the spatial Fourier transform of $G_{ij}(\tau)$, denoted $G_p(\tau)$, and its limit $\tau \to 0-$, which is the mean-particle number $<n_p>$.

The kinetic energy in the grand canonical ensemble average is given by

$$\langle T \rangle = \sum_{p,\sigma} \varepsilon_p G_\sigma(p, \tau \to 0). \tag{5.36}$$

One can also evaluate the Fourier transform of $G_p(p,\tau)$ with respect to τ at Matsubara frequencies $\omega_n = (2n+1)\pi T$:

$$G(p, i\omega_n) = \int_0^\beta d\tau e^{i\omega_n \tau} G(p, \tau) = \frac{1}{i\omega_n - (\varepsilon_p - \mu) - \sum(p, i\omega_n)}, \tag{5.37}$$

and exact information on the exact self-energy $\Sigma(p, i\omega_n)$.

Consider the evaluation of a two-particle Green's function within the Monte Carlo simulation approach. Specifically, define the s-wave pair field propagator $D(\tau)$ needed to describe superconducting correlations:

$$D(\tau) = -\left\langle T[\Delta(\tau)\Delta^+]\right\rangle, \tag{5.38}$$

where the operator Δ is defined by

$$\Delta = \left(\frac{1}{N^{1/2}}\right) \sum_l c_{i\uparrow} c_{l\downarrow}. \tag{5.39}$$

For a given Hubbard–Stratonovich configuration one calculates $D(\tau)$ by letting the up -and down-spin operator interact with the effective magnetic field given by the Hubbard–Stratonovich distribution. One finds:

$$\frac{1}{N} \sum_{ij} \int_0^\beta G_{ij\uparrow}((\tau)); \langle S \rangle) G_{ij\downarrow}(\tau; \langle S \rangle) d\tau. \tag{5.40}$$

Next, this has to be averaged over a set of Monte Carlo configurations $\{S\}$ with their appropriate weight $P(\{S\})$ to give the exact result (for a graphical representation, see Fig 5.6):

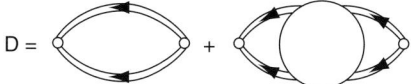

Fig. 5.6 Pair field propagator.

$$D = \frac{1}{N} \sum_{ij} \int_0^\beta \langle G_{ij\uparrow}(\tau_i \langle S \rangle) G_{ij\downarrow}(\tau_i \langle S \rangle) \rangle_{MC} d\tau. \tag{5.41}$$

Note that the first part of Fig 5.6 corresponds to the product of the average of two one-particle Green's functions, while the second part includes all effects of the pairing interaction. The two figures and eqns. (5.36)–(5.41) demonstrate a direct correspondence between the method of thermodynamic Green's functions and the Monte Carlo simulation, which allows the numerical study of these quantities in an exact form.

Having described the advantages and very attractive features of this approach, it seems fair to also discuss its drawbacks and limitations. The first is the constraint to finite and currently quite small (10×10 is a typical value) spatial lattices. The answers found need to be extrapolated to the bulk limit and, as the problem under study is a quantum-mechanical system, to zero temperatures ($\beta = L\delta\tau \to \infty$). In addition to the problems, which are similar to those arising in simulating classical systems, there are distinctly new difficulties for the quantum case, making it numerically much more extensive.

If we compare a classical problem with short-range interactions, the number of arithmetic steps for a Monte Carlo update at each site is of order N, the number of sites. For the Hubbard model, on the other hand, the non-local nature of the fermion determinant leads to $(NL)^3$ operations per site—hence a total of $(NL)^4$ operations for a spacetime sweep. It has been possible, by a refinement of algorithms, to reduce this to $N^3 L$ operations. Still, the residual increase of a factor N^2 for a typical $N \times N$ lattice contributes a factor of 10^4 to the computation for current lattice sizes.

In addition, the fermion matrices,

$$B = \prod_l \exp[-\Delta\tau h(l, \sigma)], \tag{5.42}$$

which are needed in the calculation, become ill-conditioned as the temperature is lowered, as the eigenvalues of B span a huge range between $\exp(-40)$ and $\exp(40)$ for a temperature $T = 0.1t$. Again, techniques have been developed to separate different energy scales and avoid swamping the small eigenvalues by round-off errors.

Finally, the most daunting difficulty in fermion simulations is the "sign" problem. It arises from the fact that although the partition function is positive, there is no requirement for the contributions from specific configurations also to be so. As a consequence, the product $\det M_\downarrow(\{S\}) \det M_\uparrow(\{S\})$ can fluctuate in sign. Only the presence of particle–hole symmetry for the half-filled Hubbard model, for example, guarantees the product of the two determinants for up- and down-spins to be positive definite.

If the product $\det M_\downarrow \det M_\uparrow$ changes sign, the probability distribution is taken proportional to the absolute value of the product,

$$\widetilde{P}(\{S\}) = \frac{|\det M_\uparrow(\langle S \rangle) \det M_\downarrow(\langle S \rangle)|}{\Sigma^{S'} |\det M_\uparrow(\langle S' \rangle) \det M_\downarrow(\langle S' \rangle)|}, \tag{5.43}$$

and expectation values of operators O are calculated as

$$\langle o \rangle = \frac{\left\langle o \, \mathrm{sgn}(\widetilde{P}) \right\rangle_{\widetilde{P}}}{\left\langle \mathrm{sgn}(\widetilde{P}) \right\rangle_{\widetilde{P}}}. \tag{5.44}$$

Here the subscript $\widetilde{P}$ indicates an average taken with the distribution function of eqn. (5.43). Note that if $<\mathrm{sgn}\,P>$, the average sign, becomes small, the Monte Carlo results will show large fluctuations. This is particularly troublesome for low temperature, as β becomes large and the average sign decays exponentially with increasing β. This was also found, in general, for a variety of band fillings and interaction strengths U studied. It is most unfortunate that just below half-filling the average sign decreases rapidly and makes the study of this physically very relevant region in the cuprates very difficult.

Finally, there is a difficulty in numerical simulations in extracting real frequency-domain results from the imaginary time data. Although the formal route of simply analytically continuing from $i\omega_n \to \omega + i\delta$ seems easy, the analytical continuation of numerical data is an ill-conditioned problem extremely sensitive to statistical fluctuations.

In the following, some examples of calculated quantities are given and are compared with results obtained from other studies. The local moment is defined as $m_{zi} = n_{i\uparrow} - n_{i\downarrow}$. Its average square is given by

$$\langle m_{zi}^2 \rangle = \langle n_{i\uparrow} + n_{i\downarrow} \rangle - 2\langle n_{i\uparrow} + n_{i\downarrow} \rangle. \tag{5.45}$$

At half-filling $<n_{i\uparrow} \cdot n_{i\downarrow}> = 1$, and the second term in (5.45), which measures double occupation, will decrease as U increases. For example, when $U = 8t$ (the bandwidth for two dimensions), the probability of double occupancy is 20% less than for $U = 0$.

Next consider the effective one-electron transfer term:

$$\frac{t_\mathrm{eff}}{t} = \frac{\langle c_{i\sigma}^+ c_{j\sigma} + c_{j\sigma}^+ c_{i\sigma} \rangle_U}{\langle c_{i\sigma}^+ c_{j\sigma} + c_{j\sigma}^+ c_{i\sigma} \rangle_0}. \tag{5.46}$$

For weak coupling, one finds analytically

$$\frac{t_\mathrm{eff}}{t} = 1 - \left(\frac{U}{t}\right)^2 \frac{\Delta\varepsilon_2}{\Delta\varepsilon_0}, \tag{5.47}$$

where $\Delta\varepsilon_0$ and $\Delta\varepsilon_2$ are functions of the band energy and Fermi functions. For strong coupling, on the other hand, one has

$$\frac{t_\mathrm{eff}}{t} = \frac{4(4t/U)(|\langle S_i \cdot S_j \rangle| + \frac{1}{4})}{\Delta\varepsilon_0}. \tag{5.48}$$

For the effective hopping term one can calculate the integral of the optical weight $\sigma(\omega)$,

$$8 \int_0^\infty d\omega\, \sigma(\omega) = 4\pi e^2 E_0(t_{\text{eff}}/t) \tag{5.49}$$

where E_0 is the energy per site for $U = 0$ at a given band filling $<n>$. One notes that at half-filling this will go to zero with large U. For $<n> = 1$ the integrated spectral weight increases as the vacancies created decrease the effective mass of the carriers. As an example of long-range effects, consider the equal-time magnetic moment correlation function:

$$C(l) = <m_{i+l}^z m_i^z>. \tag{5.50}$$

One finds that for the half-filled case $<n> = 1$ the antiferromagnetic correlations extend throughout the lattice, while for quarter-filling the correlations decay within one lattice spacing.

From knowledge of $C(l)$ one can calculate the magnetic structure factor $S(q)$:

$$S(q) = \sum_t \exp(iq \cdot l) <m_{i+1}^z m_i^z> \tag{5.51}$$

For the half-filled case, the peak at the antiferromagnetic wavevector (π,π) is very visible. For the case of long-range antiferromagnetic order,

$$\lim_{N \to \infty} S(\pi, \pi)/N = m^2/3, \tag{5.52}$$

where m is the antiferromagnetic order parameter. One can obtain m by extrapolating the spin–spin correlation function to infinite lattice spacing. For $U > 8t$, m is found to reach the value found for the $S = 1/2$ Heisenberg antiferromagnet (Reger and Young, 1988).

For doping off half-filling, the sign problem discussed previously presents difficulties. However, antiferromagnetic correlations are rapidly suppressed away from half-filling. The antiferromagnetic peak in the structure factor becomes smaller and seems to split into two peaks at $(\pi,\pi - \Delta_q)$ and $(\pi - \Delta_q,\pi)$. For the accessible temperature range $\beta < 6$, there is no evidence of growing incommensurate order.

Here we discuss the information content of the single-particle Green's function $G_{ij}^\sigma(\tau)$, which is related to several single particle properties:

$$G_{ij}^\sigma(\tau) = <Tc_{i\sigma}(\tau)c_{j\sigma}^+(0)>. \tag{5.53}$$

The physical quantities one may calculate are the single-particle momentum distribution $<n_{k\sigma}> = \sum_i \exp\{ik \cdot l\} G_{i+1,i}^\sigma(0^-)$, the band filling $<n>$,

$$\langle n \rangle = \frac{1}{N} \sum G_{i,j}^\sigma(0^-), \tag{5.54}$$

the chemical potential μ, and the compressibility $K = (1/n^2) dn/d\mu$.

Another very important quantity is the one-electron self-energy $\Sigma(K, i\omega_n)$, which can be calculated from the Fourier transform of G at the Matsubara frequencies ω_n:

$$\sum(k, i\omega_n) = i\omega_n - (\varepsilon_k - \mu) - G^{-1}(k, i\omega_n). \tag{5.55}$$

Experience with calculations has shown that $G_{ij}^{\sigma(\tau)}$ has the smallest statistical fluctuations in Monte Carlo simulations. It is, therefore, very helpful for obtaining one-electron results on fairly large lattices.

Several examples are given for $<n_k>$ for different sized lattices at $k = (\pi/2, \pi/2)$ $U = 4$ and $\beta = 6$; $6 \times 6 \to 14 \times 14$. The non-interacting Fermi function $f(\varepsilon_k = (\exp\{-(\varepsilon_k - \mu)\} + 1)^{-1}$ has been used with $\varepsilon_k - 2t(\cos k_x + \cos k_y)$, while the chemical potential μ has been adjusted to give $<n>$ for the half-filled and quarter-filled band respectively. We also show the mean field result:

$$\Delta = \frac{1}{N} \sum_k \frac{1}{E_k} \tag{5.56}$$

It is found that for the 16×16 lattice the mean-field gap is too large, and one needs a reduced gap near $k = (\pi/2, \pi/2)$.

The compressibility K is a good measure for the metallic or insulating character of the 2D Hubbard model. For $U = 0$,

$$K = \frac{2}{NT} \sum_k f(\varepsilon_k)(1 - f(\varepsilon_k)), \tag{5.57}$$

For small T, $K = N(\mu)$, the single-particle density of states:

$$K = \frac{\ln(16t/\mu)}{\pi^2 t}. \tag{5.57b}$$

The divergence of K as $\mu \to 0$ disappears for finite $U = 0$ as a gap opens in the single-particle spectrum.

The behavior of $\Sigma(k, i\omega_n)$ is very interesting. Mean-field theory gives for half-filling

$$\Sigma(k, i\omega_n) = \frac{\Delta^2}{i\omega_n + \varepsilon_k}, \tag{5.58}$$

For a non-interacting Fermi gas at $<n> = 1$, where $\varepsilon_k = 0$, the imaginary part of Σ has the form:

$$\text{Im } \Sigma(k, i\omega_n) = -\frac{\Delta^2}{\omega_n}. \tag{5.59}$$

This is different from the case of a Fermi liquid with

$$\text{Im } \Sigma(k, i\omega_n) = (1 - Z(k, i\omega_n))\omega_n. \tag{5.60}$$

Here $Z^{-1}(k_F, i\pi T \to 0)$ is the jump in the occupation, which defines the Fermi surface. As $Z^{-1}(k_F, i\pi T \to 0) > 1$, for a Fermi liquid Im $\Sigma(k, i\omega_n)$ has a negative slop for small ω_n different from the divergence in the presence of a gap.

From the Monte Carlo simulations, $\Sigma(k, i\omega_n)$ for the half-filled and quarter-filled cases ($k = (0, \pi)$, $k = (\pi/2, \pi/2)$) has been found and shows qualitatively quite different behavior. For $U = 4$ the quarter-filled system behaves like a Fermi liquid.

One can also extract real frequency information from the Monte Carlo data for

$$N(\omega) = -\frac{1}{4} \text{Im } G_{ii}(i\omega_n \to \omega + i\delta) \tag{5.61}$$

and for $U = 4$ and 8 find evidence for a gap. For $U < 2$, numerical difficulties make it impossible to draw unambiguous conclusions. Also, the single-particle spectral weight shows a gap, which increases with increasing U.

Of very great interest is what can be learned about superconducting correlation functions like the equal-time pair-field correlation function

$$D(l) = \langle \Delta_{l+i,i} \Delta_l^+ \rangle \tag{5.62}$$

and its structure factor

$$D(q) = \sum_l \exp(iq \cdot l) D(l). \tag{5.63}$$

The static pair field susceptibility is

$$P = \int_0^\beta d\tau \, \langle \Delta(\tau) \Delta^+(0) \rangle. \tag{5.64}$$

The pair field s-wave operator is $\Delta_l = c_{l\uparrow} c_{l\downarrow}$ and

$$\Delta = \frac{1}{N} \sum_l \Delta_l. \tag{5.65}$$

Also of interest are more general rotational symmetries like extended s-waves (S^*) and d-wave fields

$$\Delta_{s^*}(l) = \tfrac{1}{2}(c_{l\uparrow} c_{l+x\downarrow} - c_{l\downarrow} c_{l+x\uparrow}) + \tfrac{1}{2}(c_{c\uparrow} c_{l+y\downarrow} - c_{l\downarrow} c_{l+y\uparrow}),$$
$$\Delta_d(l) = \tfrac{1}{2}(c_{l\uparrow} c_{l+x\downarrow} - c_{l\downarrow} c_{l+x\uparrow}) - \tfrac{1}{2}(c_{l\uparrow} c_{l+y\downarrow} - c_{l\downarrow} c_{l+y\uparrow}), \tag{5.66}$$

which in momentum space have the form

$$\Delta_{s^*} = \left(\tfrac{1}{N}\right) \tfrac{1}{2} \sum_l \Delta = \tfrac{1}{\sqrt{N}} \sum_p (\cos p_x + \cos p_y) c_{p\uparrow} c_{-p\downarrow},$$
$$\Delta_d = \left(\tfrac{1}{N}\right) \tfrac{1}{2} \sum (\cos p_x - \cos p_y) c_{p\uparrow} c_{-p\downarrow}. \tag{5.67}$$

5.2.4 Spiral phase of a doped quantum antiferromagnet (Shraiman and Siggia, 1988–89)

A low density of vacancies in a 2D, spin-$\tfrac{1}{2}$ Heisenberg antiferromagnet leads (for a range of couplings) to a metallic phase with *incommensurate* antiferromagnetic order. This order corresponds to the staggered magnetization rotating in the plane with a wavenumber proportional to the density. It arises from the polarization of the antiferromagnetic dipole moments of the vacancies. This spiral state contains an interesting low-lying mode in its excitation spectrum, which has consequences for neutron scattering and the normal state resistivity.

CuO based compounds contain electrons with hard-core interactions on a 2D square lattice at a density of just less than one per site. Residual interactions are antiferromagnetic (AF) and are described by a *t-J* model Hamiltonian with vacancy hopping and spin-exchange terms:

$$\mathcal{H}_0 = -t \sum_{r,\sigma\hat{a}} c^+_{r+\hat{a},\sigma} c_{r,\sigma} + J \sum_{r,\hat{a}} s_r \cdot s_{r+\hat{a}}. \tag{5.68}$$

$\hat{a} = \hat{x}, \hat{y}$, the spin-$1/2$ Fermion operator $c_{r\sigma}$ is restricted to single occupancy, and $s_r = c^+_{r,\sigma} \tau_{\sigma\sigma'} c_{r\sigma}$ is the local spin operator.

In the absence of vacancies, the spin-$1/2$ Heisenberg model has long-range AF order at $T = 0$ (Chakravarty *et al.*, 1988). It can be described by the non-linear σ model for its long-wavelength properties. The ground state of a single vacancy is well understood. It forms a narrow-band ε_k of width $W \sim O(J)$ if $t \gg J$ (or $W \sim O(t^2/J)$ for $t \ll J$) with energy minima lying at $k_\nu = (\pm\pi/2, \pm\pi/2)$. The band is very anisotropic with mass $\mu_\perp \sim 1/W$, and for k perpendicular to the zone boundary is about a factor of 10 smaller than the parallel mass, $\mu_\parallel$, for $t = J$. Near the energy minima, the vacancy states involve a long-range dipolar distortion of the staggered magnetization, Q, and can be assigned an AF dipole moment $p_\alpha(k)$, which is a vector in both spin and physical space with magnitude of order min $(J^{-1}t, 1)$ and fc-dependence $p_\alpha(k) \sim \sin k_\alpha$.

Next consider the effect of a small number of vacancies, n ≪ 1. The *commensurate* Néel state is unstable (at $T = 0$ in 2D) for any n, toward a spiral state, in which the AF dipole moments of the holes order as a result of the polarization of the spins of the opposite values of the Fermi sea (located near k_v) in opposite directions. In the spiral state, the staggered magnetization, $\widehat{\Omega}$, rotates in a plane with a pitch scaling as $1/n$. There is also a *new*, low-lying excitation, the "torsion" mode, which is the collective mode associated with the transverse fluctuations of the dipole polarization, and hence with the fluctuations of the plane of the spiral in spin space. It also could be interpreted in terms of a rotation about the local $\widehat{\Omega}$ axis, which appears as an additional degree of freedom for *non-collinear* magnetic structures. We expect the AF correlation length (in some temperature range) to scale like $1/n$. The spiral phase is quite different from a $2k_F$ spin-density wave instability of the Fermi surface, and remains metallic with the vacancy Fermi surface ungapped.

These one-particle properties are included in a semiphenomenological effective Hamiltonian:

$$\mathcal{H}_{\text{eff}} = \sum_k \varepsilon_k \overline{\Psi}_k \Psi_k - g \sum_{q,a} p_a(q) \cdot j_a(q) - g' \sum_{q,k,a} \cos k_a \overline{\Psi}_{k-q/2} \hat{\tau} \Psi_{k+q/2} \cdot m(q) + \mathcal{H}_{\text{NL}\sigma}. \tag{5.69}$$

Here the vacancy is represented by a two-component spinor—the pseudospin of the vacancy arises from the sublattice structure induced by local AF correlation of the spin background. The spinor Ψ^α derived from a microscopic decomposition $c^{\alpha+}_{r,\sigma} = \Psi^{\alpha+}_r z^\alpha_{r,\sigma}$, $\Psi^{\alpha+}_r$ creates a Fermionic hole, and the spinor $z^\alpha_{r,\sigma}$ is a Schwinger spin boson. The explicit sublattice index α = A, B accommodates the staggered order (and labels the two-fold degeneracy of the vacancy ground state). The hopping part of $\mathcal{H}_0$ has the form

$$\mathcal{H}_{\text{int}} = -t \sum_{\langle r,r' \rangle} (\Psi_r^{+B} \Psi_{r'}^A \bar{z}_{r'}^A z_r^B + \text{h.c.}). \tag{5.70}$$

Define $\Psi^\alpha = h_{\alpha\alpha'} \Psi^{\alpha'}$ a $\hat{R}$ is an SU(2) rotation relating the spinor z_r, and hence the local direction of the staggered magnetization $\hat{\Omega}(r) = \bar{z}_r^A \tau z_r^A$ to a fixed basis.

The relation between $\mathcal{H}_{\text{int}}$ and eqn. (5.68) becomes more transparent in the spin-wave approximation after introducing $z_r^A = (1, a_r) z_r^B = (b_r, 1)$, transforming to the momentum representation, and identifying the staggered magnetization mode as $(a_q^+ - b_{-q})$ and the net magnetization as $(a_q^+ + b_{-q})$. **m** is the local magnetization operator, which is conjugate to $\hat{\Omega}$ and enters in the non-linear sigma model Hamiltonian,

$$\mathcal{H}_{\text{NL}\sigma} = \frac{1}{2} \Sigma_r (\chi^{-1} m^2 + \rho(\nabla_\alpha \Omega)^2), \tag{5.71}$$

where $\chi \sim 1/J$ is the susceptibility, $\rho \sim J$ is the spin-wave stiffness, and ∇_α is the lattice gradients. The second term in H_{eff} couples the background magnetization current $j = \hat{\Omega} \times \nabla_\alpha \hat{\Omega}$ with the AF dipole moment of the vacancies,

$$\hat{p}_a(q) = \sum_k \sin k_a \bar{\Psi}_{k-q/2} \hat{\tau} \Psi_{k+q/2}, \tag{5.72}$$

and gives rise to the dipolar interactions.

The phenomenological coupling constants g and g' are of order $\min(t, J)$. Note that $g = g' = t$ and $\varepsilon_k = 0$ would correspond to the bare hopping term of H_0; diagrammatically, $\varepsilon_k = 0$ emerges as the coherent part of the self-energy, while the reduction of g in the $t \gg J$ limit incorporates the downward renormalization of the coherent part of the propagator by incoherent processes. The scaling $g \sim J$ in the $t \gg J$ limit corresponds to the saturation of the single-vacancy dipole moment at $O(l)$. Note that for a low density of vacancies only states with $k = k_\nu$ are occupied; at small momentum transfers the dipolar coupling is dominant. The coupling to **m** is suppressed by an extra power of n.

Take the classical limit of eqn. (5.68):

$$\mathcal{H}_{\text{cl}} = -q\hat{p}_a(\hat{\Omega} \times \partial_a \hat{\Omega}) + \frac{1}{2}\rho(\partial_a \hat{\Omega})^2. \tag{5.73}$$

The dipoles clearly order, $\langle p_a \rangle \neq 0$, leading to a spiral AF phase with $\hat{\Omega} \times \delta\hat{\Omega} = g p_a \rho^{-1} \langle p_a \rangle$ where $\hat{\Omega}$ rotates in the plane perpendicular to the spin direction of $\langle p_a \rangle$ with a pitch along the spatial direction of $\langle p_a \rangle$. The magnitude of the inverse pitch (equivalently the incommensurability wave-number **Q**) is proportional to the total polarization and therefore to the density of holes $\mathbf{Q} = |\nabla \Omega| = g\rho^{-1}|p_a| \sim n$.

The quantum problem is more subtle, since—due to the Pauli principle—$p_a(k)$ cannot be identical for all vacancies. First consider the renormalization of the spin-wave propagator by particle–hole fluctuations as described by bubble diagrams. Only the stiffness constant is modified to lowest order in η, since the coupling to **m** vanishes at the zone center. One obtains a renormalized static stiffness $\tilde{\rho} = \rho - g^2\chi_d$ with

$$\chi_d = \frac{1}{6}\sum_{k,a}\sin^2 k_a \int dt <\bar{\Psi}_k\hat{\tau}\Psi_k(0)\bar{\Psi}_k\hat{\tau}\Psi_k(t)>, \qquad (5.74)$$

the static dipole susceptibility. The instability is signaled by $\tilde{\rho} < 0$:

$$\rho^{-1}g^2\chi_d > 1. \qquad (5.75)$$

For non-interacting particles at zero temperature $\chi_d = 4N_F<\sin^2 k_a>$, where N_F is the density of states at E_F and the angular brackets are an average over the Fermi surface. In 2D, $N_F = (\mu_\parallel/\mu_\perp)^{1/2}/2\pi$, and for t/J either large or small, $\rho^{-1}g^2\chi_d > 1$ is of order $1 \times (\mu_\parallel/\mu_\perp)^{1/2}$. Thus if the anisotropy $\mu_\parallel/\mu_\perp$ is as large as we expect, the instability occurs at arbitrarily low hole density n. The $n \to 0$ limit is, of course, singular, since the calculated stiffness renormalization applies only at wavenumbers $q^2 \ll k_F^2 \sim n$. Even in this limit, the absence of a threshold density is an artifact of 2D, and $T = 0$. In the classical limit $\chi_d \sim n/T$, while in 3D at $T = 0$, $\chi_d \sim n^{1/3}$, so that in either case the instability occurs for $n > n_c$.

A state with negative stiffness constant ρ evidently prefers to twist, and one may construct a mean-field theory for such a phase. Assume $\hat{\Omega} \times \nabla\hat{\Omega} = z\tilde{\Omega}_n \neq 0$ (in the CP' parameterization one has $<\bar{z}_{z+a}z_r> = iQ_a/2$ as the spiral order parameter). The mean field version of eqn. (5.68) reads where the $n_k^\pm$ are the occupation numbers of spin-z states. Minimizing with respect to Q_a leads to the self-consistency condition

$$\mathcal{H}_{MFT} = \sum_k \varepsilon_k n_k - gQ_a(\sin k_a(n_k^+ - n_k^-) + \frac{1}{2}\rho Q_a^2, \qquad (5.76)$$

with $n_k^\pm = (\exp\{\beta(\varepsilon_{k+} \mp gQ_a \sin k_a - \varepsilon_F)\}+1)^{-1}$, and ε_F the chemical potential. Equation (5.76) has a non-trivial solution when $g^2/\chi_d \geq \rho$ (note $\chi_d = g^{-1}(\nabla P_a/\nabla Q_a)_{|Q=0}$, hence the $\rho < 0$ instability can be identified with the onset of *incommensurate*, spiral order. For $g^2 \chi_d/2 < \rho < g^2\chi_d$ and $n \ll 1$, one finds a spatially uniform fully polarized state with $Q_a \sim n$ along the (1,0) or (0,1) directions and a positive stiffness constant. For $\rho < g^2\chi_d/2$, this model suggests phase separation, but this is unphysical, since the long-range Coulomb potential has been neglected. There may, however, also exist an intrinsically disordered phase.

It is clear from eqn. (5.76) that the dipolar polarization $<p_a> \neq 0$ is built up by populating opposite valleys of the Fermi sea with vacancies of opposite pseudo-spin. The ordering is reminiscent of Stoner ferromagnetism, except here there is no net spin polarization. In terms of the $\Psi_k^{A,B}$, fields which create spinless holes on sublattice A or B (shifting k by (π,π) sends $\Psi^A \to \Psi^A, \Psi^B \to -\Psi^B$ and is equivalent to rotating the pseudospin by π about z axis), any state with $<p_a>\perp\Omega$, as is the case for the spiral spin states, has only off-diagonal pseudospin order: that is, $<\Psi^{+A}\Psi^B> \sim Q_n \sin k_a$, but $<\Psi^{+A}\Psi^A> = <\Psi^{+B}\Psi^B>$.

The hole wave-function has equal weight on the two sublattices and a fixed phase between them. One can also see that for the spiral phase,

$$<c_\sigma^+(r')c_{\pm\sigma}> \sim \exp[iQ(r \mp r')/2]. \qquad (5.77)$$

To explore the low-energy excitations of the spiral state with given Q_a semiclassically, we examine the long-wavelength distortions of $\hat{\Omega}$, m^{-1} and $\mathbf{p}_a$ in eqn. (5.68). Define the linearized staggered magnetization operators ζ_r $[\zeta^+, \zeta] = 0$ in a rotating frame: $O_r^+\hat{\Omega} = (1 - \zeta^+\zeta/2, (\zeta - \zeta^+)/2, (\zeta + \zeta^+)/2)$ where $\tilde{O}_r$ is a uniform $O(3)$ rotation around z corresponding to the chiral state Q_a. Similarly, the long-wavelength transverse fluctuations of the dipole density are parameterized by

$$O_r^+ p_a = |<p_a>| \left(u_x, u_y, 1 - \frac{1}{2}(u_x^2 + u_y^2) \right), \tag{5.78}$$

with $u_x = \frac{1}{2}(\pi^+ + \pi - \zeta^+ - \zeta)$, $u_y = \frac{1}{2}(\pi^+ + \pi)/2i$, and $[\pi_r, \pi_{r'}^+] = \delta_{rr'}$. Introducing the magnetization operator η conjugate to ζ, $[\eta_r, \zeta_{r'}^+] = \delta_{rr'}$, substituting in eqn. (5.68), and expanding for small k and Q, one obtains

$$\mathcal{H} = \sum_k \rho Q^2 \pi_k^+ \pi_k + \frac{1}{2}\rho Q \cdot k (\pi_k - \pi_{-k}^+)(\zeta_k^+ + \zeta_{-k}) + \chi^{-1}\eta_k^+ \eta_k + \rho k^2 \zeta_k^+ \zeta_k. \tag{5.79}$$

Note that the imaginary part$(\zeta^+ - \zeta^-)$, decouples from π.

The corresponding branch (1) if the spectrum has the usual spin-wave form $\omega_k^2 = (ck)^2 (c^2 = \rho/\chi)$, with the zero mode being a global rotation of $\hat{\Omega}$ in the plane of the spiral (the phase mode of the spiral). The out-of-plane distortion of $\hat{\Omega}$, the $\zeta^+ + \zeta$ mode, mixes with the polarization fluctuations π, leading to more complex dynamics:

$$\omega_{R,T}^2 = \frac{1}{2}(c^2 k^2 + \rho^2 Q^4) \left\{ 1 \pm \left[1 - \frac{4\rho^2 c^2 Q^4 k_\perp^2}{(c^2 k^2 + \rho^2 Q^4)^2} \right]^{1/2} \right\}, \tag{5.80}$$

with $k_\perp^2 = k^2 - (k \cdot Q)^2/Q^2$. The upper branch ($R$) is spin-wave-like for $k \gg Q$; however, mixing with π introduces a gap $\omega_R = \rho Q^2$ at $k = 0$. The transverse fluctuations of the dipole polarization dominate the lower branch, the torsion mode ω_T, which lies entirely on the energy scale $\rho Q^2 \sim n^2 J$. Note that for $k \| Q$, $\omega_R = 0$. This is an artifact of the k, $Q \ll 1$ expansion. Higher-order terms in eqn. (5.71), $O((Q(k)^2))$, would induce a stiffness (or diffusivity) term for the polarization, and the corrected dispersion relation is obtained by replacing $k_\parallel^2$ by $\tilde{k}_\parallel^2 = k_\parallel^2 + \rho^{-1}D(k^2 - Q^2)^2$.

The remaining zeros of ω_T occur at $k_c = \pm Q_a$, and are associated with uniform rotation of the plane of the spiral in spin space.

The transverse fluctuations of the dipole polarization arise as a collective mode, involving slowly varying perturbations of the Fermi distributions of the vacancies and of the mode structure. The dipole (or torsion) mode is limited to small momentum transfers $q < k_F \sim Q^{1/2}$.

The behavior of the static spin-correlation function with temperature and doping should furnish a useful experimental signature of the spiral state. We expect the incommensurability $\mathbf{Q}$ to appear for $T < \varepsilon_F \sim n$ (for $n > n_c$) and remain constant. The spatial direction of $\mathbf{Q}$ is difficult to predict without a more realistic Hamiltonian; however, we note that in the $Q = (1, 1)$ state there would be a net interlayer exchange of $O(Q^2)$ (for the LaCuO-based material which may make it more favorable). The spin correlations are anisotropic because of the softness of the torsion mode in the

$k \| Q$ direction. In the classical limit, the fluctuations of Ω, parameterized by the polar angles Θ, Φ (with the uniform rotation about z taken out) have energy

$$E \sim [\rho k_\perp^2 + D(k_\|^2 - Q^2)^2]\Theta_k^2 + \rho k^2 |\Phi_k|^2, \tag{5.81}$$

where $q \cdot k_\perp = q \times k_\| = 0$.

The spin-order along $\mathbf{Q}$ can disappear for $T \sim JD^{1/2}Q \ll \varepsilon_F$ as the torsion mode "melts," leaving the correlation length $\chi \sim O(n^{-1})$. The presence of domains of different Q and topological defects in the spiral structure, which may be quenched in, will make the correlations more isotropic. However disordered the torsion mode becomes, it cannot reduce the correlation length beyond $Q^{-1} \sim n^{-1}$, since on shorter length scales the spins obey the Heisenberg model, which is ordered. These arguments favoring a spiral state do *not* require long-range order, but only $k_F \ll \zeta$, which is satisfied for $n \ll 1$.

Holes also make a potentially important non-hydrodynamic contribution to spin-wave damping via the imaginary part of the dipole susceptibility χ_d. For $k \ll k_F$ one obtains $\Gamma = \text{Im}(\omega/ck) \sim g^2 \mu c/\rho v_F$, provided $c < v_F$ or $\Gamma = g^2 \mu v_F/\rho c$ for a window around $k \sim \mu c$. For weakly localized holes when the momentum conservation constraint is absent, one finds approximately $\Gamma \sim g^2 \mu n/c$. The latter may be important at $k \ll k_F$ for low densities where $c > v_F$ and Γ is otherwise zero.

The torsion mode lying entirely at frequencies $O(n^2 J)$ is easily saturated thermally and is a possible source of the linear T resistivity of the normal state. The resistivity would arise from direct (spin flip) coupling of the holes with the collective torsion mode analogously to itinerant ferromagnets. (Scattering by a spin-diffusion m mode would give $\rho \sim T^2$ or T^l, $l > 2$.)

It has been shown that mobile vacancies in a background with at least short-range AF order have dipolar interactions, which induce their collective polarization, leading to a spiral AF phase (even if the vacancies are not strongly localized). The spiral order implies, at the one-particle level, correlations between the wavenumber and pseudospin which extend throughout the Fermi sea.

We expect that along with the ordered spiral phase there might exist (for larger values of effective coupling constant or higher vacancy density) a disordered state with local spiral twist. Dipolar interactions may also introduce pair correlations and superconductivity in singlet and triplet channels. The competition and coexistence of superconductivity and local spiral order can be explored on the mean-field level using the four-fermion Hamiltonian with dipolar interactions obtained by integrating out the spin waves.

Finally, it appears that the hopping-induced dipole moment of the vacancy persists even in the more realistic two-band model of CuO planes.

5.2.5 Slave bosons

One technique for including aspects of the charge degree of freedom in models mainly concerned with magnetic (spin) degrees of freedom is the slave boson technique (Barnes, 1976; Read and Newns, 1985; Coleman, 1984, Millis and Lee; 1987; Auerbach and Levin, 1986).

The use of slave bosons was first proposed to deal with the infinite-U Anderson model in mixed-valence systems (Read and Newns, 1985; Coleman, 1984; Auerbach and Levin, 1986), It introduces a local constraint by extending the Fock space to include auxiliary (slave) boson fields, which track the occupation of the local impurity level. The constraint is implemented by a Lagrange multiplier field λ_i. At the mean-field level, the boson fields b_i and λ_i are constant numbers b and λ; b multiplicatively renormalizes the p–d hopping yerm, while λ additively shifts the atomic levels of the d orbitals (see Eq.(5.82)).

The principal advantage of the additional boson fields, whose number operator is constrained together with the local fermion variables, is that standard field-theoretic methods can be used. Qualitatively, the splitting of the original fermion quasi-particle into a fermion-boson description allows the dynamics of the quasi-particle hopping accompanied by local spin-polarization changes to be separated into spin motion and spin and charge excitation changes.

We demonstrate in the following section how the use of the slave boson technique allows description of the quasi-particles in the cuprates in terms of renormalized parameters derived from band structure or cluster calculations. Our treatments closely follows Kotliar et al. (1988) and Millis and Lee (1987).

Consider a model for $La_{2-x}Sr_xCuO_4$ with orbitals on planar Cu and O sites, with the operator $d_{i\sigma}^+$ creating a hole in the $3d_{x^2-y^2}$ orbital with energy ε_d^0, and $c_{i\sigma}^+$ creating a hole in the $2p_{x,y}$ orbital with energy ε_p. A hopping matrix element t_{pd} is also assumed between nearest-neighbor Cu and O sites. We also include Hubbard parameters U_d, U_c for each site, and take $U_d = \infty$, $U_c = 0$. For undoped La_2CuO_4, we assume $\varepsilon_p - \varepsilon_d^0 > 0$, leading to configurations Cu^{2+} and O^{2-}. The holes introduced by doping with strontium occupy oxygen orbitals. The energy difference $D = \varepsilon_p - \varepsilon_d^0$ plays a role analogous to the Coulomb repulsion U in the Hubbard model (Zaanen, et al., 1985).

Introduce a slave boson at each copper site. As only a single linear combination of the two oxygen orbitals couples to the Cu d orbital (of b_{1g} symmetry (Zhang and Rice, 1988)), we have a two-band model with energies

$$E_{1,2}(k) = (\varepsilon_d^0 + \varepsilon_d^0 \pm R_k^0)/2,$$

$$R_k^0 = \left((\varepsilon_p - \varepsilon_d^0)^2 + 16 t_{pd}^2 \gamma_k^2\right)^{1/2}, \quad (5.82)$$

$$\gamma_k^2 = \sin^2(k_x/2) + \sin^2(k_y/2)$$

and for one hole per unit cell, the lower band E_1 is filled.

This model is formally equivalent to the Anderson lattice, and we use results from earlier work on large orbital degeneracy (N) expansion (Millis and Lee, 1987; Auerbach and Levin, 1986) Redefine the kinetic energy (hopping) term to be $t/N^{1/2}$ (for example, for $N = 2$, $t = \sqrt{2}\, t_{pd}$). Next assume that the N-fold degenerate d states can accommodate $Q = Nq_0$ holes and set $q_0 = 1/2$. The deviation from half-filling, δ, is defined by $H = Q(1 + \delta)$ for the total number of holes per unit cell. The salve bosons are introduced to enforce the constraint.

$$b_i^+ b_i + d_i^+ d_i = Q. \quad (5.83)$$

$N = 2$ represents the spin degeneracy. Only if $N = 2$ and $Q = 1$ is the constraint equivalent to $U_d = \infty$.

From the $1/N$ expansion we know that the lowest order gives mean-field theory (Hartree–Fock approximation) with $<b> = b_0 = \sqrt{N}r_0$, and the effective hopping matrix element is $<\sigma_0> = (t/\sqrt{N})b_0 = tr_0$. In addition, the position of the d-level is renormalized, $\varepsilon_d^0 \to \varepsilon_d$, and the renormalized band structure obeys eqn. (5.1) with the replacements $t_{pd} \to \sigma_0$ and $\varepsilon_d^0 \to \varepsilon_d$. These bands are the actual quasi-particle bands, and one finds the chemical potential by filling the lower band with H/N holes. The Fermi surface contains H holes, and the Luttinger theorem is obeyed.

We have to solve for the mean field parameters r_0, ε_d.

$$r_0^2 + \sum_k u_k^2 f(E_1(k)) = q, \quad \varepsilon_d - \varepsilon_d^0 = \frac{t}{r_0} \sum_k u_k v_k \gamma_k f(E_1(k)), \quad (5.84)$$

where

$$\left.\begin{array}{c} u_k^2 \\ v_k^2 \end{array}\right\} = (1 \pm (\varepsilon_p - \varepsilon_d)/R_k) \quad (5.85)$$

are the weights of the Cu and O orbital in the lower band.

The limit $4t \ll D$ can be studied analytically in terms of the two dimensionless parameters D/t and δ. For $\delta < 0$ the shift in ε_d can be found perturbatively from

$$r_0^2 = -\frac{\delta}{2} + \sum_k (1 - u_k^2) f((E_1(k))) \quad (5.86)$$

for $t \ll D$, $u_k^2 \approx 1$ and $r_0^2 = -\delta/2$. The bandwidths of the two bands are given by $8r_0^2 t^2/D$; that is, they are narrowed by a factor $|\delta|$ compared to the bare bandwidth $8t_{pd}^2/D$. This is quite analogous to the band narrowing in the Hubbard model for $U/t \gg 1$.

The physical situation described is a Fermi liquid obeying Luttinger's theorem, in which δ holes in the lower "Hubbard" band correspond to $H = N/2(1 + \delta)$ holes, which have a mass-enhancement $m^*/m \approx 1/\delta$. We also see by inspection of eqn. (5.86) that $r_0 = 0$ at $\delta = 0$, so that the perturbative solution discussed so far disappears for $\delta > 0$. There exists another solution, in which ε_d is renormalized almost up to ε_p, and therefore $\varepsilon_p - \varepsilon_d = 4t^2/D \sum_k \gamma_k^2 f(E_1)$. If $\sigma_0 \ll \varepsilon_p - \varepsilon_d$ (which can be seen to hold), we can expand $\Sigma_k(1 - u_k^2) f(E_1)$ to be $\gamma_0 4t^2/D \Sigma_k \gamma_k^2 (E_1) \gg \gamma_0^2$. For that case we have the solution

$$r_0^2 \approx \frac{\delta}{2} \left(\frac{2t}{D}\right)^2 \sum_k \gamma_k^2 f(E_1(k)) \quad (5.87)$$

and we find the bandwidths are still given by $8\delta t_{pd}^2/D$, as was the case for $\delta < 0$ (see Fig 5.7).

The discontinuous jump in the position of the bands relative to the ε_p, ε_d^0 for δ changing sign is due to the fact that the chemical potential must be near ε_p for $\delta > 0$, because the additional holes occupy oxygen sites. The holes on oxygen can delocalize

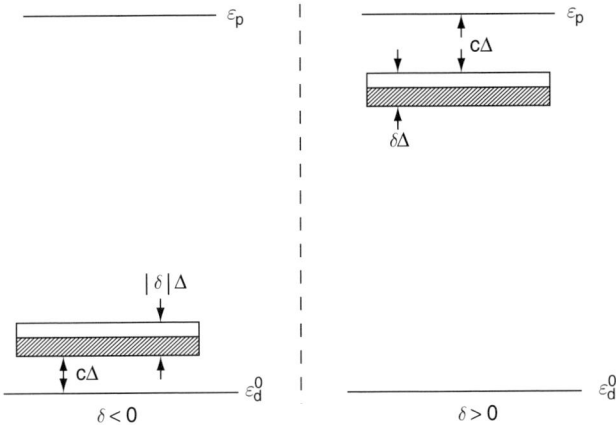

Fig. 5.7 Narrow band.

by virtual hopping into Cu sites with an energy gain of $4t_{pd}^2/D$ from perturbation theory, explaining the narrow band below ε_p (see Fig 5.7).

Physically, we do not expect band narrowing to continue for arbitrarily small $|\delta|$, as we anticipate an antiferromagnetic (Néel) state at and near half-filling. We should compare J with the delocalization energy $4t_{pd}^2/D$ and expect the Fermi liquid picture to hold for $\delta > t_{pd}^2 D(D+Uc)$.

We note that it is possible to go beyond the present mean-field approximation and include fluctuations in the slave bosons (Kotliar et al., 1988).

Extensions to multiband models with several slave bosons can be found in Grilli et al. (1990), Castro et al. (1991), and Feiner et al. (1992), which show encouraging agreement with experimental data.

5.3 Non-Fermi-liquid models

5.3.1 The resonant valence bond (RVB) model and its evolution

It was proposed by Anderson (Anderson et al., 1987) at the very beginning of the discovery of superconductivity in the cuprates (Bednorz and Mueller, 1986) that magnetic fluctuations were responsible for the pairing. In addition, Anderson suggested (Anderson et al., 1987) that a novel type of ground state consisting of singlet pairs may form a quantum spin liquid which could move on the underlying square lattice by resonant tunneling once the filling moved away from half, in which all the spins are paired. Such a RVB state was proposed by Anderson for a triangular lattice (Anderson, 1987) a long time ago. This fascinating proposal was rapidly taken up and expanded upon by many others—in particular, by Kivelson, Rokhsar, and Sethna (1987).

Because of the large degeneracy of spin configurations near the half-filled state, and also because of the possibility of quantum fluctuations introducing additional novel possibilities for the $s = 1/2$ state of the carriers, it took some time before it was demonstrated explicitly by calculation (Reger and Young, 1988) that in fact, the spin–$1/2$ Heisenberg model in two dimensions has an antiferromagnetic ground state with an effective moment of $0.4\mu_B$ reduced by transverse quantum fluctuations from the classical value of $1/2\mu_B$.

One of the central points stressed by Anderson (1987–88) is the nature of the normal metallic state and, in particular, the temperature dependence of the resistivity, proportional to the temperature T. This empirical law seems to be obeyed by almost all the newly discovered cuprate superconductors with high superconducting transition temperatures.

Furthermore, it was emphasized in the evolution of these ideas (Anderson, 1987–88) that the single-band Hubbard model contains the physics essential for the understanding of the low-energy (<1 eV) behavior of the quasi-particles. The $U \gg 4t$ limit of the Hubbard model is assumed to be the relevant parameter regime, and some of the exact results found for the 1D Hubbard model such as charge and spin separation (Haldane, 1981) are postulated to also hold in higher dimensions.

A particularly successful application of these ideas was made to experimental determination (Ong et al., 1991) of the temperature dependence of the Hall effect (Anderson, 1991).

We refer the interested reader to the literature on this topic (Anderson, 1987–88), and have also listed several of the proceedings of recent conferences and workshops on the superconducting cuprates (Bedell *et al.*, 1990; Baskaran *et al.*, 1991; Mueller and Olsen, 1988; Shelton *et al.*, 1989; Tachiki *et al.*, 1991).

5.3.2 Anyon models and fractional statistics

The second proposal for a non-Fermi-liquid origin of superconductivity in the cuprates (Bednorz and Mueller, 1986) arose from the striking success of the ideas of fractional statistics (Wilczek, 1982; Laughlin, 1983; Halperin, 1984; Arovas *et al.*, 1984) in explaining the fractional quantum Hall effect (Laughlin, 1986).

In its application to the cuprates, it is argued that spin-charge separation may occur as it does for the one-dimensional Hubbard chain (Haldane, 1981), which has been named the "Luttinger liquid." The separate spin- and charge-carrying entities are shown to obey Fermi and Bose statistics respectively. This break-up of the hole carrier created by doping into separate excitations—"spinons" and "holons"—was also an early part of the RVB description of the cuprates.

It was proposed by Kalmeyer and Laughlin (1987) that both the chargeless, spin–$1/2$ excitations ("spinons") and the charged, spinless excitations ("holons") obey 1/2-fractional statistics. The basic analogy is that as charge is fractionalized in the fractional quantum Hall effect, here the spin–$1/2$ is also fractional.

Although the arguments for such a model to apply for the cuprates are very appealing on aesthetic grounds and have found strong resonances with other theorists (Laughlin, 1989; Wen et al., 1989; Laughlin, 1988), it was soon realized that it made some striking predictions for experimental observation, such as its breaking of parity P(reflection in the plane) and time-reversal symmetry T.

Experiments on muon-spin precession (Kiefl et al., 1990) placed stringent upper limits on local magnetic fields estimated by Halperin (1991), and experimental tests of optical activity and dichroism showed conflicting results. In fact, the most sensitive experiment (Spielman et al., 1990) did not show any effect.

We refer the interested reader to the literature on this topic, and have listed several of the proceedings of recent conferences and workshops on the superconducting cuprates (Bedell et al., 1990; Baskaran et al., 1991; Mueller and Olsen, 1988; Shelton et al., 1989; Tachiki et al., 1991).

5.4 Conclusions

In this chapter we have considered a very tiny subset of the huge number of papers on spin-fluctuation effects in the cuprates. As the major emphasis of the rest of the book is on other mechanisms and our own view of these materials, it seemed necessary for balance to mention this very active field.

The relevant regime for superconductivity starts at small values of the carrier concentration n_c, near the half-filled band, which characterizes the antiferromagnetic starting materials La_2CuO_4 and $YBa_2Cu_3O_6$. The central idea of the models we have described is that the motion of the carriers introduced by doping is dominated by the proximity to antiferromagnetic order in the CuO_2 planes. The intra-atomic Coulomb repulsion U between two holes on the same site is assumed to remain the dominant energy scale ($U > 4t$). Superconductivity within the Fermi-liquid-based models such as the spin-bag and the two-dimensional Hubbard model occurs via the exchange of paramagnons, and leads to a d-wave order parameter for the superconducting gap function.

In the non-Fermi-liquid models the superconducting state is not yet expressed in a generally accepted form: in Anderson's view, interlayer hopping of the Luttinger liquid is invoked, and a gage force arising from the fractional 1/2 statistics of the carriers leads to pairing in Laughlin's anyon model.

Applicability of the Hubbard model to pairing in cuprates requires an experimental manifestation of the d-wave pairing.

6
Experimental methods: Spectroscopic

In this chapter we will describe several of the most important experimental techniques that have been used to probe the most fundamental properties of the superconducting state, the energy gap, and the pairing interaction. The techniques that will be described are all spectroscopic: they involve the tunneling of quasiparticles through an insulating barrier or through a point contact, the interaction of electromagnetic waves or photons with a superconducting film or surface, the attenuation of ultrasonic sound waves, and the relaxation and or resonance of muons interacting with a superconducting compound.

6.1 Tunneling spectroscopy

The BCS theory was very successful in explaining many of the fundamental properties of the elemental superconductors. However, the mechanism of the pairing in many of the more exotic materials was uncertain. Tunneling spectroscopy has proved to be a very powerful tool in probing both the nature of the interaction as well as verifying the validity of the strong coupling theory of Eliashberg (1960). In this section we will describe the experimental techniques that are used to obtain the tunneling spectra, present the relevant equations for the tunneling density of states, and describe the beautiful inversion method developed by McMillan and Rowell (1969).

6.1.1 Experimental method

There are several experimental methods that have been used to generate tunneling spectra. The most widely used are methods based on the deposition of a barrier and a counter-electrode. The simplest manifestation of this method requires the deposition of the superconducting electrode, the formation of the barrier, either by oxidation of the superconductor, or by a deposited insulating layer, and the final deposition of another metallic or superconducting electrode at right angles to the original film. This simple cross-stripe structure can be made using contact masks, and does not require sophisticated lithography. An example of such a structure is illustrated in Fig. 6.1. Subsequently, junctions have been made by a trilayer process where the electrodes and the barrier are deposited sequentially in the same deposition run and the junction is defined photolithographically afterward. This method allows the preparation of small and very uniform tunnel structures. Point contacts have also been used very

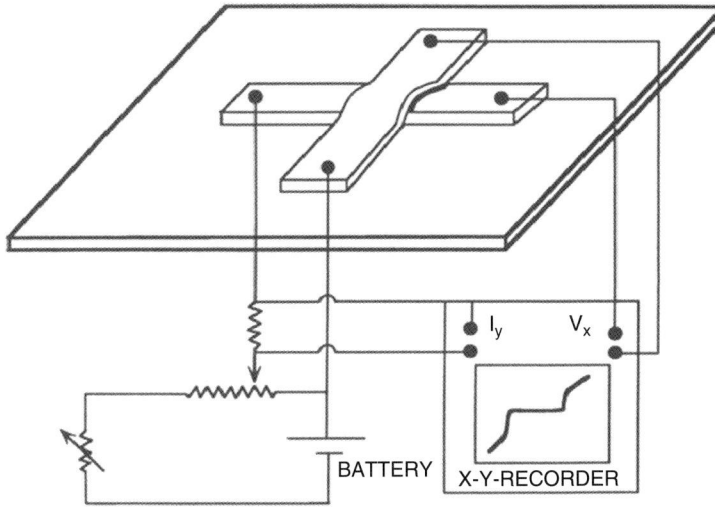

Fig. 6.1 A simple schematic of a cross-stripe junction and the circuitry necessary to measure an I–V characteristic.

successfully to make tunneling measurements. In this case a sharpened and sometimes oxidized needle with a small voltage applied to it is slowly moved into the proximity of a superconducting surface. The needle position is adjusted until the desired current starts to flow. Current–voltage characteristics can then be measured. This technique has been reasonably successful in probing oxide superconductors, since it has been very difficult to prepare thin insulating barriers by conventional deposition methods.

The important quantities that need to be measured are the direct current voltage characteristic, I–V, the derivative of the I–V, dI/dV, and the second derivative, d^2I/dV^2. These data must be taken with the sample in both the normal and superconducting state and with the Josephson or pair tunneling contribution quenched by a small magnetic field. These measurements are somewhat tricky and require ac lock-in techniques (see McMillan and Rowell, 1969; Wolf, 1985). The I–V characteristic and its first derivative, dI/dV, are essential for determining the tunneling density of states $N_T(\omega)$, and is therefore very important in making the connection to the theory. The first derivative data must be taken to very high precision, because only the deviations from the "BCS" (or weak coupling) density of states are significant, and these deviations should be measured to about 1%. Indeed, from these measurements one can find the energy gap as well as detailed information about the phonon modes (or other excitations) responsible for the attractive pairing interaction. Just as importantly, this technique may be unique in determining whether the excitations responsible for the superconductivity are not phononic. Details of these analyses will be described here.

6.1.2 Energy gap and transition temperature

In any complete analysis of tunneling data, it is important to know the T_c for the junction. Since the tunneling process involves a very thin region at the surface of the superconductor, the T_c of the junction may be slightly different then the "bulk" T_c for the rest of the film. Since T_c is an important parameter in the rest of the analysis, it is important to know it quite accurately. The simplest method for determining the T_c of the junctions is to plot the ratio of the conductances in the superconducting and normal state for various temperatures and then extrapolate to a ratio of 1. This is the best estimate of the transition temperature of the superconducting material.

Determining the gap from the conductance is in principle simple, but in practice is complicated by the non-idealities of real junctions.

6.1.2.1 MIS junctions. If the tunnel junction counter-electrode is a normal metal, then the normalized conductance of the junction is given simply by

$$\sigma = (dJ/dV)_s/(dJ/dV)_n = \int N_T(\omega')[-df(\omega' + eV)/d\omega'']d\omega' \qquad (6.1)$$

where $N_T(\omega) = \mathrm{Re}[|\omega|/(\omega^2 - \varepsilon^2)^{1/2}]$, f is sharply peaked and positive at $E' = -eV$ with a half width of order T, and ε is the energy gap. Thus, at $T = 0$, $\sigma = N_T(\omega)$. Direct information about the gap is contained in the normalized conductance. In Fig. 6.2 we show the current voltage characteristics of Al–Pb junctions taken by

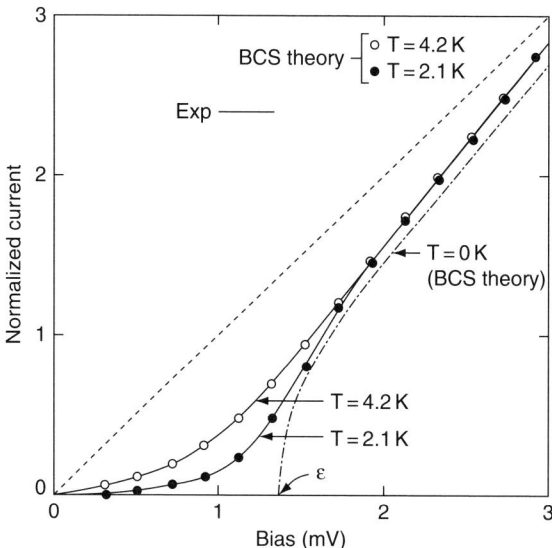

Fig. 6.2 Comparison between experimental and theoretical I–V characteristics of an Al–Pb junction. This figure demonstrates the excellent agreement with BCS theory. (Figure reproduced from Nicol et al., 1960. © 1960 by the American Physical Society.)

Nicol, Shapiro, and Smith (Nicol et al., 1960) above the transition temperature of aluminum These were the first data where a quantitative comparison between the experiment and the BCS theory were made.

In order to determine the gap from arbitrary tunneling data, a simple method is to measure the normalized conductance σ at zero bias and compare with the tabulated data of σ versus ε/T (Bermon, 1964). Of course, the complete conductance curve can be fitted to the functional form of eqn. 6.1, which has been fully calculated and tabulated by Bermon. Of course, these results are in the weak coupling limit and will lose some accuracy as the coupling becomes stronger.

6.1.2.2 SIS junctions. For junctions with two superconducting electrodes the determination of the gap is more straightforward. If both electrodes are equivalent than for $T \ll T_c$, there should be a sharp discontinuity in the conductance at 2ε. Of course, for non-ideal (real) junctions, the rise is not abrupt but is typically broadened by lifetime effects or by a distribution of gaps. In this case the point of maximum slope, dI/dV, should be used. In the case of a nearly linear rise, the midpoint should be used (McMillan and Rowell, 1969).

6.1.2.3 S_1IS_2 junctions. For junctions with inequivalent superconducting electrodes then there is a cusp in the conductance at $\varepsilon_1 - \varepsilon_2$ and an abrupt rise in the conductance at $\varepsilon_1 + \varepsilon_2$. Thus, by clearly locating the position of the cusp and using the same technique for the steep rise that has been described above for the SIS junctions, one can determine the gaps for both superconducting electrodes.

If $T \ll T_c$ then the value of the junction conductivity is an exponentially decreasing function of the smaller gap divided by temperature, and in fact obeys the relation (see Giaever and Megerle, 1961)

$$\sigma(V=0) = (2\pi\varepsilon_1/T)^{1/2} \, exp(-\varepsilon_1/T) \qquad (6.2)$$

Thus the value of the gap for the lower T_c electrode can be estimated from the temperature dependence of the conductivity at very low bias.

6.1.3 Inversion of the gap equation and $\alpha^2 F(\Omega)$

The Eliashberg (1960, 1961) gap equations describe the direct influence of the phonon density of states, $F(\omega)$, on the energy-dependent order parameter $\Delta(\omega)$. In turn, the order parameter modifies the electronic density of states that can be directly determined by tunneling conductance measurements. Just how the phonon spectrum influences $\Delta(\omega)$ and the normalized conductance was shown by Scalapino, Schrieffer, and Wilkins (1966). They calculated the effect of a single peak in $F(\Omega)$ on $\Delta(\omega)/\Delta_o(\omega)$ and on $N_T(\omega)/N(0)$. The results are shown in Fig. 6.3. It is clear from this calculation that strong peaks in the phonon density of states show up as dips in the normalized tunneling conductance.

All the information about the phonon density of states and the strength of the electron phonon interaction are contained in the tunneling measurements. The method by which the information is extracted is by a numerical inversion of the gap equations.

Thus, inverting the Eliashberg equations to determine the coupled phonon density of states $\alpha^2 F(\Omega)$ and μ^* (the Coulomb pseudopotential) is an extremely powerful

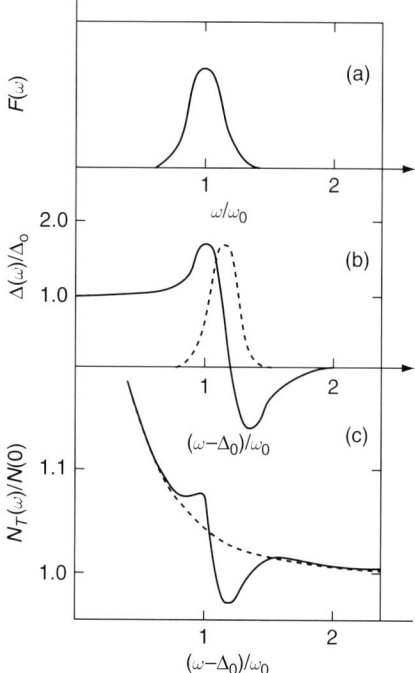

Fig. 6.3 The effect of a single peak in the phonon spectrum (a) on the order parameter (b) and on the tunneling density of states. (Figure reproduced from Scalapino et al., 1966. © 1966 by the American Physical Society.)

method for determining the degree to which any superconductor can be described in the framework of BCS superconductivity, and, of course, was initially used to determine the accuracy of the Eliashberg theory itself, which was found to be accurate to within several percent. The method developed by McMillan and Rowell (1965) involves solving numerically the integral Eliashberg equations for a given set of parameters, calculating the density of states, comparing the calculated values to the measured values, adjusting the input parameters, and iterating the procedure until the calculated density of states matches the measured one. A detailed description of the procedure and the application to Pb can be found in McMillan and Rowell (1969), but the highlights of this treatment are included here.

The integral equations for the normal and pairing self-energies for a dirty superconductor are:

$$\xi(\omega) = [1-Z(\omega)]\omega = \int_{\varepsilon_0}^{\infty} d\omega' \, \mathrm{Re}[\omega'/(\omega^2-\Delta^2)^{1/2}]$$
$$\int d\Omega \alpha^2(\Omega) F(\Omega)[D_q(\omega'+\omega)-D_q(\omega'-\omega)]$$
(6.3)

$$\phi(\omega) = \int_{\varepsilon_0}^{\omega_c} d\omega' \, \mathrm{Re}[\Delta'/(\omega'^2-\Delta'^2)^{1/2}] \int d\Omega \alpha^2(\Omega) F(\Omega)[D_q(\omega'+\omega)-D_q(\omega'-\omega)-\mu^*]$$
(6.4)

where $D_q(\omega) = (\omega + \Omega + i0^+)^{-1}$, $\Delta(\omega) = \phi(\omega)/Z(\omega)$, and $\Delta_o = \Delta(\Delta_o)$. $F(\Omega)$ is the phonon density of states, and $\alpha^2(\Omega)$ is an effective electron–phonon coupling function for phonons of energy ω.

For an isotropic superconductor the energy dependent normalized tunneling conductance (see Eq.(6.1)) gives the superconducting density of states

$$N(\omega) = \sigma(\omega) = Re[\omega/(\omega^2 - \Delta(\omega)^2)^{1/2}] \qquad (6.5)$$

Eqns. (6.3) and (6.4) are solved by iterating them together. We start with a guess for $\alpha^2(\Omega)F(\Omega)$ (based on the measured differential conductivity) and μ^* and a zeroth-order guess for the order parameter; that is, $\Delta^{(o)} = \Delta_o$ for $\Omega < \Omega_0$ and $\Delta^{(o)} = 0$ for $\Omega < \Omega$, where Δ_0 is the measured energy gap, ε, (see Sec. 6.1.2.1) and ω_o is the maximum phonon frequency. Using these parameters we find $\xi^{(1)}$, $\phi^{(1)}$, $Z^{(1)}$, and hence $\Delta^{(1)}$. The iteration is continued until $\Delta^{(n)}$ converges to three decimal places. At this point, $N(\omega)$ is computed and compared to the experimental result.

Now the task is to adjust the zeroth-order $\alpha^2F(\Omega)$ and μ^* to provide convergence of the calculated density of states to the measured density of states. To do this it is necessary to calculate the linear response of the density of states to a small change in $\alpha^2F(\Omega)$. This amounts to calculating the derivative $\delta N_c(\Omega)/\delta \alpha^2 F(\Omega)$. This allows us to estimate the change in the zeroth-order guess for $\alpha^2F(\Omega)$ necessary for the first-order iteration. The change is given by

$$\delta \alpha^2 F(\Omega) = \int d\omega' [\delta N_c(\Omega)/\delta \alpha^2 F(\Omega)]^{-1} [N_c(\Omega') - N_c^{(o)}(\Omega')] \qquad (6.6)$$

Thus

$$[\alpha^2 F(\Omega)]^{(1)} = [\alpha^2 F(\Omega)]^{(o)} + \delta \alpha^2 F(\Omega) \qquad (6.7)$$

Since the gap equations are not linear, the iteration needs to be continued until the calculated density of states reproduces the measured density of states. During the iteration process, μ^* is adjusted so that the calculated energy gap agrees with the measured energy gap. For many conventional superconductors, μ^* is found to be approximately 0.1, though by virtue of the role it plays in the inversion procedure it can be zero or even negative. If some spectral weight is missed in the tunneling data because it occurs at energies beyond the voltage range of the measurements, then the inversion procedure reflects this in a reduced value of μ^* A recent confirmation of this fact was shown by Schneider et al. (1987) for the cluster compound YB_6.

This procedure has been used to show that most of the conventional superconductors are indeed very well described by the Eliashberg equations. For example, the best results have been obtained for lead (see McMillan and Rowell, 1969), where the Eliashberg theory has been fully tested and found to be very accurate. Fig. 6.4 shows the normalized density of states from the tunneling conductance measurements, and Fig. 6.5 shows the calculated $\alpha^2 F(\Omega)$ resulting from the inversion procedure described previously (McMillan and Rowell, 1969).

94 *Experimental methods: Spectroscopic*

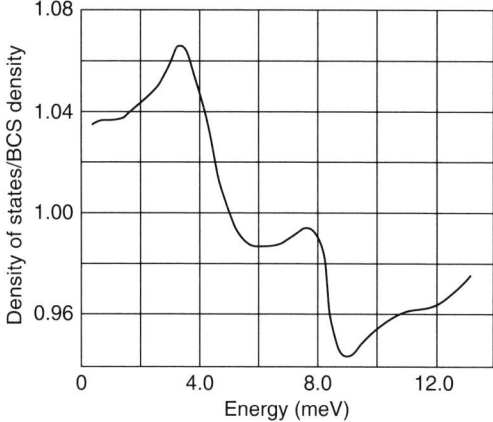

Fig. 6.4 The electronic density of states of Pb divided by the BCS density of states versus $E - \varepsilon_0$. (Figure reproduced from McMillian and Rowell, 1969.)

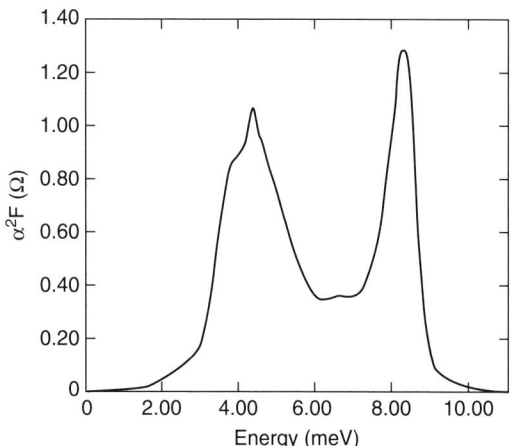

Fig. 6.5 The function $\alpha^2 F(\Omega)$ for Pb, found by fitting the data of Fig. 3.4. (After McMillian and Rowell, 1969).

6.1.4 Electron-phonon coupling parameter λ

The Eliashberg equations can be solved for T_c, and the solution can be simplified with the use of the parameter λ, which is related simply to $\alpha^2 F(\Omega)$ by the following equation (see eqn. (2.29)):

$$\lambda = \int d\Omega 2\alpha^2 F(\Omega)/\Omega \qquad (6.8)$$

We can show (eqn. (3.35)) that for an isotropic superconductor for any value of λ,

$$T_c = 0.25\Omega/(e^{2/\lambda_{eff}} - 1)^{1/2} \tag{6.9}$$

where $\lambda_{eff} = (\lambda - \mu^*)[1 + 2\mu^* + \lambda\mu^* t(\lambda)]^{-1}$ and Ω is the average coupled phonon energy. λ is considered to be the most direct measure of the strength of the electron–phonon interaction which strongly effects the nature of the superconductivity; that is, if $\lambda \ll 1$ then we have weak coupling or "BCS" superconductivity, whereas if $\lambda \gg 1$ we have strong-coupled superconductors.

It is also important to note that normal electrons near the Fermi surface are dressed with a cloud of "virtual" phonons. This dressing of the electron properties shows up as an enhancement of the cyclotron mass, the Fermi velocity, and the electronic heat capacity. The enhancement over the band value is given by exactly the same parameter λ that is implicit in the Eliashberg gap equations. Thus detailed Fermiology provides an independent method for determining the strength of the electron–phonon interaction as well as the accuracy of the inversion.

In fact, the degree to which phonons are responsible for the superconductivity can be tested by detailed comparisons of calculated specific heat and measured specific heat (Section 2.4). The method depends on measurements of the temperature dependence of the specific heat, determinations of $\alpha^2 F(\Omega)$ from tunneling spectroscopy, and values of $F(\Omega)$ from neutron scattering experiments. The main idea of this procedure is to use the manner in which the phonon dressing of the electronic part of the Sommerfeld constant is unrenormalized at high temperatures. By comparing the high-temperature and low-temperature values of the Sommerfeld constant, one can determine λ, the electron–phonon coupling strength. The electronic specific heat is separated from the total specific heat by integrating the measured $F(\omega)$ to determine the lattice specific heat and then subtracting it. Tunneling spectroscopy, through its determination of $\alpha^2 F(\Omega)$, is also a measure of λ, provided that phonons are fully responsible for the superconductivity. If some other boson contributes strongly to the pairing, then the $\alpha^2 F(\Omega)$ determined by tunneling will not have the correct magnitude, since the inversion procedure, which assumes that phonons are fully responsible, will improperly weight the phonon peaks. A simple integration of the tunneling $\alpha^2 F(\Omega)$ will give a value of λ which can be compared to the value determined by the analysis of the specific heat. If they agree, then phonons account for the superconductivity, and if they disagree, then some other pairing interaction is present. This procedure has been carried out by Kihlstrom et al. (1987) for several A-15 structure superconductors. It was determined that V_3Si was completely phononic, whereas Nb_3Ge was not! It would be very useful to carry out this procedure for the cuprate superconductors. Unfortunately, the state of the art in tunneling spectroscopy at the present time for these compounds is not adequate for a very reliable determination of $\alpha^2 F(\Omega)$ Some of the more interesting data on tunneling into the cuprate superconductors will be presented in Chapter 9, which treats the cuprates in great detail.

6.2 Scanning tunneling microscopy and spectroscopy

Scanning tunneling microscopy and spectroscopy has become an extremely valuable tool for the understanding of properties of superconducting compounds—especially the cuprates and the Fe-based compounds. The scanning tunneling microscope was developed by Binnig and Rohrer (1982), and revolutionized the study of surfaces, as it was able to image single atoms on the surface (see Fig. 6.6). It was soon realized that this tool could also measure locally the electronic structure of the surface with atomic resolution by varying the tip voltage and measuring the current at a fixed distance from the surface.

The derivative of the I–V curve—namely, dI/dV—can easily be shown to be proportional to the density of electronic states at the surface being explored (Tersoff and Hamann, 1985). By scanning the tip over a surface, a very-high-resolution map of the density of states, and hence if the surface is superconducting, a map of the energy gap, can be obtained. Since the spatial resolution of the microscope is much higher than any of the relevant superconducting lengths, these maps can provide information on the effect of local perturbations, such as magnetic or non-magnetic impurities, as was first demonstrated by Ji *et al.* (2008), and as illustrated in Fig. 6.7 What is clearly observed in the figure is the effect of a magnetic impurity (Mn or Cr) on the density of states in the superconducting gap of Pb, indicating the local destruction of superconductivity by magnetic impurity. This destruction is very localized around the

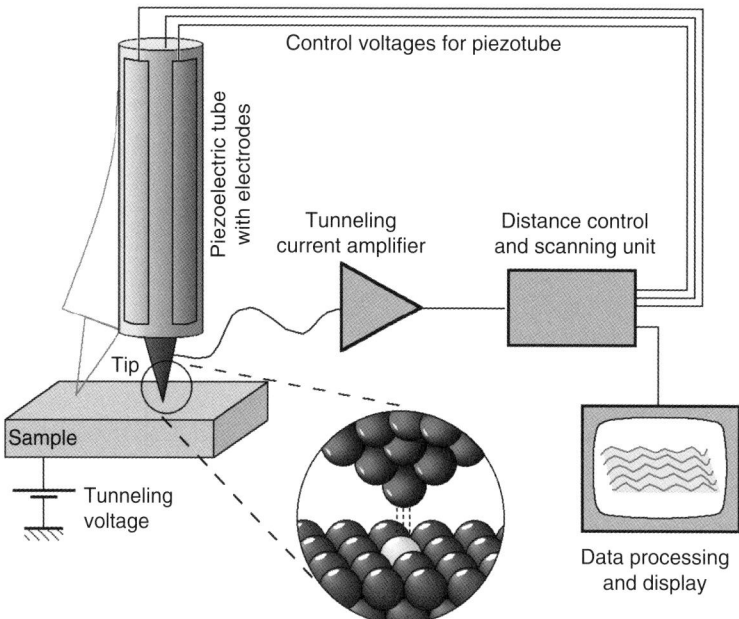

Fig. 6.6 A schematic of a scanning tunneling microscope. (Figure: Michael Schmid, TU Wien.)

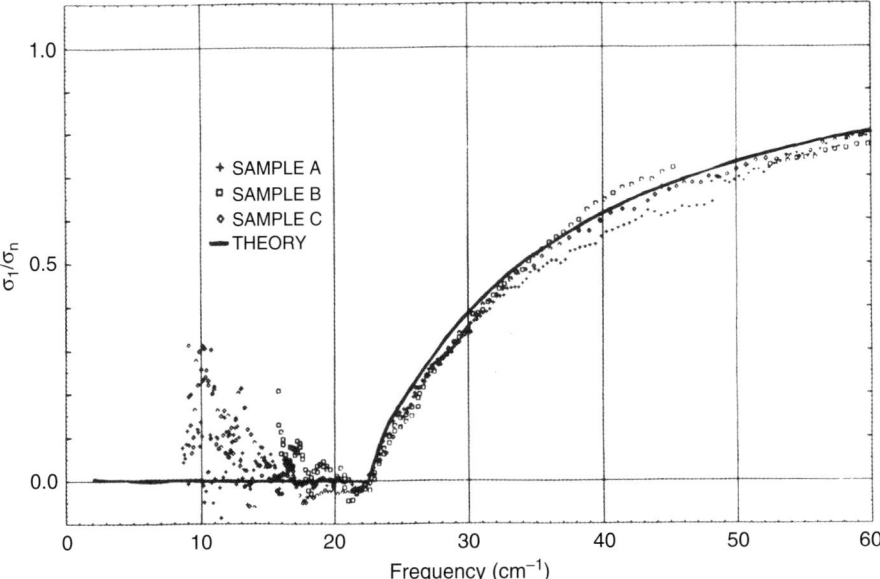

Fig. 6.7 Measured and calculated value of the conductivity ratio for three Pb films as a function of the phonon frequency. (Figure reproduced from Ginsberg and Hebel, 1969.)

impurity, and covers a region that is approximately the coherence length of Pb. This ability to look at the spectrum reflecting the local density of states for the cuprates in the pseudogap region is crucial for understanding the nature of this region, as will be described and illustrated in detail in Chapter 11.

6.3 Infrared spectroscopy

Study of the absorption of electromagnetic energy has played a very important role in the history of superconductivity. It provided some very early confirmation of the BCS theory and provides a simple method for determining the energy gap, especially for thin films where the transmission, the reflection, and the absorption can be estimated. Mattis and Bardeen (1958), and independently, Abrikosov, Gor'kov, and Khalatnikov (1959), derived the general equations for the complex conductivity. For the convention where the incoming radiation is in a plane wave $\exp[i(\omega t - \mathbf{q}\mathbf{r})]$, then $\sigma = \sigma_1 - i\sigma_2$. The results of the Mattis–Bardeen theory for σ_1/σ_n and σ_2/σ_n are given by the following two equations (Ginsberg and Hebel, 1969):

$$\frac{\sigma_1}{\sigma_2} = \frac{2}{\omega} \int_\varepsilon^\infty [f(E) - f(E+\omega)]g(E)dE + \frac{1}{\omega}$$
$$\times \int_{\varepsilon-\omega}^{-\varepsilon} [1 - 2f(E+\omega)]g(E)dE$$
(6.10)

$$\frac{\sigma_2}{\sigma_n} = \frac{1}{\omega} \int_{\varepsilon-\omega,-\varepsilon}^{\varepsilon} \frac{[1 - 2f(E+\omega)] \left(E^2 + \varepsilon^2 + \omega E\right)}{(\varepsilon^2 - E^2)^{\frac{1}{2}} [(E+\omega)^2 - \varepsilon^2]^{\frac{1}{2}}} dE \qquad (6.11)$$

If $\omega > 2\varepsilon$, then the second term in eqn. (6.10) is included and the lower limit of integration in eqn. (6.11) should be $-\varepsilon$ instead of $\varepsilon - \omega$.

Measuring two of the following three rates of energy propagation, reflection, transmission, or absorption for a plane wave normal to the surface of a thin film whose thickness is less than the penetration depth and the coherence length allows a direct determination of σ_1/σ_n and σ_2/σ_n. These results can then be compared directly to the theoretical expressions, eqns. (6.10) and (6.11).

The earliest experiments by Glover and Tinkham (1957), however, measured only the transmitted energy, so that unambiguous values of σ_1/σ_n and σ_2/σ_n could not be obtained. However, they could estimate these quantities using the Kramers–Kronig transform relation between σ_1 and σ_2, as well as the sum rule which requires that

$$\int \sigma_{1s}(\omega) d\omega = \int \sigma_{1n}(\omega) d\omega \qquad (6.12)$$

Eqns. (6.10) and (6.11) predict that for $T = 0$, σ_1/σ_n is zero for frequencies up to the gap frequency 2ε and finite above this frequency, whereas for finite temperatures, σ_1/σ_n will be finite but vanishingly small if $(-\varepsilon/T) \ll 1$. The Kramers–Kronig transformation requires that σ_2/σ_n also be small near the gap frequency, which means that both the reflection and absorption will be small, the result being that the transmission has a sharp peak at the gap frequency. Thus a rather unambiguous determination of the gap was made with these early infrared measurements.

Later measurements were made of both transmission and reflection. In these cases, rather accurate determinations of σ_1/σ_n and σ_2/σ_n could be made and compared to the theory (eqns. (6.10) and (6.11)). An example of this type of result is illustrated in Fig. 6.7, which shows both measured curves of σ_1/σ_n as well as the theoretical curve based on the Mattis–Bardeen theory.

Direct measurements of the absorption of electromagnetic energy as a function of frequency can be used to accurately determine the energy gap. If the temperature is low enough, so that $\exp[-\varepsilon/T] \ll 1$, then there is a well-defined absorption edge at 2ε Measuring the absorption is done typically by bolometric methods, by which the sample is thermally isolated from its surroundings and the absorption is measured by the temperature rise.

Precise infrared spectroscopy can also be used to reconstruct the function $\alpha^2(\Omega)\Phi(\Omega)$. This method was proposed by Little and collaborators (see Holcomb et al.,1996; Little et al., 1999), and is based on the thermal-difference-reflectance spectroscopy technique. This technique enables the determination of $\alpha^2(\Omega)\Phi(\Omega)$ for an energy interval that is much larger than when using the tunneling method (see Section 6.1). The reflectivity of the sample is measured with a high degree of precision at different temperatures, and the ratio of the difference relative to their sum is determined. Using a theoretical method developed by Shaw and Swihart (1968), an inversion of the data is performed and the function $\alpha^2(\Omega)\Phi(\Omega)$ can be obtained. The large range of energies allowed using this method enables higher-energy electronic modes to be revealed if they are present. This method has been applied successfully to the cuprates (see Chapter 10).

6.4 Ultrasonic attenuation

Historically, the first method used to measure the energy gap and to check the validity of the temperature dependence of the BCS gap was ultrasonic attenuation. The experimental method involves mounting transducers on opposite sides of a crystal of the superconducting material, launching a sound wave into the sample, and measuring the ratio of the attenuation below the transition temperature to its value just above the transition. The simplest case to consider is for the propagation of longitudinal sound waves where the interaction of the sound wave with the crystal is due to the change in the crystal potential which accompanies the wave. If the phonons associated with the sound wave have energies (ω) less than the gap energy, 2ε, then the attenuation at very low temperatures (T$\ll T_c$) should be proportional to $\exp[-\varepsilon/T]$, since only the thermally excited quasiparticles can absorb the phonons.

In their landmark paper, Bardeen, Cooper, and Schrieffer (1957) calculated the ratio of the ultrasonic attenuation in the superconducting state α_s divided by the attenuation in the normal state, α_n, for the case when $ql \gg 1$, where q is the wavenumber of the sound wave, l is the electron mean free path, and $\omega < 2\varepsilon$. In this case, the ratio α_1/α_n is equal to $2f(\varepsilon(T), T)$, where f is the Fermi function. This expression

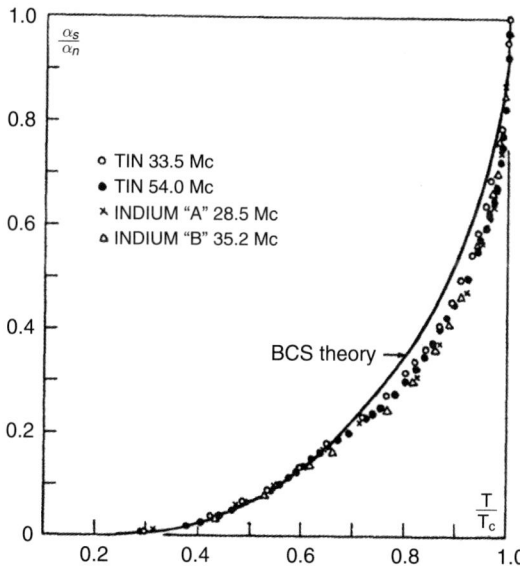

Fig. 6.8 The attenuation ratio as a function of reduced temperature, compared to the BCS variation. (Figure reproduced from Morse and Bohm, 1957. © 1957 by the American Physical Society.)

is valid even for $ql \ll 1$. Thus, by fitting α_s/α_n as a function of temperature to the Fermi function, one can determine the value of the energy gap and its temperature dependence. Fig. 6.8 is the result from the paper by Morse and Bohm (1957), where the measurements on tin and indium are compared with the BCS theory.

6.5 Angle-resolved photoemission

Angle-resolved photoemission has emerged as one of the best tools for experimentally determining the electronic structure of highly correlated superconductors—especially the superconducting oxides such as the cuprates and ruthenates. By measuring the energy and angular distribution of electrons that are photo-emitted from the material in response to a beam of high-energy photons, one can determine much about the energy and the wavevector (momentum) of the electrons that are moving in the material—particularly those materials that cannot be explained by using an independent particle approximation. Over the last score of years, this technique has reached extraordinary resolutions in energy and angular distribution, and better than 2 meV in energy and 0.2 degrees of angular resolution are now a reality. For a detailed review of the state of art for ARPES, see Damascelli (2004). ARPES can provide information on the band structure and band dispersion, determine the complicated Fermi surface, and provide information on the strength and nature of many-body correlations that are important for the single-electron excitation spectrum and how they are responsible for many of the important physical properties. A key feature of the cuprates in the pseudogap (mainly underdoped) region are Fermi arcs that are incomplete Fermi contours that change in length with temperature and doping. These structures reflect the unusual properties of the pseudogap region that may reflect the underlying inhomogeneity of these compounds (see Chapter 11). Fermi arcs were first discovered using ARPES, and they provide important information about the unusual electronic structure of the cuprates, especially in the underdoped pseudogap part of the phase diagram.

A typical ARPES system resides at the end of a synchrotron beamline, and such a system is shown schematically in Fig. 6.9.

In addition to the dispersion curves and the construction of the band structure and the Fermi surface, ARPES can determine the k-dependence of the superconducting energy gap. The symmetry of the gap, as well as strong anisotropies, can be determined.

6.6 Muon spin resonance (μSR)

Muon spin relaxation or resonance (μSR) has become an important technique (see the reviews by Keller, 1989; Schenck, 1985) for studying the magnetic properties of novel materials and in particular the magnetic behavior of novel superconductors, including many described in Chapters 10 (cuprates) and 13 (novel superconducting materials). A muon is an unstable particle with either positive charge or negative charge (μ^+ or μ^-) that lives for about 2 microseconds and spontaneously decays into a positron (electron) and a neutrino–antineutrino pair. Muons are produced in an accelerator that uses high-energy protons that are accelerated to more than 500 MeV and then

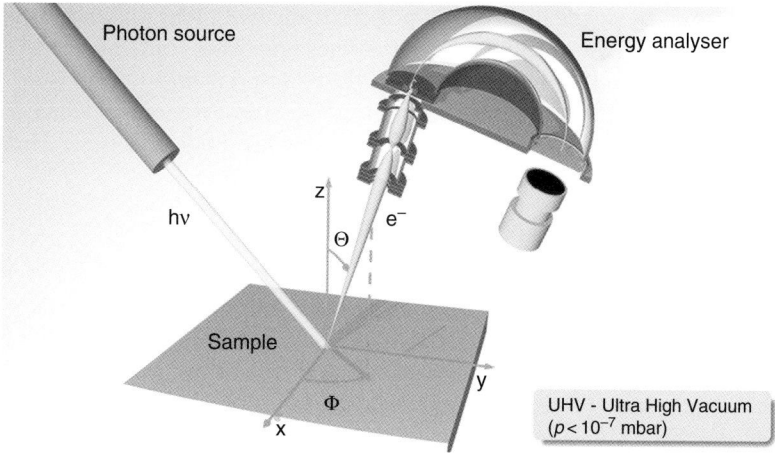

Fig. 6.9 A highly focused photon source from a synchrotron impinges on the sample, and the electrons emitted are analyzed using a double hemispherical analyzer that can measure the dispersion (E versus k) along an angular cut in k-space.

collide with a light-element target such as beryllium or carbon. This nuclear reaction produces positively or negatively charged pions. The pions are very unstable, and in about 25 nanoseconds decay into muons and muon neutrinos or antineutrinos. The muons of interest in studies of most solids are the positively charged muons (μ^+). Beams of these muons are produced from pions decaying at rest in the surface layer of the primary light element target, and these muons are called surface muons. Negative muon beams cannot be produced in this manner, though they are not of significant interest for studies of the solid state, for reasons that will not be discussed here. The surface muon (μ^+) beam has a momentum of 2.98 MeV/c (in the rest frame of the pion), and an energy of 4.119 MeV. This beam emerges from the target isotropically from the decaying pion at rest.

Serendipitously, the decay of the pion exhibits maximal parity violation for this weak interaction, and thus the surface muons are perfectly spin-polarized *opposite* to their momentum. Thus, when the beam hits the material being studied, the muons arrive nearly 100% spin-polarized. This makes this technique in many respects superior to other magnetic resonance probes that rely on obtaining a high spin polarization in a magnetic field and utilizing the equilibrium spin-polarization. High spin-polarization can thus be achieved only at very high field and very low temperatures.

There are only a very few meson facilities in the world that can make high precision (μSR) measurements: the Paul Scherrer Institute (PSI) in Villigen, Switzerland, which also currently has a very-low-energy muon beam; the Tri-University Meson Facility in Vancouver, Canada (TRIUMF); the Booster Muon facility at the high-energy research organization (KEK) in Japan (soon to be replaced by J-PARC in Tokai Japan); and ISIS, at the Rutherford Appleton Laboratory in the UK. The experimental setup (continuous beam or pulsed, transverse, longitudinal, or zero field) and the nature of the beams differ at each of these facilities, but in general they can all study the interaction of muons in their host material and obtain very detailed information about magnetism in the host compounds. For a more detailed discussion about muon physics and (μSR) specifically, the following document is very instructive: http://muon.neutron-eu.net/muon/files/muSRBrochure.pdf

6.6.1 μSR studies of superconductivity

The discovery of the high critical temperature cuprate superconductors led to an explosion of research aimed at discovering the mechanism and understanding the properties of these novel materials. μSR has provided some key information about the magnetic properties of the cuprates and other novel superconducting systems. In particular, being able to map out the antiferromagnetic phase diagram for both LaSrCuO and YCaBaCuO has been extremely important for understanding the interplay of magnetism and superconductivity in the cuprates (Niedermayer *et al.*, 1998). More recently, μSR studies of very weak magnetism in YBaCuO (Sonier *et al.*, 2001) have highlighted the role of spatial inhomogeneities in the cuprates, and strongly supports our understanding of the role of the inhomogeneous structure of the cuprates detailed in Chapter 10. Finally, information about the symmetry of the order parameter, especially where it is suspected that there is time reversal symmetry-breaking—for example, in the case of p-wave superconductivity—can be obtained uniquely by μSR measurements because of their exquisite sensitivity to very weak magnetism (Luke *et al.*, 1998).

7
Multigap superconductivity

7.1 Multigap superconductivity: general picture

Consider a superconductor containing several different groups of electrons occupying distinct quantum states. The most typical example is a material with several overlapping energy bands (Fig. 7.1).

One can expect that each band will possess its own energy gap. Let us clarify this point and introduce the necessary definition. Of course, if the energy gap were defined as the smallest quantum of energy which can be absorbed by the material, then only the smallest gap of the system would satisfy this definition; it would be impossible for the system to have several energy gaps. To avoid misunderstanding, when talking about multi-gap structure of a spectrum we will mean explicitly that the density of states of the superconducting electrons contains several peaks. The magnitudes of these gaps will differ according to the variations in the densities of states and coupling constants across the bands.

The two-gap model was first considered by Suhl et al. (1958) and independently by Moskalenko (1959). A general treatment based on quantum field-theoretical techniques, together with an analysis of various thermodynamic, magnetic, and transport properties, was presented by Geilikman et al. (1967).

Each band contains its own set of Cooper pairs. Since, generally speaking, $p_{F;i} \neq p_{F;l}$ (here $p_{F;i}$ and $p_{F;l}$ are the Fermi momenta for different bands), there is negligible pairing of electrons belonging to different energy bands. Indeed, only electrons with equal and opposite momenta form the pairs. This does not mean, however, that within each band the pairing is completely insensitive to the presence of the other. On the contrary, a peculiar interband interaction and the appearance of non-diagonal (interband) coupling constants are fundamental properties of the multigap model.

Indeed, consider two electrons belonging to band i. They exchange phonons, and as a result form a pair. There exist two pairing scenarios. In one of them the first electron emits a virtual phonon and makes a transition into a state within the same

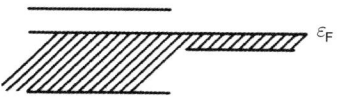

Fig. 7.1 Overlapping bands.

energy band. The second electron from the same band absorbs the phonon and also remains in the band, forming a bound pair with the first one. This is the usual pairing picture of the BCS theory, described by a coupling constant λ_i. However, the presence of the other energy band gives rise to an additional channel. Namely, the first electron, originally located in the i band, can emit a virtual phonon and make a transition into the l band. The phonon is absorbed by the second i-band electron, which also is scattered into the l band, where it pairs up with the first electron. Thus the initial state had two electrons in the i band, while the final state finds a pair in the l band. Of course, the initial and final states must obey conservation of energy (both must be located at the Fermi level) and conservation of momentum (total momentum must be equal to zero) laws. Charge transfer processes like this are described by non-diagonal coupling constants λ_{il}, and because of them the system is characterized by a single critical temperature. Otherwise, each set of electrons would have its own T_c.

The Hamiltonian describing the pairing interaction has a form (in a weak coupling approximation):

$$\hat{H}_{int} = \sum_{\substack{i,l \\ k,p}} \lambda_{il} a^+_{i;k,+} a^+_{i;-k,-} a_{l;-p,-} a_{l;p,+} \tag{7.1}$$

The terms with $i = l$ (then $\lambda_{il} = \lambda_{li} = \lambda_i$) correspond to the intraband transition, whereas the terms with $i \neq l$ describe the interband transitions; the indices "+" and "−" correspond to various spin projections.

7.2 Critical temperature

We introduce self-energy parts $\Delta_i(\mathbf{p}, \omega_n)$, which describe electron pairing in the ith band. They satisfy the system of equations shown in Fig. 7.2, or, in analytical form,

$$\Delta_i(\mathbf{p}, \omega_n) = \sum_{i=1}^{n} \sum_{\omega_{n'}} \int \lambda_{il} D(\mathbf{p} - \mathbf{p}', \omega_n - \omega_{n'}) F^+_{ll}(\mathbf{p}', \omega_{n'}) d\mathbf{p}',$$

where

$$F^+_{ll}(\mathbf{p}, \omega_n) = \frac{\Delta_l}{\omega_n^2 + \xi^2 + \Delta_l^2}. \tag{7.2}$$

D is the phonon Green's function, λ_{il} describes the transition of an electron pair from the ith to the lth band accompanied by phonon absorption or emission, and Δ_i is the order parameter.

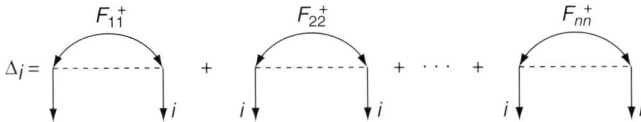

Fig. 7.2 Diagrammatic equations for the order parameters.

Let us consider the case of weak coupling. Introducing the effective coupling constants in the usual way and summing over $\omega_{n'}$, we write eqn. (7.2) in the form:

$$\varepsilon_i = \sum_{l=1}^{n} g_{il}\nu_l\varepsilon_l \int d\xi_l \frac{\tanh(E_l/2T)}{E_l}$$
$$E_l = \left(\xi_l^2 + |\varepsilon_l|^2\right)^{1/2} \tag{7.3}$$

One can see from eqn. (7.3) that the gaps ε_i in different bands appear at the same temperature T_c. In the most interesting two-gap case, the critical temperature turns out to be

$$T_c = 1.14\tilde{\Omega} e^{-1/\tilde{\lambda}}$$
$$\tilde{\lambda} = \frac{1}{2}\left[\lambda_1 + \lambda_2 + \sqrt{(\lambda_2 - \lambda_1)^2 + 4\lambda_{12}\lambda_{21}}\right] \tag{7.4}$$

This result can be obtained easily from eqn. (7.3) ($i,l = 1,2$) taken at $T = T_c$.

One can see directly from eqn. (7.4) that the presence of the second gap is a favorable factor for T_c, that is, its value is larger relative to the one-gap case. This statement is valid even if $\lambda_2 = 0$ (see Section 7.4)—that is, for the case when the pairing in the second band is not intrinsic—but is induced by the interband transition. One can say that this feature is caused by an increase in a phase space for virtual transitions.

The presence of overlapping bands is typical for many superconductors. Nevertheless, it was not possible to observe multigap structure in conventional superconductors. This is due to their large coherence lengths: the inequality $\xi \gg l$ (l is the mean free path), which holds for most conventional superconductors, leads to averaging because of interband scattering. As a result, the multigap structure is washed out and the one-gap model is applicable.

The situation in novel superconductors—for example, the high-T_c oxides, MgB_2—is entirely different. The coherence length is small, and this leads to the observation of a multigap structure (see Chapters 10 and 13).

A two-gap spectrum was first observed in the exotic system, Nb-doped $SrTiO_3$ compound (Binnig et al., 1980). The STM technique (Section 6.2) was employed, and the two-gap structure of the spectrum was observed. The structure appears as a result of doping and corresponding filling of the second band. As noted previously, this was the first observation of the two-gap structure. According to theoretical analysis by Fernandes et al. (2013), the two-gap structure can be observed also for the Nb-doped $SrTiO_3$ films and interface; that is, for the quasi-2D case.

7.3 Energy spectrum

Here we focus mainly on the two-gap case, because this case is relatively simple and is the most important one. The temperature dependences of the energy gaps can be evaluated from eqns. (7.3) and, generally speaking, are very different from that for the single-band BCS case; the typical dependences can be seen in Fig. 7.3.

The energy spectrum and properties of two-gap superconductor are described by three coupling constants, $\lambda_1, \lambda_2, \lambda_{12}$, which are entering the Hamiltonian (7.1).

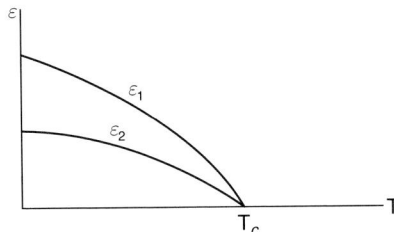

Fig. 7.3 The dependences $\varepsilon_1(T)$, $\varepsilon_2(T)$.

Let us express the energy gaps ε_i in terms of these coupling constants. We can employ the method similar to that used by Pokrovsky (1961) to study anisotropy of the energy gap. The solution of eqn. (7.3) at $T = 0$ K can be sought in the form:

$$\varepsilon_i = 2\tilde{\Omega}\phi_i \exp(-\alpha); \qquad (7.5)$$

then $\phi_i = QX_i$, where X_i is the normalized solution of the following system of linear equations:

$$\phi_i = \alpha \sum_l \lambda_{il} \phi_l \qquad (7.6)$$

corresponding to the smallest eigenvalue α. The calculation leads to the following expression for coefficient Q:

$$Q = \exp\left[-\sum_i \nu_i |X_i|^2 \ln |X_i| \bigg/ \sum_i \nu_i |X_i|^2\right] \qquad (7.6a)$$

For the two-gap case:

$$X_1 = \lambda_{12} S^{-1}, \; X_2 = \left(\tilde{\lambda} - \lambda_1\right) S^{-1}$$

$$S = \left[\lambda_{12}^2 + \left(\tilde{\lambda} - \lambda_1\right)^2\right]^{1/2};$$

$\tilde{\lambda}$ is defined by eqn. (7.4).

From these results one can show (Kresin and Wolf, 1990b) that one of the gaps always exceeds the BCS value, while the other is less than this value. Indeed, we can write

$$\frac{2\varepsilon_1(0)}{T_c} = 3.52 e^{A_1}, \quad \frac{2\varepsilon_2(0)}{T_c} = 3.52 e^{A_2}, \qquad (7.7)$$

where

$$A_1 = \nu_2 \tau^2 R \ln\left(\tau^{-1}\right), \quad A_2 = -\nu_1 R \ln\left(\tau^{-1}\right)$$
$$R = (\nu_1 + \nu_2 \tau^2)^{-1}, \quad \tau = (\tilde{\lambda} - \lambda_1)/\lambda_{12}$$

or

$$\tau = \frac{1}{2}\left(\lambda_2 - \lambda_1 + \left[(\lambda_2 - \lambda_1)^2 + 4\lambda_{12}\lambda_{21}\right]^{1/2}\right)\lambda_{12}^{-1}$$

One can see that regardless of the value of τ we always have $A_1 A_2 < 0$. Therefore (see eqn. (7.7)), one can state the following theorem: one of the gaps in a two-gap system is always smaller than the BCS value, whereas the other gap is greater than this value.

Note that qualitatively this result is related to the previously mentioned (Section 7.1) impact of impurities on the multigap structure: the impurity scattering leads to averaging of the gaps and to the one-gap picture which is observed for conventional superconductors. Such an averaging would lead to the BCS relation $2\varepsilon_i = 3.52 T_c$ if, indeed, the value of one gap exceeds the BCS value, and for the second gap $2\varepsilon(0)_{min} < 3.52 T_c$. Otherwise, the average value would be large (or smaller) than that following from the BCS theory.

Therefore, the inequality $2\varepsilon(0)/T_c > 3.52$ can be caused not only by strong coupling effects (see Chapter 3), but by the presence of the multi-gap structure. That is why an analysis of the experimental data when $2\varepsilon(0)/T_c > 3.52$ should be performed with considerable care (see, for example, Santiago and de Menezes, 1989).

Temperature dependences of the gaps for some specific two-gap superconductor are determined by the values of the coupling constants $\lambda_1, \lambda_2, \lambda_{12}$. Nevertheless, one can formulate the general feature; namely, the ratios $\eta = \varepsilon_2/\varepsilon_1$ are equal at $T = 0$ K and near $T = T_c$, that is, $\eta(0) = \eta(T_c)$ (Kresin, 1973). Indeed, the gaps at $T = 0$ K and $T \to T_c$ are the solutions of the following set of equations which follow directly from eqn. (7.3) ($i,k = a,b$):

$$T = 0 \text{ K}: \varepsilon_i = \sum_{i,k} \lambda_{ik} \ln \frac{2\tilde{\Omega}}{|\varepsilon_k|} \qquad (7.3a)$$

$$T \to T_c: \varepsilon_i = \ln \frac{\tilde{\Omega}}{T_c} \sum_{i,k} \lambda_{ik}\varepsilon_k \qquad (7.3b)$$

As noted previously, the solution of eqn. (7.3a) can be sought in the form: $\varepsilon_i = 2\tilde{\Omega}\varphi_i \exp(-\alpha)$. As a result, we obtain eqn. (7.6), allowing us to determine $\eta(0)$.

The ratio $\eta(T_c)$ can be determined from eqn. (7.3b), and is the solution of the equation $\lambda_{12}\eta^2 + (\lambda_1 - \lambda_2)\eta - \lambda_{12} = 0$. This equation permits us to express the ratio $\eta = \eta(T_c)$ in terms of the coupling constants $\lambda_1, \lambda_2, \lambda_{12} = \lambda_{21}$. As for the ratio $\eta(0)$, it can be obtained from eqns. (7.5) and (7.6). One can see that $\eta(0)$ and $\eta(T_c)$ are described by similar equations, and therefore, indeed, the ratios of the gaps at $T = 0$ K and $T = T_c$ are equal.

Let us stress again that the deviations from the BCS behavior and the inequality $2\varepsilon_i \neq 3.52 T_c$ here are caused not by strong coupling (see Section 3.5), but by the presence of a multigap structure.

7.4 Properties of two-gap superconductors

Let us, at first, stress one more important feature of the two-gap picture. In the usual theory of superconductivity, in the weak coupling approximation, the energy gap is related to T_c by the universal relation $2\varepsilon(0) = 3.52\,T_c$ (see Section 3.2). This is due to the fact that for such a case the system is characterized by a single parameter; namely, by one coupling constant, λ. All other parameters can be expressed in terms of λ. Instead of λ one can select the critical temperature T_c as a such parameter. As a result, all other characteristic quantities, including the energy gap at $T = 0$ K, $\varepsilon(0)$, can be expressed in terms of T_c by universal relations. In a multi-band model with a large number of coupling constants there is no such universal relation between the energy gaps $\varepsilon(0)$, and T_c.

For the two-gap superconductors one can select as the independent and experimentally measured parameters the value T_c and the values of the energy gaps $\varepsilon_1(0)$ and $\varepsilon_2(0)$.

The temperature dependences of the thermodynamic, electromagnetic, and transport parameters differ from those obtained in the usual one-gap BCS theory. For example, for $T \to 0$ the heat capacity is the sum of two exponents; the band with the smaller superconducting gap makes the dominant contribution, and as a result $C_{el.|T\to 0}$ decreases slower than in the one-band case.

7.4.1 Penetration depth; surface resistance

The superconducting current for the multigap superconductor can be written in the form:

$$\boldsymbol{j}(\boldsymbol{q}) = \sum_i Q_i(\boldsymbol{q})\, \boldsymbol{A}(\boldsymbol{q}) \tag{7.8}$$

Here $\boldsymbol{A}$ is a vector potential, summation is over all overlapping energy bands, and the kernel $Q_i(\boldsymbol{q})$ is (see, for example, Abrikosov et al., 1975)

$$Q_i(\boldsymbol{q}) = \frac{3\pi}{4}\frac{n_i e^2}{m_i} T \sum_n \int_{-1}^{1} \frac{(1-\mu^2)\,d\mu}{\left(\omega_n^2 + |\varepsilon_i|^2\right)^{1/2}} \frac{\varepsilon_i^2}{\left(\omega_n^2 + \varepsilon_i^2 + \left(\frac{v_i q \mu}{2}\right)^2\right)} \tag{7.9}$$

Here N_i and m_i are the electron concentration and effective mass for the ith band, and ε_i and v_i are the corresponding energy gap and Fermi velocity, $\omega_n = (2n+1)\pi T$.

The penetration depth λ can be calculated from the expression (see, for example, Abrikosov and Khalatnikov, 1959); we consider the case of diffuse reflection:

$$\delta = \pi \left(\int_0^\infty dq \ln\left[1 + \frac{Q(q)}{q^2}\right] \right)^{-1} \tag{7.10}$$

One can consider various limiting cases. If $v_i q \ll T_c$, $\varepsilon_i(0)$ (London case), then

$$Q_i(q) = 4\pi e^2 \sum_i n_{i;s}/m_i \tag{7.11}$$

where $n_{i;s} = \pi n_i \varepsilon_i^2 \, T \sum_{\omega_n} \left(\omega_n^2 + \varepsilon_i^2\right)^{-3/2}$ is the concentration of "paired" electrons in the ith band.

Let us consider the low-temperature region. Based on eqns. (7.10) and (7.11) one can obtain for the two-gap picture (this is the most interesting case) that the quantity $\Delta \delta = \delta(T) - \delta(0)|_{T \to 0}$ is a sum of two exponential terms; that is,

$$\Delta \delta = a e^{-\frac{\varepsilon_1(0)}{T}} + b e^{-\frac{\varepsilon_2(0)}{T}} \tag{7.12}$$

This expression is valid for the usual s-wave pairing. According to eqn. (7.12), at $T \to 0$ the temperature dependence of $\Delta \delta$ is determined by the smaller gap; that is, the penetration depth decreases slower than in the BCS one-band case.

Based on eqns. (7.10) and (7.11), one can evaluate the dependence of $\Delta \delta(T)$ for a whole temperature range up to $T = T_c$ (see Fig. 7.4 and the discussion in Section 7.4.4). One can also study the opposite limiting case, $\nu_i q \gg T_c$, $\varepsilon_i(0)$ (Pippard case).

One can have peculiar situations in which $\xi_2 \ll \delta \ll \xi_1$ (δ is the penetration depth, ξ_1, ξ_2 are the coherence lengths in the two bands). In this case a single sample will contain two groups of electrons, one of which is Pippard-like, while the other is London-like (Geilkman et al., 1967b, 1968).

For the novel systems, such as MgB$_2$, the penetration depth and its anisotropy were studied by Kogan and Zhelezina (2004). The treatment was based on the Eilenberger theory (1968).

The surface resistance at $T \to 0$ can also be written as a sum of two exponents.

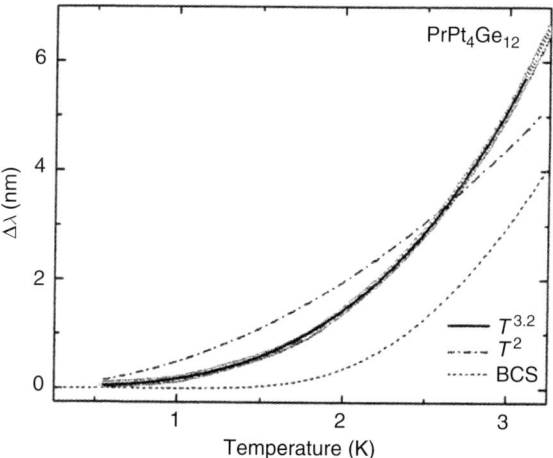

Fig. 7.4 Temperature dependence of the penetration depth $\Delta \delta(T)$ for PrPt$_4$Ge$_{12}$. Solid, dash-dotted, and dashed lines represent the fit of $\Delta \delta \sim T^{3.2}$, $\Delta \delta \sim T^2$ (point node), and single-gap BCS model, respectively. (Figure reproduced from Zhang et al., 2013. © 2013 by the American Physical Society.)

7.4.2 Strong magnetic field: Ginzburg–Landau equations for a multigap superconductor

One can study also the behavior of a multigap superconductor in an arbitrary magnetic field near T_c. In other words, we are talking about the generalization of the Ginzburg–Landau equations for the multigap case (Geilikman et al., 1967). For the usual one-gap case, the Ginzburg–Landau equations were derived from the microscopic theory by Gor'kov (1959); see Abrikosov et al. (1975). For the multigap case one can use the expansion near T_c and the expression for the current:

$$\Delta_i^*(\mathbf{r}) = \ln\left(\frac{2\gamma\tilde{\Omega}}{\pi T}\right)\sum_k \lambda_{ik}\Delta_k^*(\mathbf{r}) + \frac{7\zeta(3)}{48(\pi T_c)^2}$$
$$\times\left[\sum_k \lambda_{ik} v_k \left(\frac{\partial}{\partial \mathbf{r}} - 2ie\mathbf{A}\right)^2 \Delta_k^*(\mathbf{r}) - 6\sum_k \lambda_{ik}\Delta_k^*(\mathbf{r})|\Delta_k^*(\mathbf{r})|^2\right] \quad (7.13)$$

$$\mathbf{j}(\mathbf{r}) = \sum_k \left[\frac{ie}{m_k}\left(\Delta_k \frac{\partial \Delta_k^*}{\partial \mathbf{r}} - \Delta_k^* \frac{\partial \Delta_k}{\partial \mathbf{r}}\right) - \frac{4e^2}{m_k}|\Delta_k|^2 \mathbf{A}\right]\frac{7\zeta(3)n_k}{16(\pi T_c)^2}$$

Using the method proposed by Gor'kov and Melik-Barkhudarov (1963), one can seek the solution of eqn. (7.13) in the form:

$$\Delta_i(\mathbf{r}) = X_i Q^*(\mathbf{r}) + \varphi_i^*(T) \quad (7.14)$$

where $Q(\mathbf{r}) \propto \left[\frac{T_c - T}{T_c}\right]^{\frac{1}{2}}$ and slowly depends on coordinates, φ_i is a small correction, $\varphi_i \propto \left[\frac{T_c - T}{T_c}\right]^{\frac{3}{2}}$, X_i are defined by eqns. (7.5)–(7.6a). Substituting eqn. (7.4) in (7.9), we can arrive at the following equations:

$$a\left(\frac{\partial}{\partial \mathbf{r}} - 2ie\mathbf{A}\right)^2 Q^*(\mathbf{r}) + b\left[\frac{T_c - T}{T_c}\right] Q^*(\mathbf{r}) - cQ^*(\mathbf{r})|Q(\mathbf{r})|^2 = 0$$
$$\mathbf{j} = \left[ie\left(Q(\mathbf{r})\frac{\partial Q^*(\mathbf{r})}{\partial \mathbf{r}} - Q^*(\mathbf{r})\frac{\partial Q(\mathbf{r})}{\partial \mathbf{r}}\right) - 4e^2|Q(\mathbf{r})|^2 \mathbf{A}(\mathbf{r})\right]d \quad (7.15)$$

where

$$a = \rho\sum_i v_i |X_i|^2 \,;\, b = \sum_i v_i |X_i|^2\,;$$
$$c = 6\rho\sum_i v_i |X_i|^4\,;$$
$$d = 3\rho\sum_i \frac{n_i}{m_i}|X_i|^2 \,;\, \rho = \frac{7\zeta(3)}{48(\pi T_c)^2}$$

We can see from eqn. (7.15) that the Ginzburg–Landau theory can be used for the multigap superconductor; the function $Q(\mathbf{r})$ is playing role of the order parameter, and the values of coefficients are different from those for the usual one-gap case. It is essential that the Ginzburg–Landau theory contains a single order parameter, $Q(\mathbf{r})$, even for the multigap case.

7.4.3 Heat capacity

In the low-temperature region ($T \to 0$) the heat capacity is a sum of two exponents; the band with the smaller energy gap makes the dominant contribution, and as a result, $C_{el.|T\to 0}$ decreases slower than in the one-gap case. The dependence is similar to eqn. (7.12).

In order to evaluate the jump in heat capacity at $T = T_c$ one can use the expression for the entropy at $T \to T_c$ (see, for example, Lifshitz and Pitaevsky, 2002):

$$S = \frac{2\pi^2}{3} T \sum_i \nu_i - T \sum_i \nu_i \left(\frac{\varepsilon_i}{T}\right)^2 \qquad (7.16)$$

Based on eqns. (7.15) and (7.16), we can obtain

$$\frac{\Delta C}{C_n(T_c)} = \left(\frac{\Delta C}{C_n(T_c)}\right)_{BCS} \cdot \rho \qquad (7.17)$$

where $\left(\frac{\Delta C}{C_n(T_c)}\right)_{BCS} \approx 1.4$ and $\rho = 1 - \frac{\sum_{i,k} \nu_i \nu_k (|X_i|^2 - |X_k|^2)^2}{2 \sum_{i,k} \nu_i \nu_k |X_i|^4}$ (see eqns. (7.6) and (7.6a)). It is essential that $\rho < 1$ and, therefore, the multigap situation leads to the jump in heat capacity to be smaller than for the usual one-gap case. The impact is similar to that for the energy gap anisotropy (Pokrovsky, 1961). For some superconductors the jump in heat capacity is larger than that in the BCS theory. This phenomenon is caused by the strong coupling effects (see Section 3.5 and eqn. (3.40)).

7.4.4 Experimental data

As was stressed at the beginning of this chapter, the term "two-gap superconductivity" refers to the situation when the electronic density of states displays two distinct peaks. As a consequence, the two-gap structure can be detected by tunneling spectroscopy. In this way, a two-gap structure in the tunneling spectrum of the high-T_c compound (YBCO; see Chapter 10) was observed by Geerk et al. (1988).

As noted in the previous section, the presence of the two-gap structure greatly affects the temperature dependences of various quantities. In Chapters 10 and 13 we will discuss such manifestations for the high-T_c cuprates, MgB$_2$ and other compounds. Let us discuss here, as an example, the data obtained for a relatively new superconducting material, PrPt$_4$Ge$_{12}$, which belongs to the family of so-called filled-skutterudite compounds. It has a relatively high value of T_c ($T_c \approx 8$ K; see Gumenik et al., 2008). Various manifestations of the multigap structure of this superconductor were observed by Zhang et al. (2013). They study a high-quality single crystal of the compound, so that the mean free path ($l \approx 10^2$ nm) greatly exceeds the value of the coherence length ($\xi \approx 13.5$ nm). Because $l >> \xi$, there is no averaging of the gaps, and it is natural to expect the presence of several gaps for such a complex material.

Figure 7.4 shows the measured dependence of $\Delta\delta = \delta(T) - \delta(0)$ on temperature (solid line). The observed dependence ($\propto T^{3.2}$) cannot be fitted by the usual BCS exponential curve or by the power law ($\propto T^2$) valid in the presence of the nodes. One can provide an excellent fit with the use of the two-gap model (see eqn. (7.12)) with the values of the energy gaps: $\varepsilon_1(0) = 0.8 T_c$ and $\varepsilon_2(0) = 2 T_c$. Note that, according to Zhang et al. (2013), these values are in agreement with the criterion (7.7). Indeed, $(\varepsilon_1(0)/T_c) < (\varepsilon(0)/T_c)_{BCS}$, whereas $(\varepsilon_2(0)/T_c) > (\varepsilon(0)/T_c)_{BCS}$. The two-gap picture with these values of the gaps allows Zhang et al. (2013) to describe with great accuracy the data on heat capacity.

7.5 Induced two-band superconductivity

Let us consider the two-band model and the special case of $\lambda_2 = 0$; that is, let us suppose that one of the bands is intrinsically normal. Below the critical temperature, both bands are in the superconducting state. In other words, superconductivity in the second band is induced by the interband transitions. The critical temperature is given by eqn. (7.4) with

$$\tilde{\lambda} = \frac{1}{2}\left[\lambda_1 + \left(\lambda_1^2 + 4\lambda_{12}\lambda_{21}\right)^{1/2}\right] \tag{7.4a}$$

In the present case the bands are not spatially separated. Instead, near the Fermi surface the states belonging to different bands are separated in momentum space. The induced superconducting state arises due to phonon exchange. Note that other excitations such as excitons, plasmons, and so on, are also capable of providing interband transitions.

It is important to note that $\tilde{\lambda} > \lambda_1$, which is due to an effective increase in the phonon phase space. As noted previously, this means that the presence of the second band is favorable for superconductivity. As a result of charge transfer, the second band acquires an induced energy gap which depends on the constant λ_{12}.

As remarked previously, the ratios $2\varepsilon_1(0)/T_c$ and $2\varepsilon_2(0)/T_c$ differ from the value $2\varepsilon(0)/T_c = 3.52$ of the BCS theory. Namely, $2\varepsilon_1(0)/T_c > 3.52$ and $2\varepsilon_1(0)/T_c < 3.52$. In the case of induced two-gap superconductivity, the ratio λ_{21}/λ_1 could be very small. As a result, the values of the energy gaps may differ very noticeably from each other.

Now consider the case of strong coupling. We can write:

$$\Delta_1(\omega_n)Z_1 \cong \lambda_1 \sum_{\omega_{n'}} k_{\omega_n,\omega_{n'}} \Delta_1(\omega_{n'}), \tag{7.18}$$

$$\Delta_2(\omega_n) \cong \lambda_{21} \sum_{\omega_{n'}} k_{\omega_n,\omega_{n'}} \Delta_1(\omega_{n'}), \tag{7.18a}$$

where

$$k_{\omega_n,\omega_{n'}} = \tilde{\Omega}^2[\tilde{\Omega}^2 + (\omega_n - \omega_{n'})^2]^{-1}(\omega_{n'}^2 + \Delta_1^2)^{-1/2},$$

The energy gap is the solution of the equation $\omega = \Delta(-i\omega)$, where $\Delta(z)$ is the analytical continuation of the function $\Delta(\omega_n)$.

We obtain for the ratio of the gaps:

$$\eta = \frac{\varepsilon_2(0)}{\varepsilon_1(0)} \cong \frac{\lambda_{21}(1+\lambda_1)}{\lambda_1} \tag{7.9}$$

The presence of strong coupling leads to the appearance of the factor $(1 + \lambda_1)$. We see that strong coupling tends to decrease the relative difference in the values of the gaps.

7.6 Symmetry of the order parameter and multiband superconductor

Consider the case when the Fermi surface, in addition to the main ("large") part, contains small pockets. Such complex structure is not uncommon for many novel systems, such as heavy fermions materials, organics, borocarbides, and the high-T_c cuprates. If, in addition, the d-wave component of the order parameter dominates, then the value of the induced order parameter for the pocket may end up being close to zero (Gor'kov, 2012). In other words, we are dealing with the coexistence of the d-wave superconductivity in one band and the "unpaired" carriers in the pocket. Indeed, the induced order parameter is determined by eqn. (7.3) with $\lambda_2 = 0$ (see eqn. (7.18a)). Because of the d-wave symmetry, the order parameter in the "large" band has parts with opposite signs, and their compensations leads to the absence of a pairing amplitude in the pocket.

Two-band induced superconductivity is caused by phonon-mediated exchange, but does not require a spatial separation of the intrinsically superconducting and normal subsystems. Another type of induced superconductivity, the proximity effect, is based on such a separation, and will be considered in the next chapter.

8
Induced superconductivity: proximity effect

8.1 Proximity "sandwich"

The proximity effect was discovered by Meissner (1960) and involves spatially separated normal and superconducting subsystems. The simplest type of proximity system is shown schematically in Fig. 8.1.

A superconducting state is induced in the normal film N under the influence of the neighboring superconducting film S. Indeed, experimentally one finds that film N begins to exhibit the Meissner effect; see, for example, Simon and Chaikin (1981). Qualitatively, such an effect means that two electrons from the N film can tunnel into S film, forming the Cooper pair. This leads to an appearance of the pairing function F_N^+ (see Figs. 3.1 and 3.2). One can introduce also the order parameter Δ_N.

An important parameter characterizing the proximity system is the coherence length in the normal film, ξ_N. It describes the scale of penetration of the superconducting state into N film. This quantity, introduced by Clarke (1969) is temperature dependent and equals $\xi_N = v_F/2\pi T$. Thus, lowering the temperature benefits the proximity effect. For example, it has been observed (Mota et al., 1989) that for NbTe–Cu(Ag) systems, large scale (up to 2.10^5 A) of induced superconductivity could be achieved at ultra-low temperatures.

There exist different approaches to describe the proximity effect, and among them, two are particularly prominent. One is based on the Ginzburg–Landau theory and makes use of the boundary conditions at the S–N interface (the "Orsay school;" see, for example, de Gennes (1964, 1966), and Deutcher and de Gennes (1969)). This approach takes into account directly the spatial variation of the order parameter $\Delta(\boldsymbol{r})$.

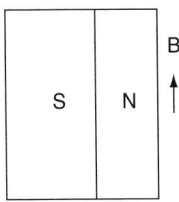

Fig. 8.1 Proximity S–N system.

The other approach is the tunneling model developed by McMillan (1968b). The superconducting state in the N film arises due to Cooper pair tunneling from the superconducting film into the normal one. In the tunneling model it is assumed that the order parameters in the S and N films are constant throughout the films. In other words, it is assumed, for example, that the thickness of the normal film is much smaller than the coherence length, ξ_N; that is, $L_N \ll \xi_N$. An attractive feature of this model, in addition to its elegance and simplicity, is that it is applicable not only near T_c but at any temperature.

8.2 Critical temperature

Let us calculate the critical temperature of an S–N proximity sandwich within the framework of the McMillan model. We will employ the method of thermodynamic Green's functions. Let us introduce the self-energy parts, Σ_2^α and Σ_2^β ($\alpha(\beta) \equiv S(N)$), describing pairing in the films. They satisfy the following equations:

$$\Sigma_2^\alpha = \Sigma_{2,ph}^\alpha + \tilde{T}^2 \int d\boldsymbol{p}' F^\beta(\boldsymbol{p}', \omega_n),$$
$$\Sigma_2^\beta = \tilde{T}^2 \int d\boldsymbol{p}' F^\alpha(\boldsymbol{p}', \omega_n), \qquad (8.1)$$
$$\Sigma_{2,ph}^\alpha = T \sum_{\omega_{n'}} \int d\boldsymbol{p}' \zeta_\alpha^2(\boldsymbol{p}, \boldsymbol{p}') D(\omega_n - \omega_{n'}, \Omega(q)) F^\alpha(\boldsymbol{p}', \omega_{n'})$$

Here, D is the phonon Green's function (see eqn. 3.5), and F^α and F^β are the pairing Green's functions; for example,

$$F^\alpha(\omega_n, \boldsymbol{p}) = -\frac{\Sigma_2^\alpha(\omega_n, \boldsymbol{p})}{\omega_n^2 Z_\alpha^2 + \xi_\alpha^2(\boldsymbol{p}) + \Sigma_2^{(\alpha)^2}(\omega_n, \boldsymbol{p})} \qquad (8.1a)$$

A similar expression can be written for F^β. Here, ξ_α is the energy of an ordinary electron referred to the Fermi level, Z is the renormalization function, and $\tilde{T}$ is the averaged tunneling matrix element.

Equations (8.1) are written for the case when the existence of the pair condensate in the β film is due totally to the proximity effect.

In the weak coupling approximation the renormalization functions at $T = T_c$ are equal to

$$Z_i(\omega_n) = 1 + \frac{\Gamma^{ik}}{|\omega_n|}; \quad \Gamma^{ik} = \pi \tilde{T}^2 \nu_k V_K; \quad i, k = \alpha, \beta; \ i \neq k \qquad (8.2)$$

ν_k and V_K are the density of states and volume. The parameters Γ^{ik} were introduced by McMillan (1968), and, for example, $\Gamma^{\beta\alpha}$ can be written in the following form:

$$\Gamma^{\beta\alpha} = \frac{v_{F\perp} \sigma}{2 B L_\beta} \qquad (8.3)$$

where $v_{F\perp}$ is the Fermi velocity, σ is the barrier penetration probability, and B is a function of the ratio of the mean free path to the film thickness. If $T = T_c$, we should put $\Sigma_2^{(\alpha)} = 0$ in the denominator of eqn. (8.1a).

Making use of eqns. (8.1) and (8.2), we arrive, after some manipulation, at an equation for T_c (Kresin, 1982):

$$\ln \frac{T_c}{T_c^\alpha} = -\frac{\Gamma_{\alpha\beta}}{\Gamma}\frac{1}{\lambda_\alpha}\int d\Omega g(\Omega) \\ \times \left\{\left[\psi\left(\frac{1}{2}+\frac{\Gamma}{2\pi T_c}\right)-\psi\left(\frac{1}{2}\right)\right]\frac{\Omega^2}{\Omega^2+\Gamma^2}+\frac{\Gamma^2}{\Omega^2+\Gamma^2}\ln\frac{2\Omega Y}{\pi T_c}\right\} \quad (8.4)$$

Here, T_c^α is the critical temperature in an isolated α film, $\Gamma = \Gamma_{\alpha\beta}+\Gamma_{\beta\alpha}$, and ψ is the digamma function. It should be noted that in general, $g(\Omega) \neq g^a(\Omega)$; in other words, the presence of the interface may alter the function $g(\Omega)$.

Equation (8.4) allows us to evaluate T_c for any proximity system, and is valid for any relation between Γ and T_c^α.

Let us consider some special cases. Assume that $\Gamma \gg T_c$. Making use of the asymptotic form of the digamma function, we find, after simple substitutions, that

$$T_c = T_c^\alpha \left(\frac{\pi T_c^\alpha}{2\langle u\rangle \gamma}\right)^\rho, \quad (8.5)$$

where

$$\rho = \Gamma_{\alpha\beta}/\Gamma_{\beta\alpha};\ \Gamma = \Gamma_{\beta\alpha}(1+\rho); \\ u = \Gamma^{1-\delta}\Omega^\delta;\ \delta = \Gamma^2(\Omega^2+\Gamma^2)^{-1} \quad (8.5\text{a})$$

or

$$\rho = (\nu_\beta/\nu_\alpha)(L_\beta/L_\alpha) \quad (8.5\text{b})$$

Recall that it is assumed that $L_i \ll \xi_i (i \equiv \alpha, \beta)$.

The mean value is to be understood in the following sense:

$$\ln\frac{\langle u(\Omega)\rangle}{T_c} = \frac{1}{\lambda_\alpha}\int d\Omega\, g_\alpha(\Omega)\ln\frac{u(\Omega)}{T_c}. \quad (8.6)$$

Equation (8.5) is valid for any relation between Γ and $\langle\Omega\rangle$ (keeping in mind that $\Gamma \gg T_c$). If $\Gamma \ll \langle\Omega\rangle$ we arrive at the expression obtained by McMillan (1968):

$$T_c = T_c^\alpha \left(\frac{\pi T_c^\alpha}{2\,\Gamma\gamma}\right)^\rho \quad (8.7)$$

In the opposite limit, $(\Gamma \gg \tilde\Omega)$ (Cooper, 1961), we obtain the result

$$T_c = T_c^\alpha \left(\frac{\pi T_c^\alpha}{2\langle\Omega\rangle\gamma}\right)^\rho \quad (8.8)$$

Finally, if $\Gamma \lesssim T_c$, $\tilde\Omega \gg T_c$, $\rho \ll 1$, we find

$$T_c = T_c^\alpha \exp(-F) \quad (8.9)$$

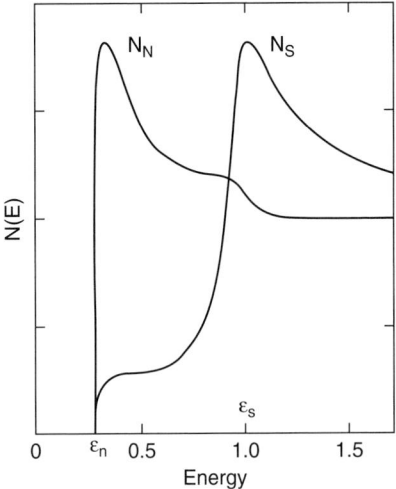

Fig. 8.2 Densities of states for an S–N system ("two-gap structure").

where

$$F = (\Gamma_{\alpha\beta}/\Gamma)\left[\psi\left(\frac{1}{2} + \frac{\Gamma}{2\pi T_c}\right) - \psi\left(\frac{1}{2}\right)\right] \tag{8.10}$$

Another interesting quantity to determine is the induced energy gap $\varepsilon_\beta(0)$. It can be found from eqn. (8.1), and is given by the simple expression (McMillan, 1968b)

$$\varepsilon_\beta(0) = \Gamma^{\alpha\beta} \tag{8.11}$$

Figure 8.2 displays the density of states of the superconducting electrons. The density of states is characterized by two distinct peaks, corresponding to two gaps: $\varepsilon_\alpha(0)$ and $\varepsilon_\beta(0)$.

Clearly, the behavior of an isolated metallic film is strongly affected by the boundary conditions. Indeed, electrons are scattered by the surface, and, for example, one has to distinguish the cased of specular or diffuse scattering. However, in the case of a proximity system—for example, an N–S sandwich, the scattering picture becomes quite peculiar. Indeed, two N-electrons can tunnel into the S film (we assume that $E_N < \varepsilon_S$), making a transition into the Cooper pair condensate. That is to say, the presence of the N–S boundary leads to the "disappearance" of two N-electrons. At the same time, we know that the disappearance of an electron can be regarded as the appearance of a hole. As a result, the transfer of two N-electrons into a Cooper pair on the S-side can be described as the disappearance of one electron with simultaneous creation of a hole (envisioning this creation instead of the disappearance of the second electron). Therefore, in terms of the scattering picture, the process at the N–S boundary can be described as the disappearance of an electron and the creation of a hole (Andreev scattering; Andreev, 1964; see also Blonder et al., 1982, and the reviews by Tinkham,

1996; Waldram, 1996; and Deutscher, 2005). Generally speaking, this corresponds to the Cooper diagram in Figs. 3.1 and 3.2. The pairing can be described as an electron → hole transition.

We focused on the S–N proximity systems. One can also study the contact of two superconductors S_α – S_β with different values of the critical temperature. In this case the proximity sandwich has T_c, which value is intermediate between $T_{c;\alpha}$ and $T_{c;\beta}$.

The physics of proximity systems is an interesting and broad area. For example, the study of Josephson junctions S–N–I–S and S–N–S is concerned with Josephson tunneling in proximity systems. Other active areas include the electrodynamics of proximity systems, as well as the properties of S_α – S_β contacts ($T_c^\alpha \neq T_c^\beta$). If, for example, $T_c^\alpha > T_c^\beta$, then at $T > T_c^\beta$ the superconducting state penetrates into β film, but the decay of the pairing function is slower than for the SN contact (Covaci and Marsiglio, 2006).

Note that for the S_α – S_β system at $T_c^\alpha > T > T_c^\beta$ the McMillan tunneling model is applicable, and equations similar to eqn. (8.1) can be used to calculate, for example, the temperature dependence of the energy gaps. For example, the data by Gilabert et al. (1971) for the Pb–Sn system are in a good agreement with the tunneling model (Fig. 8.3).

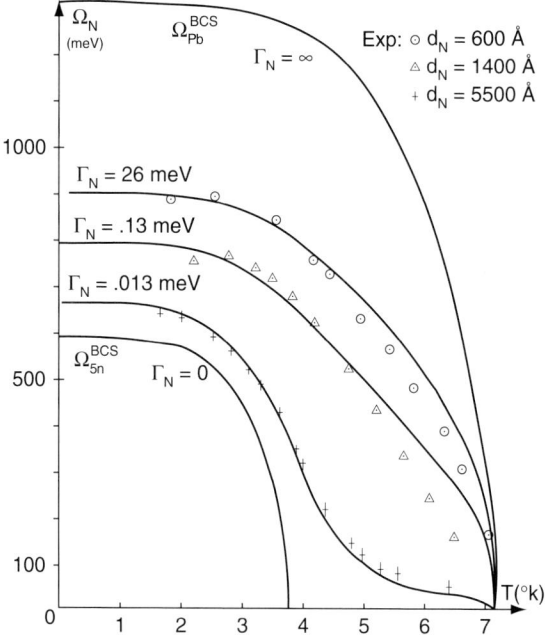

Fig. 8.3 Temperature variation of the energy gap $\Omega_N \equiv \varepsilon_N$ of tin films backed by 5000Å of lead. Experimental points are compared with theoretical curves from the MacMillan tunneling model: $\Gamma_N \equiv \Gamma^{\alpha\beta}$. (Figure reprinted from Gilabert et al., 1971. © 1971, with permission from Elsevier.)

As noted previously, the described tunneling model is valid if the thickness of the normal (N) film is smaller than the coherence length ξ_N. If we are concerned with the spatial dependence of the order parameter, one should use the approach based on the Ginzburg–Landau theory with corresponding boundary conditions (see, for example, Deutcher and de Gennes, 1969). A very efficient method based on the Green's functions, averaged over energy (Eilenberger, 1968; Larkin and Ovchinnikov, 1969, 1986) was developed by Usadel (1970); see, for example, Gueron et al. (1996), and Martinis et al. (2000). These questions, however, are beyond the scope of the present book.

8.3 Proximity effect versus the two-gap model

At this point we may compare the results found for the proximity system with those of the two-gap model. There is a profound analogy between these two situations. In both cases we are dealing with induced superconductivity (we are assuming that in the second band, as in the N film, there is no intrinsic pairing). In both cases an induced energy gap appears. However, the mechanisms giving rise to induced superconductivity are very different. In the two-band model the system's bands are "separated" in momentum space, and the second band acquires an order parameter due to phonon exchange. The phase space is effectively increased. In the proximity effect, on the other hand, the systems are spatially separated, and superconductivity is induced by the tunneling of Cooper pairs.

In both cases we end up with an induced energy gap, but the physical differences manifest themselves in the behavior of the critical temperature. The two-band picture turns out to be favorable for T_c, which becomes higher than in the single-band case (eqn. (7.4a)). On the contrary, the proximity effect depresses T_c. The critical temperature of a proximity sandwich is lower than that of an isolated S film. This can be seen directly from eqns. (8.7) and (8.8).

8.4 Pair-breaking: gapless superconductivity

As discussed in previous section, the proximity effect results in a decrease in T_c relative to its value for the superconducting film; that is, $T_c < T_c^S$. Qualitatively, this is a consequence of the concurrent diffusion of Cooper pairs from the superconducting film α into the normal film β and the "normal" electrons into α film. As a result, the density of the "superconducting" component—that is, of the number of Cooper pairs in the α film—decreases relative to their density in an isolated α film. Another way to express this is that the proximity effect leads to the break-up of pairs in the superconducting film. As a result, the critical temperature decreases relative to T_c^α in the isolated α film. This decrease is described by eqns. (8.5)–(8.8).

The proximity contact (for example, the S–N "sandwich") is an example of a spatially inhomogeneous system, and the pair-breaking effect is a consequence of the inhomogeneity. However, this effect was introduced initially by Abrikosov and Gor'kov (1961), and the pair-breaking appeared as a result of violation of time-reverse symmetry. More specifically, they focused on the impact of magnetic impurities. The localized magnetic moment (magnetic impurity) is trying to align the magnetic moments of two electrons forming a Cooper pair, and therefore to destroy the bound state of the pair. Abrikosov and Gor'kov generalized the so-called "cross"-diagrammatic technique developed by them (1959) to study the impact of non-magnetic impurities and defects (for the description of this technique, see, for example, Abrikosov et al. (1975)) for the case of magnetic moments. It has been shown that the value of the critical temperature T_c is depressed relative to its value T_c^0 in the absence of the pair-breakers, so that

$$\ln\left(\frac{T_c^0}{T_c}\right) = \psi(0.5 + \gamma_S) - \psi(0.5) \tag{8.12}$$

Here, ψ is the digamma function, $\gamma_S = (2\pi T_c^0 \tau_S)^{-1}$, and τ_s is the scattering relaxation time.

Equation (8.12) is similar to eqns. (8.4) and (8.10), describing the impact of the proximity effect on the value of T_c. This is not occasional, since in both cases we are dealing with the pair-breaking phenomenon.

The pair-breaking effect has also been studied by de Gennes (1966) and by Skalski et al. (1964); see also the review by Maki (1969). Note also that pair-breaking can be also caused by an external magnetic field (Maki and Tsuneto, 1964).

Experimentally, one can observe a marked change in the critical temperature. Even small impurity concentration (a few percent) leads to complete destruction of the superconducting state.

The presence of magnetic impurities affects not only the critical temperature but also the energy gap. It turns out that the value of the energy gap is affected even stronger than T_c. As a result, one can observe a remarkable phenomenon: the gapless superconducting state (Abrikosov and Gor'kov, 1961). One can show that in the usual case the energy gap vanishes at concentration $n \cong 0.9 n_{cr}$, where n_{cr} is the concentration at which the superconductivity disappears ($T_c = 0$). The superconducting state persists at $n_{cr} > n > 0.9 n_{cr}$, though the energy gap is equal to zero. This result is of fundamental importance. It illustrates the fact that the presence of the energy gap is an important feature but not a condition necessary for the existence of superconductivity. The order parameter may exist even in the absence of the energy gap.

The gapless state can be visualized qualitatively as being described by the "two-liquid" model. According to this model, the superconductor contains two "components." The presence of a "normal" component leads to the absence of the energy gap, whereas the "superfluid" component is responsible for zero resistance, the Meissner effect, and so on.

Let us consider the impact of magnetic impurities on the induced superconducting state. More specifically, one can consider the proximity system (S–N contact), so that the energy gap ε_N is induced by interaction with the intrinsically superconducting film S. Another example is the two-band system when the pairing in one of the bands is induced by the interband transitions (Section 7.5.). It turns out that the region of the gaplessness can be greatly extended (Kresin and Wolf, 1995); that is, it might appear at concentrations much smaller than in the usual isolated one-gap superconductor.

Consider the α–β system, where α is an intrinsically superconducting part. As for the β subsystem, its superconducting state is induced by the interaction with the α part; as mentioned previously, it could be a normal film for the S–N proximity contact, or the energy band with the pairing induced by the interband transitions only.

Consider, at first, the S–N proximity system and assume that the N film contains magnetic impurities. To describe the magnetic scattering one should add an additional term Γ_M^β (Kaizer and Zuckermann, 1970) into the expression for the renormalization term Z_β (see eqn. (8.2)). Assume also that $\Gamma^{\alpha\beta} \ll \Delta_\alpha^{ph} \cong \varepsilon_\alpha; \Delta_i^{ph} = \sum_i^{ph} Z_i$, where ε_α is the energy gap for the α film. One can prove that the induced energy gap ε_β becomes equal to zero at $\Gamma_{\beta M}^C \cong \Gamma_M^{\beta\alpha} \cong \varepsilon_\beta$. Indeed, in this case, the total density of states becomes different from zero at $\omega \to 0$. We can estimate the corresponding concentration of magnetic impurities. The mean free path for such a gapless state is equal to $l_m = v_F^\beta \tau_M^C \approx v_F^\beta \varepsilon_\beta^{-1} \cong \xi_\beta$, where ξ_β is the coherence length for the N film.

It is essential that the transition to such a gapless state is not accompanied by a noticeable shift in T_c. The value of this shift depends on the parameters of the α film and $\Gamma^{\beta\alpha}$. For realistic values of the quantities, the shift in T_c appears to be small. For example, for the value of the induced gap $\varepsilon_\beta \cong 0.8 T_c$ and $\Gamma^{\beta\alpha} = (\pi/4)T_c$, we obtain $\Delta T_c/T_c \cong 10^{-2}$. Therefore, indeed, gaplessness does not lead to a noticeable shift in T_c. The picture differs drastically from that for usual superconductors, where $(\Delta T_c/T_c) \cong 0.9$.

The case with two-band superconductor could be analyzed in a similar way. In this case it is impossible to put magnetic impurities in the β subsystem, because, unlike the proximity system, there is no special separation of both subsystems. However, one can consider the realistic case when $\Gamma_M^\beta \lesssim \varepsilon_\beta$, whereas $\Gamma_M^\beta \ll \varepsilon_\alpha$. The energy gap ε_β vanishes (gapless state) at the critical value of the scattering amplitude $\Gamma_M^C \cong \varepsilon_\beta = \left(\frac{\lambda_{\beta\alpha}}{\lambda_\alpha}\right)\varepsilon_\alpha$. In this case there is no drastic change in T_c, despite an appearance of the gapless state.

9
Isotope effect

9.1 General remarks

The isotope effect occupies a special place in the physics of superconductivity. Its discovery was the major step preceding the BCS theory by demonstrating that lattice is involved in the formation of the superconducting state. The effect was studied theoretically by Frohlich (1950) and was discovered by studying various isotopes of mercury (Maxwell, 1950; Reynolds, et al., 1950). As the mass M was varied between 199.5 and 203.4, the value of T_c changed from 4.185K to 4.140K It was established that the following relation holds:

$$T_c = (\text{const.}) \cdot M^{-\alpha}; \alpha \approx 0.5 \tag{9.1}$$

The dependence (9.1) and value $\alpha \approx 0.5$ are in a total agreement with the BCS expression (3.13) for the critical temperature, since the vibrational frequency $\tilde{\Omega} \propto M^{-1/2}$.

At first sight, the isotope effect seems to be a rather straightforward phenomenon, but such an impression is misleading. In reality, it is a complex effect controlled by many factors. Indeed, in many cases one can observe large derivations from the canonical value $\alpha = 0.5$. As a result, the presence of the isotopic dependence means that the lattice is involved in the pairing, but it is difficult to provide a quantitave estimation of the degree of this involvement.

In the following we discuss various factors affecting the value of the isotope coefficient.

9.2 Coulomb pseudopotential

The BCS expression (3.28) contains the Coulomb pseudopotential μ^* defined by eqn. (3.27). One can see directly from eqn. (3.27) that the value of μ^* depends on phonon frequency, and its change upon the isotope substitution also affects the value of T_c and, correspondingly, the value of the isotope coefficient α.

Starting from the expression (9.1), defining the isotope coefficient, we can obtain the following general expression:

$$\alpha = -(M/T_c)(\partial T_c/\partial M) \tag{9.2}$$

Then we can write:

$$\alpha = -(M/T_c)(\partial T_c/\partial \Omega)(\partial \Omega/\partial M) \tag{9.3}$$

Based on eqn. (3.27), (3.28), and (9.3), we obtain:

$$\alpha = \alpha_{|\mu^*=0} \left[1 - \left(\frac{\mu^*}{\lambda - \mu^*} \right)^2 \right] \tag{9.4}$$

One can see that the presence of the Coulomb factor leads to a decrease in the value of the isotope coefficient. For example, if $\lambda = 0.5$ and $\mu^* = 0.15$, we obtain $\alpha = 0.4$. For $\lambda = 0.4$ and $\mu^* = 0.15$, the decrease is even more noticeable: $\alpha \approx 0.3$.

Eqn. (3.28) is valid for the case of weak coupling. One can see from eqns. (3.22) and (3.33) that for the intermediate and strong coupling (see also eqn. (3.35)), the presence of the Coulomb pseudopotential also leads to a noticeable impact on the isotope effect.

9.3 Multi-component lattice

As shown previously, the Coulomb pseudopotential noticeably affects the value of the isotope coefficient. If the lattice contains several atoms per unit cell, the picture becomes more complicated (Geilikman, 1976). Indeed, even for a two-atomic lattice we can obtain any value of α. This can be seen directly from the expression for the two phonon modes (Maradudin et al., 1963):

$$\omega_{ac}^2 = \sum_{\alpha=1}^{3} \gamma_\alpha \left(\frac{1}{M_1} + \frac{1}{M_2} \right) - \left[\left(\sum_{\alpha=1}^{3} \gamma_\alpha \right)^2 \left(\frac{1}{M_1} - \frac{1}{M_2} \right)^2 + \frac{4}{M_1 M_2} \left(\sum_{\alpha=1}^{3} \gamma_\alpha \cos q_0 d \right)^2 \right]^{1/2}$$

$$\omega_{op}^2 = \sum_{\alpha=1}^{3} \gamma_\alpha \left(\frac{1}{M_1} + \frac{1}{M_2} \right) + \left[\left(\sum_{\alpha=1}^{3} \gamma_\alpha \right)^2 \left(\frac{1}{M_1} - \frac{1}{M_2} \right)^2 + \frac{4}{M_1 M_2} \left(\sum_{\alpha=1}^{3} \gamma_\alpha \cos q_0 d \right)^2 \right]^{1/2}$$
(9.5)

The substitution $M_1 \to M_1^*$ can lead to any value of α, depending on M_2, force constants γ_α, and so on.

9.4 Anharmonicity

Anharmonicity of the lattice can strongly affect the value of the isotope coefficient. The well-known superconductor displaying such an effect is the Pd-H compound. The isotope substitution $H \to D$ leads to an increase in T_c,; that is, the isotope coefficient α (see eqn. (9.1)) appears to be negative. The idea that such an unusual behavior of the isotopic dependence is caused by anharmonicity was proposed by Ganzuly (1973). A detailed analysis, based on the strong coupling theory (see Chapter 3) was carried out by Klein and Cohen (1992). They demonstrated that anharmonicity of the lattice leads to the dependence of vibrational frequency on mass which drastically different from the $\tilde{\Omega} \propto M^{-\frac{1}{2}}$ law. The calculation for the Pd-H system performed up to sixth order of the ionic displacement show that the coupling constant λ (see eqn. (3.23)) increases upon the isotope substitution. This increase is caused by the effective decrease in the characteristic frequency; the Hopfield factor $\eta = <I>^2 v_F$, see eqn. (3.23), was assumed being constant, so that the increase in λ is caused by lattice dynamics only. The value of T_c was determined from eqn. (3.35). The value of μ^* for the Pd-H system was chosen to be $\mu^* = 0.115$, since it provides the best agreement with the data (Schirber and Northrop, 1974). The value of the isotope coefficient obtained by Klein and Cohen (1992) appears to be negative, in agreement with experimental observation.

Therefore, anharmonicity can drastically change the isotopic dependence of the critical temperature; it may lead even to a negative sign of the isotope coefficient.

9.5 Isotope effect in proximity systems

In this and the following sections we consider the factors which are not related directly to the pairing mechanism, but nevertheless contribute to the isotope effect. Indeed, the value of the critical temperature could be affected by an external factor ($T_{c;0} \to T_c$; $T_{c;0}$ is the intrinsic value of the critical temperature), which is not related to the lattice dynamics.

The most interesting case corresponds to the situation when the relation between $T_{c;0}$ and T_c is not linear. Let us be more specific. As an example, consider the isotope effect in a proximity system $S-N$ (where S and N are superconducting and normal films; see Chapter 8 and Fig. 8.1). It turns out that the value of α depends on the relative thicknesses of the films. Indeed, assume that the thickness $L_N \ll \xi_N$, where $\xi_N = h v_{F;N}/2\pi T$ is the coherence length for the N film. We can then use a well-known McMillan tunneling model (1968b). According to this model, the proximity effect is described by the parameter $\Gamma = \Gamma_{SN} + \Gamma_{NS}$, where $\Gamma_{ik} = \tilde{T}_{ik}^2 \nu_k V_k$, $\tilde{T}_{ik}$ is the tunneling matrix element, ν_k is the density of states (per unit of volume V), $i,k = \{S,N\}$, $i \neq k$. We assume that $\Gamma \ll \langle\Omega\rangle$. One can see that, indeed, the relation between T_c and T_{c0} is non-linear. The critical temperature of the whole system T_c differs from T_{c0} (T_{c0} is the critical temperature of the isolated S film), and is described by eqn. (8.7).

If we make an isotope substitution $M \to M^*$ for the isolated S film, one can measure the shift in T_{c0} and determine the isotope coefficient α_0 which is described by eqn. (9.2). The presence of the N film leads to a change in T_c and in the value of the isotope coefficient. One can see directly from eqns. (9.2) and (8.7) that the shift in T_c and the new value of the isotope coefficient $\alpha = -(M/\Delta M)(\Delta T_c/T_c)$ differ from $\Delta T_{c0}/T_{c0}$ and the value of α_0. Indeed, the value of T_c for the whole sandwich is determined by eqn. (8.7). Then the value of the shift ΔT_c is determined not only by ΔT_{c0} but also by the value of the parameter ρ defined by eqn. (8.5b); it reflects the presence of the N film. More specifically, based on eqn. (8.7), we obtain:

$$\Delta T_c/T_c = (\Delta T_{c0}/T_{c0})(1+\rho)$$

As for the value of the isotope coefficient, we obtain

$$\alpha = \alpha_o[1 + (\nu_N L_N/\nu_S L_S)] \tag{9.6}$$

One can see that, indeed, the value of the isotope coefficient is modified by the proximity effect. Moreover, $\alpha > \alpha_o$. Therefore, a decrease in T_c, which is a well-known feature of the proximity effect, and is described by eqn. (8.7), is accompanied by an increase in the isotope coefficient. It is interesting that one can modify the value of α by changing the thicknesses of the films. The increase in the thickness of the normal film L_N leads to decrease in T_c, but the value of α increases. For example, if $\alpha_0 = 0.2$, $\nu_N/\nu_s = 0.8$. and $L_N/L_s = 0.5$, then $\alpha = 0.28$. If we increase the thickness of the normal film, so that $L_N = L_s$, then $\alpha = 0.36$.

Note that the increase of the isotope coefficient discussed in this section is not related to lattice dynamics; as a result, the value of α can, in principle, exceed, the value of $\alpha_{0;\text{ph, max}} = 0.5$.

9.6 Magnetic impurities and isotope effect

In this section we focus on another isotope effect which is also not related to lattice dynamics: we consider a superconductor which contains magnetic impurities.

The presence of magnetic impurities leads to decrease of the critical temperature, T_c, relative to the intrinsic value T_{c0}, because of the pair-breaking effect described in Ch. 8. This depression of T_c is described by eqn. (8.12): $\ln(T_{c0}/T_c) = \Psi[0.5 + \gamma_s] - \Psi(0.5); \gamma_s = (2\pi T_c^0 \tau_s)^{-1}$; Ψ is the digamma function. In this case also (c.f. eqn. (8.7)) we are dealing with a non-linear relation between T_{c0} and T_c.

The isotope substitution $M \to M^*$ for the sample without magnetic impurities allows one to observe the shift in T_{c0} and measure the isotope coefficient. One can see directly from eqn. (8.12) that the shift in T_c and the value of the isotope coefficient α in the presence of magnetic impurities differ from $\Delta T_{c0}/T_{c0}$ and the value of α_0.

Based on eqns. (9.2) and (8.12), one can arrive at the following equation (Carbotte et al., 1991; Kresin et al., 1997):

$$\alpha = \alpha_0 [1 - \Psi'(0.5 + \gamma_s)\gamma_s]^{-1} \qquad (9.7)$$

Note that an increase in the concentration of magnetic impurities n_M leads to increase in γ_s. Eqn. (9.7) is valid in a broad range except a very small region near n_{cr} (then T_c is close to $T = 0$ K and the condition $\Delta T_c/T_c \ll 1$ is not satisfied).

The presence of magnetic impurities leads to an increase of the isotope coefficient ($\alpha > \alpha_0$), since $\Psi' > 0$; this can be seen directly from the expression

$$\Psi'(0.5 + x) = \sum_{k=0}^{\infty} (k + 0.5 + x)^{-2}$$

One can study the dependence α on the concentration of magnetic impurities. For small γ_s (small values of n_M), $\Delta \alpha \propto n_M$. Therefore, near T_{c0} the critical temperature displays a linear decrease with increasing n_M, whereas the isotope coefficient increases linearly as a function of n_M. In the region $\gamma_s \gg 1$ one can use an asymptotic expression for the digamma function, and we obtain $\Delta \alpha \propto n^2{}_M$; that is, the dependence becomes strongly non-linear.

We can see directly from eqns. (9.6) and (9.7) that the impact of the proximity effect, as well as the magnetic scattering on the value of the isotope coefficient, is caused by the non-linear relation between T_c and T_{c0}. Qualitatively, both factors provide the pair-breaking, and this leads to an effective increase in the normal electronic component.

9.7 Polaronic effect and isotope substitution

Some novel materials, including the high-T_c cuprates, display dynamic polaronic effect (see Sections 2.6.2 and 3.9). These materials contain sub-groups of ions (for example, oxygen ions in the cuprates), and each of them is characterized not by one but two close equilibrium positions. In addition, the carriers in materials of interest (see Chapter 10) are provided by doping. Then one can observe a novel isotope effect (Kresin and Wolf, 1994). The point is that the doping in such materials represents a charge transfer process. In this case the isotope substitution affects the doping, and therefore the carrier concentration n. For doped materials the critical temperature depends strongly on n, and this leads to isotopic dependence of T_c. In the following we will discuss all implications of such an isotope effect for the physics of high-T_c cuprates (Chapter 10) and manganites (Chapter 12), but here we describe the basic scenario of the polaronic isotope effect.

If the charge transfer occurs in the framework of the usual adiabatic picture (Chapter 2), so that only the carrier motion is involved, then the isotope substitution does not affect the forces and therefore does not change the charge transfer dynamics. However, the situation of strong non-adiabaticity (polaronic effect; see Section 2.6.2) is different, and does not allow the separation of electronic and nuclear motions. In this case, charge transfer appears as a more complex phenomenon, which involves nuclear motion, and this leads to a dependence of the doping on isotopic mass.

Let us consider the case when the lattice configuration—that is, the positions of some definite ions (such as axial oxygen for YBCO, see Chapter 10)—corresponds to a degenerate or near-degenerate state. This means that the degree of freedom describing its motion corresponds to electronic terms crossing (see Fig. 2.4). Then the ion has two close equilibrium positions.

The charge transfer in this picture is accompanied by the transition to another electronic term. Such a process is analogous to the Landau–Zener effect (see, for example, Landau and Lifshitz, 1977). The charge transfer corresponds to the transitions between the first and second terms.

The total wavefunction (see Section 2.6.2) can be written in the form

$$\Psi_i(\mathbf{r}, \mathbf{R}, t) = a(t)\Psi_1(\mathbf{r}, \mathbf{R}) + b(t)\Psi_2(\mathbf{r}, \mathbf{R}) \tag{9.8}$$

Here

$$\Psi_i(\mathbf{r}, \mathbf{R}) = \Psi_i(\mathbf{r}, \mathbf{R})\Phi_i(\mathbf{R}), \; i = \{1, 2\}$$

$\Psi_i(\vec{r}, \vec{R}), \Phi_i(\vec{R})$ are the electronic and vibrational wavefunctions that correspond to two different electronic terms (see Fig. 2.4).

Qualitatively, the charge transfer for such non-adiabaticity can be visualized as a multistep process. First, the carrier makes a transition from the reservoir site to the ion; then the ion transfers to another term, and this is finally followed by the transition of the carriers to the plane (for the cuprates). The second step is affected by the isotope substitution.

Let us separate the Z-coordinate of the ion (the axis Z has been chosen to be parallel to the ionic motion between the minima, so that $\Phi_i(\mathbf{R}) = \Phi_i(\boldsymbol{\rho}_i)\varphi_i(Z)$, $\mathbf{R}_i$

corresponds to the other degrees of freedom. We do not assume the electronic terms to be similar and, as a result, they differ on the energy level spacing. This leads to some splitting between the vibrational levels (see Fig. 2.4), and the transition of the ion is not a resonant one. In the harmonic approximation,

$$\varphi_1(Z) = (\pi a_1)^{-1/2} \exp\left(-\frac{(z-z_{10})^2}{2a_1^2}\right) \tag{9.9}$$

$$\varphi_2(Z) = (\pi a_2)^{-1/2} \exp\left[-\frac{(z-z_{20})^2}{2a_2^2}\right] H_\nu[(z-z_{20})a_2^{-1}] \tag{9.9a}$$

Here, $a_i = (M\,\Omega_i)^{-1}$ are the vibrational amplitudes, H_ν is the Hermite polynomial, M is the ionic mass, and Z_{i0} are the equilibrium positions. We assume that the ionic motion corresponds to the zero vibrational state of the first term, and ν th level for the second term. Assume also that $b(0) = 0$. The dependence $b(t)$ describes the dynamics of the charge transfer $(1 \to 2)$.

As noted previously (Section 2.6.2), it is convenient to employ the diabatic representation, and for the average value, $\tilde{b}^2$, we obtain the expression (2.40a). The asymmetry of the potential for the axial oxygen and correspondingly the inequality $\varepsilon \neq 0$ are playing key roles. Assume a large asymmetry (see Fig. 2.4), so that $|L_0 F_{12}| \ll \varepsilon$. Then the carrier concentration

$$n_{pl} \propto \overline{\tilde{b}^2} = \frac{2L_0^2 |F_{12}|^2}{\varepsilon^2} \tag{9.10}$$

Since the Franck–Condon factor F_{12} and the value of ε depends directly on the mass M, the carrier concentration also depends on M. Its presence reflects the fact that the electronic and vibrational states are not separate (polaronic effect).

Let us focus now on the isotope coefficient $\alpha = (-M/T_c)(\partial T_c/\partial M)$. We should distinguish two contributions, so that $\alpha = \alpha_0 + \alpha_{na}$, where $\alpha_0 = -(M/T_c)(dT_c/d\Omega)$ $(d\Omega/dM)$ describes the usual isotope effect caused by change in the phonon spectrum (Ω is a characteristic phonon frequency for that mass). If the corresponding mode in the polyatomic system does not contribute noticeably to the pairing, then the coefficient α_0 is small.

The term α_{na} corresponds to a different isotope effect which is due to the dependence $n(M)$ described previously, and also to the dependence $T_c(n)$. Namely:

$$\alpha_{na} = -\frac{M}{T_c}\frac{\partial T_c}{\partial n}\frac{\partial n}{\partial M} \tag{9.11}$$

According to eqns. (9.10) and (9.11), the dependence $n(M)$ is determined mainly by the Franck–Condon factor. Based on eqns. (9.9)–(9.11), and the dependence $\varepsilon^2 \propto M^{-1}$ (therefore $n \propto Me^{-sM^{0.5}}$, $s = \frac{k^{0.5}d^2}{2\hbar}$, $k^{-0.5} = k_1^{-0.5} + k_2^{-0.5}$, where k_i are the elastic constants and d is the distance between the minima) we obtain the following expression for the isotope coefficient:

$$\alpha_{na} = \gamma\frac{n}{T_c}\frac{\partial T_c}{\partial n} \tag{9.12}$$

where $\gamma \cong$ constant (it has weak logarithmic dependence on M). Expression (9.12) can be compared directly with experimental data (see Chapter 10).

Equation (9.12) contains the derivative $\frac{\partial T_c}{\partial n}$. Therefore, the isotope coefficient α_{na} is equal to zero if $T_c = T_{c,max}$. For the cuprates, T_c has a maximum at some value $n = n_{max}$. Therefore, the described mechanism leads to the situation that the maximum value of T_c corresponds to a minimum in the value of the isotope coefficient. This phenomenon, indeed, has been observed experimentally (see Chapter 10).

Let us make several comments. Continuous doping leads not only to an increase in concentration n, but to a change in electronic terms; the structure of the terms becomes less asymmetric. For example, for the cuprates (see Chapter 10) this is reflected in the movement of the average position of the axial oxygen toward the plane. This motion has been observed experimentally (Jorgensen et al., 1990). Such an evolution leads to an additional decrease in the value of α.

The second comment is related to the overdoped region. Our analysis is applicable to a single-phase system, that is, for the underdoped region of the phase diagram. Structural transitions can drastically modify this picture. If the overdoped region corresponds to the same phase as the underdoped region (this is probably the case for BiPbCaSrCuO), then in the region $n > n_{max}$, the isotope coefficient should become negative; such an effect has indeed been observed by Bornemann et al. (1991). However, if the region $n > n_{max}$ is a multiphase compound (as is the case for La$_{2-x}$Sr$_x$CuO$_4$; see Bozin et al., 2000), then the picture described above can be used for $n < n_{max}$ only. Finally, one should note that according to the described model the values of α are not limited by a maximum of 0.5, and may exceed this value.

Therefore, a strong non-adiabaticity of the ions in the doped superconductors, such as the high-T_c oxides, leads to the dependence of the carrier concentration on M, and this factor affects the value of T_c.

9.8 Penetration depth: isotopic dependence

The polaronic isotope effect described previously is a consequence of the dependence of the carrier concentration on the ionic mass $M: n = n(M)$. Since not only T_c but other properties are also affected by the value of the carrier concentration, one can expect the isotopic dependence of various quantities. As an important example, let us discuss the isotopic dependence of the penetration depth, δ. This dependence was introduced theoretically by Kresin and Wolf (1995) and then analyzed in detail by Bill et al. (1998). The effect has been observed experimentally for the high-T_c oxides, and we will discuss it in Chapter 10.

By analogy with T_c, let us define a new isotope coefficient β by the dependence

$$\delta \sim M^{-\beta} \qquad (9.13)$$

As a result, β is determined by the expression (see eqn. (9.2))

$$\beta = -(M/\delta)(\partial \delta/\partial M). \qquad (9.14)$$

describing the change in the value of the penetration depth induced by the isotope substitution: $M \to M^* = M + \Delta M$.

Let us consider the London limit. Then the penetration depth is given by the well-known relation:

$$\delta^2 = \frac{mc^2}{4\pi n_s e^2} = \frac{mc^2}{4\pi n\varphi(\frac{T}{T_c})e^2}, \qquad (9.15)$$

where m is the effective mass. n_s is the superconducting density of charge carriers, related to the normal density n_s through $n_s = n\varphi(T/T_c)$. The function $\varphi = (T/T_c)$ is a universal function of (T/T_c). For example, for conventional superconductors, $\varphi \simeq 1 - (T/T_c)^4$ near T_c, whereas $\varphi \simeq 1$ near $T = 0$. We can now determine the isotope coefficient β of the penetration depth from the relation which follows from eqns. (9.14) and (9.15):

$$\beta \equiv -\frac{M}{\delta}\frac{\partial \delta}{\partial n_s}\frac{\partial n_s}{\partial M} = \frac{M}{2n_s}\frac{\partial n_s}{\partial M}. \qquad (9.16)$$

Because of the relation $n_s = n\varphi(T)$, one has to distinguish two contributions to β. There is the usual (BCS) contribution, β_{ph}, arising from the fact that $\varphi(T/T_c)$ depends on ionic mass through the dependency of T_c on the characteristic phonon frequency. Indeed, isotopic substitution leads to a shift in T_c and thus in β, which might be noticeable near T_c.

The presence of the factor n (see eqn. (9.15)) leads to a peculiar isotope effect for the penetration depth with $\beta \equiv \beta_{na}$ (β_{na} corresponds to the non-adiabatic contribution.)

Let us consider the region near $T = 0$ K. Then

$$\beta_{na} \simeq \left(\frac{M}{2n}\right) \cdot \left(\frac{dn}{dM}\right) \qquad (9.17)$$

Comparing eqns. (9.11), (9.12), and (9.17), one infers that $\beta_{na} = -\gamma/2$ and thus establishes a relation between the non-adiabatic isotope coefficient of T_c, α_{na}, and β_{na}:

$$\alpha_{na} = -2\beta_{na}\frac{n}{T_c}\frac{\partial T_c}{\partial n}. \qquad (9.18)$$

The equation contains only measurable quantities, and can thus be verified experimentally. It is interesting to note that β_{na} and α_{na} have opposite signs when $\partial T_c/\partial n > 0$ (which corresponds to the underdoped region in high-T_c cuprates). In Chapter 10 we will discuss the isotopic dependence of T_c and δ for the cuprates.

Let us consider another case when the penetration depth is affected by isotopic substitution, and again the dependence is not related directly to the lattice dynamics. Namely, consider the $S-N$ proximity system (Bill et al., 1998), where S is a superconductor and N is a metal or a semiconductor (see Section 8.5.1). Assume that $\delta < L_N \ll \xi_N$ (this is certainly satisfied in the low-temperature regime, since $\xi_N = \hbar v_F/2\pi T$), where L_N and ξ_N are the thickness and the coherence length of the normal film, respectively. One can show that in this case the penetration depth is described by the expression:

$$\delta^{-3} = a_N \Phi \qquad (9.19)$$

where $a_N = $ constant; its value depends only on the material properties of the normal film (it is independent of the ionic mass), and

$$\Phi = \pi T \sum_{n \geq 0} \frac{1}{x_n^2 p_n^2 + 1}, \qquad p_n = 1 + t\sqrt{x_n^2 + 1} \qquad (9.19\text{a})$$

$x_n = \omega_n/\varepsilon_S(T)$, $\omega_n = (2n+1)\pi T$, and $\varepsilon_S(T)$ is the superconducting energy gap of S film. In the weak-coupling limit considered here, $\varepsilon_S(0) = rT_{c0}$, with $r = r_{BCS} = 1.72$, t is the dimensionless parameter, $t = r(L_N/L_S)(T_{c;S}/\Gamma_0)$ ($\Gamma_0 \equiv \Gamma^{\alpha\beta} \propto L_S^{-1}$ is the McMillan parameter; see Chapter 8), and L_N and L_S are the thicknesses of the normal and superconducting films.

Since δ depends non-linearly on T_{c0} (through the proximity parameter t), the penetration depth will display an isotope shift due to the proximity effect. From eqn. (9.19) we can calculate the isotope coefficient β_{prox} of the penetration depth due to the proximity effect:

$$\beta_{prox} = -\frac{2\alpha_0}{3\Phi} \sum_{n>0} \frac{x_n^2 p_n^2}{x_n^2 p_n^2 + 1}\left(1 - \frac{t}{p_n\sqrt{x_n^2 + 1}}\right), \qquad (9.20)$$

where α_0 is the isotope coefficient of T_{c0} for the superconducting film S alone.

Note that the isotope coefficients β_{prox} and α_0 generally have opposite signs. In addition, the isotope coefficient depends on the proximity parameter t; that is, on the thickness ratio $l = L_N/L_S$ of the normal and superconducting films, and on the MacMillan tunneling parameter Γ_0. Note also that the isotopic shift of the penetration depth is *temperature dependent*: $|\beta_{prox}|$ increases with increasing temperature.

Therefore, the presence of a normal layer on a superconductor induces an isotope shift of the penetration depth.

Note that, similarly to the proximity effect, the isotopic dependence of the penetration depth is also affected by another pair-breaking effects: by magnetic scattering (Bill et al., 1998). A detailed description of this effect, however, is beyond the scope of this book.

10
Cuprate superconductors

The cuprates are uniquely interesting superconducting compounds due to their record transition temperature, their peculiar properties, and their potential for applications. This chapter will discuss the properties and some of the key theoretical concepts that can be used to understand their superconducting and normal behavior. It will also provide an experimental and theoretical description of some very key aspects of these materials: for example, the role of magnetic impurities and the role of phonons for the mechanism.

10.1 History

Superconductivity above 35 K—a spectacular record transition temperature—was discovered by Bednorz and Mueller (1986) in the compound $Sr_{(2-x)}Ba_xCuO_4$, which we will abbreviate as the (214) compound, reflecting the ratio of rare earth to copper to oxygen in the parent compound, Sr_2CuO_4 (Fig. 10.1). This was not a totally serendipitous discovery, as Bednorz and Mueller were searching for superconductivity in compounds that exhibited strong polarizability and polaronic effects (Bednorz and Mueller, 1987; review: Bussmann-Holder and Mueller, 2012). These compounds (metallic ceramics) were synthesized initially by the Raveau group (Michel and Raveau, 1984; see also Raveau et al., 1991). The original figure reproduced from the Bednorz and Mueller is shown as Fig. 10.1.

The discovery by Bednorz and Mueller resulted in a worldwide flurry of activity to search for other superconductors in this family. Early in 1987, Paul Chu's group (Wu et al., 1987) discovered superconductivity at around 95 K in another member of this family: $YBa_2Cu_3O_7$ (1237)—or, if we omit the number of oxygens from the abbreviation, (123). This was a major discovery, since it meant that superconductivity could be observed above the liquidus temperature of nitrogen, and again, every laboratory could demonstrate superconductivity without having to deal with liquid helium! Soon afterward there were further discoveries that raised the maximum T_c, including compounds of bismuth and/or thallium, strontium, calcium, copper, and oxygen (1201), (1112), and (2223), which raised the transition temperature to 127 K (Kaneko et al., 1991), and finally the highest T_c compound consisting of mercury, barium, calcium, copper, and oxygen (1223), with a T_c around 133 K at ambient pressure (Schilling et al., 1993), and over 160 K under high pressure (Gao et al., 1994).

One of the key structural features of all of these compounds in this large family is the very well defined copper oxygen planes where the charge carriers reside and consequently where the superconductivity occurs.

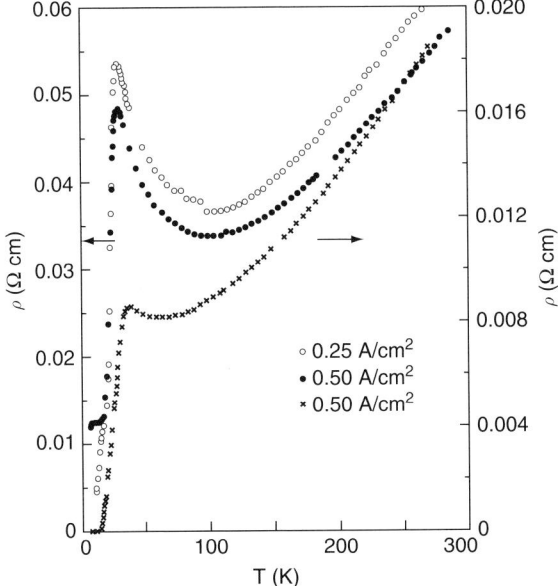

Fig. 10.1 Temperature dependence of the resistivity in $Ba_xLa_{5-x}Cu_5O_{5(3-y)}$ for samples with $x = 1$ (upper curves, left scale) and $x = 0.75$ (lower curve, right scale). (Figure reproduced from Bednorz and Mueller, 1986, with kind permission from Springer Science and Business Media.)

10.2 Structure of the cuprates

The cuprates generally are perovskites that all contain copper oxygen planes that form a checkerboard lattice with alternating coppers and oxygens where the oxygen atoms form squares with a copper at the center. The critical temperature is related to the number of copper oxygen planes in a unit cell peaking out at three for the Hg (2223) compound. One of these compounds, Y (123), is unique in that it contains not only two copper oxygen planes but a chain of alternating copper and oxygen ions. This structure is shown in Fig. 10.2, where the top and bottom layers are clearly missing oxygen between the copper atoms in the left-to-right direction. Between these chain layers are two complete copper–oxygen planes. Thus, single crystals of this compound are very anisotropic in the x–y direction, since the chains are also conductive but are more one-dimensional compared to the two-dimensional planes. This compound seems to be the most robust for applications, since it is doped by the missing oxygen in the chain layer, and the compound is stoichiometric rather than having random substitutions in the thallium, mercury, or bismuth layer.

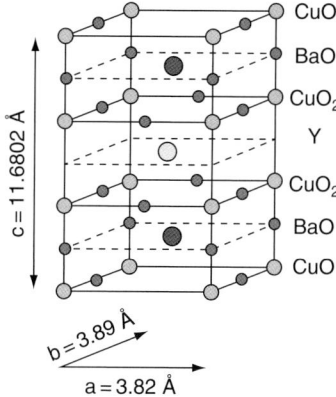

Fig. 10.2 The crystal structure of YBa$_2$Cu$_3$O$_7$, showing the copper oxygen planes and the chain layer.

10.3 Preparation of bulk and film cuprates

The cuprates are traditional ceramic oxide materials, and bulk and thick films can be prepared using ceramic processing techniques. Typically, high purity and very fine powders of the constituent oxides or carbonates are mixed together and calcined at around 1000° C for about an hour. They are then mixed again, pressed into pellets, and sintered for about 24 hours—again at around 1000° C in oxygen. For example, for the Y(123) compound the powders that are mixed are yttrium oxide (Y$_2$O$_3$), barium carbonate (BaCO$_3$), and copper oxide (CuO). This process for the Y(123) compound will yield a 94 K superconductor (Qadri *et al.*, 1987). Thick films can be prepared in a very similar fashion by using a binder and a doctor-blade technique. Thick ribbons have also been prepared in this fashion.

Thin films can be prepared using almost any of the physical and chemical vapor deposition techniques that are used for preparing other ceramic materials. These methods include sputtering, ion beam deposition, molecular beam epitaxy, chemical vapor deposition, pulsed laser deposition, and flash evaporation. The key to high-quality films is in finding the right deposition conditions (atmosphere, temperature, and so on) that maintain the stoichiometry, especially with regard to the oxygen content. All of the aforementioned techniques can be used to prepare high-quality films on insulating substrates. For deposition of textured films on metallic substrates for conductor applications, the use of an ion beam textured buffer layer has proven quite successful (Groves *et al.*, 1999). For understanding the physics of the cuprates, the deposition technique of preference is oxide molecular beam deposition, pioneered by Bozovic and Eckstein (Bozovic *et al.*, 1994; see also Bozovic and Eckstein, 1998).

10.4 Properties of the cuprates

10.4.1 Phase diagram

The majority of the cuprates, being chemically doped compounds, have a very similar phase diagram. The parent compound is typically an antiferromagnetic insulator that is doped by substituting, for example, strontium or barium for lanthanum, bismuth, thallium, or mercury. As the number of dopants is increased, the Neel temperature decreases until the sample becomes conducting and then superconducting with a nearly parabolic shape transition temperature versus doping, with a maximum transition temperature at a doping concentration x of around 0.15 for the (214) and related compounds. Y(123)—perhaps one of the most interesting cuprates—is doped in another way by adding oxygen to the chain layer. The optimum doping occurs when the chain is complete and the oxygen content is 7. The region of the superconducting part of the phase diagram below the maximum T_c (optimal doping) is called the underdoped region. Above T_c one can observe a peculiar so-called "pseudogap" state consisting of mixed superconducting and normal "islands." This state will be discussed in Chapter 11. The part of the phase diagram above the maximum T_c is the overdoped region. The phase diagram is quite similar for the electron-doped compounds and the hole-doped compounds, though the transition temperature is typically much higher for the hole-doped cuprates. A schematic of the phase diagram is illustrated in Fig. 10.3.

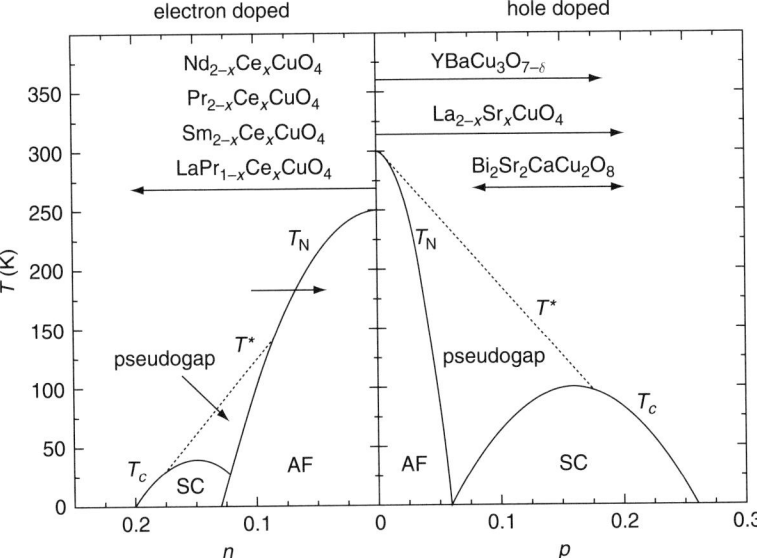

Fig. 10.3 The doping phase diagram for most of the cuprates, showing the antiferromagnetic region, the superconducting region, and the pseudogap region.

10.4.2 Critical field H_{c2}

One of the most interesting and novel behaviors exhibited by the cuprates is the novel temperature dependence of the upper critical field H_{c2} for the cuprates, and in particular the bismuth (2201) compound in the overdoped region. As we know, for the conventional superconductors the dependence $H_{c2}(T)$ has almost linear increase near T_c and then saturates at $T \to 0$ to some value $H_{c2}(0)$; see Helfand and Werthamer (1964, 1966). What is remarkable is that for the cuprates the upper critical field has a positive curvature at all temperatures that is exaggerated at low temperatures (Osofsky et al., 1993; Mackenzie et al., 1993; Fig. 10.4). The data show a sharp and almost linear increase toward $H_{c2}(0)$, this value greatly exceeds that for the usual systems. This is clearly shown in Fig. 10.4. This unusual behavior eluded explanation until it was realized that the dopants (in this case, oxygen ions probably on apical sites) are also pair breakers with magnetic moments, and at temperatures close to T_c suppress the critical temperature and the critical

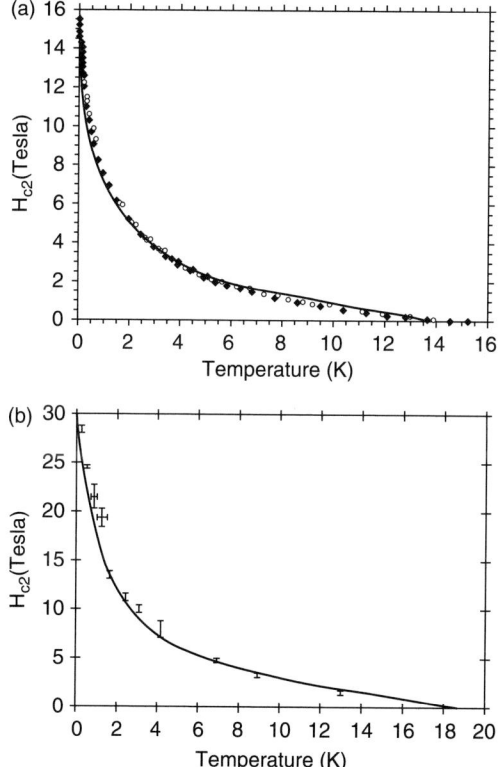

Fig. 10.4 Dependence $H_{c2}(T)$ for a) $Tl_2Ba_2CuO_6$ and b) $Bi_2Sr_2CuO_6$. Solid points, experimental data (Mackenzie et al., 1993; Osofsky et al., 1993); solid lines, theory. (Figure reproduced from Ovchinnikov and Kresin, 1996.)

field so that the increase of H_{c2} with decreasing temperature is much smaller than in a conventional superconductor without the pair breaking. The pair-breaking effect is caused by spin–flip scattering, because it is necessary to satisfy the spin conservation law. However, at low temperatures the picture becomes different (Ovchinnikov and Kresin, 1996). The correlation between the magnetic moments becomes important, and the spin-flip scattering is frustrated. The spin-flip relaxation time becomes temperature dependent, and increases as $T \to 0$. As a result, the superconducting state becomes less depressed ("recovery"), leading to the observed increase in the value of the critical field.

10.4.3 Two-gap spectrum

Chapter 7 contains a description of the two-gap scenario which is caused by the presence of two groups of electronic states (for example, two overlapping energy bands). This phenomenon has never been observed in conventional superconductors. Indeed, because of the inequality $l < \xi$ (l is a mean free path, and ξ is the coherence length) the scattering by impurities mixes the electronic states, and we are dealing with an usual one-gap picture. However, the coherence length is short for the high-T_c cuprates, and this leads to the possibility of observing the two-gap spectrum.

A remarkable system which displays the two-gap structure of the spectrum is the YBCO compound which has two conducting subsystems: planes and chains. The most direct observation of two gaps was performed by Geerk et al. (1988) using tunneling spectroscopy. Indeed, two peaks in the density of states have been observed.

It is interesting that the second (smallest) gap is induced in the chains through two channels (Kresin et al., 1992; Kresin and Wolf, 1992): 1) the usual two-band scenario (see Chapter 7) when the second gap is induced by the phonon-mediated plane-chain transitions, and 2) the proximity effect caused by spatial separation of these units. These two channels have an opposite impact on T_c (see Section 8.3.), and as a result, the value of T_c is determined, mainly, by the intra-plane pairing.

10.4.4 Symmetry of the order parameter

The concept of pairing with finite value of the orbital momentum l is not new. It was introduced by Landau in 1959, shortly after the appearance of the BCS theory (see Lifshitz and Pitaevsky, 2002). Later it was discussed for various systems (see, for example, Geshkenbein et al., 1987).

The case of d-wave symmetry ($l = 2$; we consider the singlet pairing) corresponds to the $d_{x^2-y^2}$ structure of the gap parameter. It is essential that we are dealing with the d-wave symmetry of the gap parameter in momentum, not real space; this is different from the usual quantum-mechanical picture of the electronic wavefunction with $l = 2$.

The symmetry of the order parameter in the cuprates has been the subject of intense study since the original discovery in 1986 (see, for example, reviews by Mannhart and Chaudhari, 2001; Van Harlingen, 2012; Tsuei and Kirtley, 2000). Although the symmetry does not provide the key to the mechanism it was still heavily studied as it is an important feature of these materials. Especially, one should mention the experiments by Tsuei et al. (1994) and by Kouznetsov et al.(1997).

Experiments on tricrystal rings by Tsuei *et al.* provided very strong evidence that the order parameter has d-wave symmetry, especially in nearly optimally doped $YBa_2Cu_3O_{7-\delta}$ (1997). This experiment used flux quantization in rings with 0, 2, and 3 junctions that showed $1/2$ flux quanta only in rings with three junctions. This result implied an order parameter that had a large d-wave component. A d-wave order parameter (in **k** space) has a zero in the k_x and k_y directions and a maximum at 45 degrees to these axes.

One important consequence of pure d-wave symmetry is that there would be no Josephson tunneling in the c direction between a purely d-wave superconductor and an s-wave superconductor, because the net Josephson current from all of the lobes would sum to zero. A very important experiment was carried out by Kouznetsov *et al.*, who measured a finite Josephson tunneling in the c direction between Pb, which is a pure s-wave superconductor, and a twinned sample of $YBa_2Cu_3O_{7-\delta}$. These results proved unambiguously that the symmetry of the order parameter was mixed with a clear s-wave component that changes sign across a twin boundary. The coexistence of s- and d-components has also been discussed by Mueller (1995).

The d-wave symmetry and corresponding presence of the nodes of the gap parameter means that we are dealing with the absence of the gap in the density of states, though the gaplessness can also be observed as the impact of the pair-breaking.

Experimentally, the d-wave symmetry is manifested in measurements of the temperature dependence of the penetration depth (Hardy *et al.*, 1993); the data show linear (non-exponential) behavior at $T -> 0\,K$. The angular photoemission data (ARPES) also demonstrated the d-wave nature of the pairing (Shen *et al.*, 1993; Ding *et al.*, 1996).

Note that the d-wave structure of the gap parameter is very beneficial for many novel superconductors, including the cuprates. Indeed, as we know, the attraction between the carriers should overcome the Coulomb repulsion, and for conventional systems this process is helped greatly by the logarithmic weakening of this repulsion (see eqn.(3.27)). This factor is caused by the large value of the Fermi energy and, correspondingly, by large scale for the coherence length. However, for the cuprates the coherence length is rather short, and this factor is not so effective. The d-wave order parameter is changing its sign. Because the equation for T_c has the structure $\Delta = \Sigma K \Delta$, it is clear that this feature allows us to obtain the solution, even if in some region the kernel K has a negative sign (the contribution of the attraction is smaller relative to the repulsion in this region).

An interesting experiment was performed by Razzoli *et al.* (2013). With the use of photoemission (ARPES) technique they observed the s–d evolution of the pairing gap as a function of doping. More specifically, the highly underdoped $La_{2-x}Sr_xCuO_4$ with $x \approx 0.08$ ($T_c \approx 20$ K) displays an s-wave gap, and the ensuing increase in doping leads to the s–d evolution, so that at the optimum doping ($x = 0.145, T_c = 33$ K) the sample displays the d-wave order parameter. A similar observation was reported for the underdoped $Bi_2Sr_2CaCu_2O_{8+\delta}$ compound (Vishik *et al.*, 2012). One can see that the observed s–d evolution correlates with the doping level, and, correspondingly, with the value of T_c and the decrease in the value of the coherence length.

10.5 Isotope effect

The isotope effect described previously and defined by eqn. (9.1) is a rather complex phenomenon which can be affected by a number of factors (see Chapter 9). The effect has been observed in the cuprates (see, for example, Crawford et al., 1990; Franck et al., 1991; Babushkina, 1991; and reviews by Franck, 1994; Keller, 2005); its dependence on the doping level appears to be very peculiar. More specifically, the value of the oxygen isotopic coefficient α (see eqn. (9.1)) is relatively small at optimum doping, but increases rather rapidly upon decreasing doping up to values that are even larger than that in the BCS theory.

10.5.1 Polaronic state

As we know, the value of the critical temperature depends strongly on the carrier concentration n, that is, on the doping level, $T_c \equiv T_c(n)$. As described previously (Section 9.7), the value of the carrier concentration is affected by the isotope substitution and, more specifically, by the dynamic polaronic effect which occurs if some ion is characterized not by one but by two close potential minima. Such a polaronic effect, indeed, leads to the observed isotopic dependence (Kresin and Wolf, 1994).

Let us start with an apical oxygen ion. The dynamics of the apical oxygen plays an important role in the high-T_c oxides (Mueller, 1990). The cuprates are doped materials, and the charge transfer occurs through this ion. The key observation was reported by Haskel et al. (1997); namely, the "double well" structure of the apical oxygen ion (Fig. 10.5) has been observed with the use of X-ray absorption techniques.

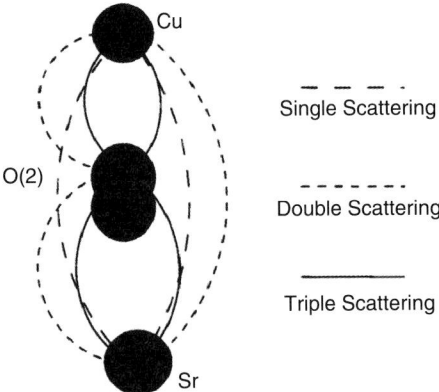

Fig. 10.5 "Double-well" structure for the apical oxygen. (Figure reproduced from Haskel et al., 1997. © 1997 by the American Physical Society.)

The in-plane oxygen ions can also display the polaronic effect (the site selective isotope substitution was described by Zeck et al., 1994). It is important for an analysis of the electron–lattice interaction (see Section 10.6), but it also affects the dynamics of the doping. This is observed for $Y_{1-x}Pr_xBa_2Cu_3O_7$ samples studied by Keller (2003) and Khasanov et al. (2003), and is caused by the mixed valence state of the Pr ions.

The polaronic isotope effect is described by eqn. (9.12). One can see directly from this equation that $\alpha_{na} \propto (\partial T_c/\partial n)$, and therefore, at the optimum doping $(T_c \equiv T_{c;max})$, the isotope coefficient should have a minimum value; this, indeed, corresponds to the observed picture. A decrease in the doping level leads to the finite value of the derivative and to a rapid increase in the value of the isotope coefficient; its value can, in principle, exceed that for the BCS theory.

Equation (9.12) contains experimentally measured quantities, and can be compared with the experimental data. Such an analysis for a whole temperature range has been performed by Weyeneth and Mueller (2011; Fig. 10.6). One can see remarkable agreement between the theory and experimental data for several cuprates, and such an agreement provides a key support for the polaronic concept.

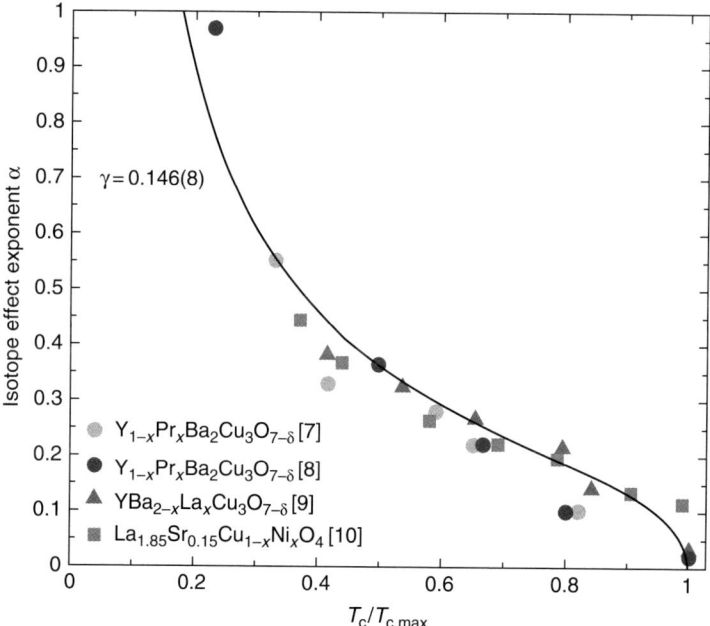

Fig. 10.6 The oxygen isotope effect exponent α as a function of $\frac{T_c}{T_{c,max}}$ for several cuprates. Solid line, theory, eqn. (12). (Figure reproduced from Weyeneth and Mueller, 2011, with kind permission from Springer Science and Business Media.)

10.5.2 Isotopic dependence of the penetration depth

The carrier concentration affects not only T_c, but also the value of the penetration depth (see Section 9.8). That is why one can expect that the penetration depth should also display the isotopic dependence. This is an unusual effect which has never been observed for conventional superconductors. This effect is specific for the doped materials only, and is due to the dynamic polaronic effect. The cuprates represent the most important family of such superconductors.

The isotope effect for the penetration depth has indeed been observed (Zhao et al., 1997; review by Keller, 2003, 2005).

The isotopic dependence of the penetration depth δ at $T = 0$ K is described by eqns. (9.17) and (9.18). Let us note first that unlike the isotopic dependence of T_c (eqn. (9.11)), the isotope coefficient β is not small at $T = T_{\max}$.

Equation (9.18) establishes the relation between the isotope coefficients describing the impacts of the isotope substitution on the T_c and the penetration depth. Since in the underdoped region the dependence $T_c(n)$ is close to linear, one can expect, based on eqn. (9.18), that in this region, $|\alpha| \simeq 2|\beta|$. Such a dependence is in good agreement with the empirical Uemura plot (Uemura et al., 1989; see also the discussion by Khasanov et al., 2006).

10.6 Mechanism of high T_c

Since the discovery of high-T_c oxides there has been an intensive study of these novel materials. However, despite such research, the question of the pairing mechanism is still controversial. Various mechanisms—electron–lattice interactions, and electronic and magnetic contributions (see Chapters 3–5)—have been analyzed, and as a result we have witnessed an intensive development of the theory of superconductivity.

One can state that the final determination of the origin of the high-T_c state in the cuprates requires performing some critical experiments. Indeed, theoretically one can demonstrate that there are a number of mechanisms which can produce a high-temperature superconducting state. Only a few critical experiments can allow us to select the proper mechanism and rule out those which are not effective for the cuprates. The situation is similar to that of the physics of conventional superconductivity. The discovery of the isotope effect provided an initial indication of the importance of the ionic subsystem, but tunneling spectroscopy and, more specifically, the method developed by McMillian and Rowell (Section 6.1), has been a key and decisive tool in the final determination of electron–phonon interaction as the major mechanism. Similarly, some key experiments will determine the origin of high T_c in the cuprates.

Experimental and theoretical studies of the isotope effect (see Section 10.5 and Fig. 10.6) provide a key support for the presence of the polaronic states in the cuprates. The wavefunction has the form (2.39), and is not just a product of electronic and ionic terms. This means that electronic and lattice degrees of freedom are not separable, and this imposes strong restrictions on the theory.

The following analysis focuses mainly on the electron–lattice interaction and its contribution to the pairing (see the review by Kresin and Wolf, 2009). As shown in Chapter 3, the electron–lattice interaction can, in principle, lead to high values of T_c.

Of course, this statement alone does not provide the answer to the question about the nature of the superconducting state in cuprates, but it means that electron–lattice interaction cannot be summarily ruled out as a potential mechanism.

Note that immediately after the discovery of high-T_c oxides, many researchers excluded the electron–lattice interaction from the list of potential mechanisms. For the most part, this was due to the natural temptation to introduce something new and exciting into the field, as opposed to relying on the important principle of Occam's razor: *pluralitas non est ponenda sine neccesitate* (one should not increase, beyond what is necessary, the number of entities required to explain anything). An additional key factor was the conviction that despite the electron–phonon interaction being successful as an explanation for superconductivity in conventional materials, this mechanism is not sufficient to explain the observed high values of T_c. This question was addressed in Section 3.4, and it is clear that, in principle, electron-lattice interaction can provide high values of T_c.

Note that, as described previously (Chapter 6), tunneling spectroscopy is a powerful tool which can provide key information about the pairing mechanism, and it is very tempting to use it to study the nature of high-T_c superconductivity in the cuprates. However, there is a serious challenge. As we know, the coherence length, which is defined as $\xi = v_F/2\pi T_c$, where v_F is the Fermi velocity, is an important parameter; its value characterizes the scale of pairing, and can be visualized as the size of the pair. For usual superconductors, the value of ξ is rather large $(\sim 10^3 - 10^4$ Å$)$, whereas for the cuprates it is quite small: $\xi \approx 15$–20 Å. The length scale for providing the tunneling current at the interface between the superconductor and the insulator—the depth over which the tunneling current originates—is the pairing coherence length, and, as noted above for conventional superconductors, this is a large quantity that greatly exceeds the thickness of the surface layer. In the cuprates the coherence length is very short, and this makes the measurements difficult.

Nevertheless, such experiments have been performed. One of the first tunneling experiments (Dynes *et al.*, 1992) was carried out to study the yttrium–barium–copper-oxide (YBCO) compound. The inversion procedure carried out in this paper resulted in the dependence $\alpha^2(\Omega)F(\Omega)$, shown in Fig. 10.7. The calculated value of the critical temperature was $T_c \approx 60$ K. This value lies below the experimental one, but is still quite high.

Break-junction tunneling spectroscopy, which provides a high-quality contact, was employed by Aminov *et al.* (1994) and Ponomarev *et al.* (1999). They demonstrated (Fig. 10.8) that the current–voltage characteristic for the $Bi_2Sr_2CaCa_2O_8$ compound contains an additional substructure which correlates strongly with the phonon density of states (PDS); the PDS was obtained by Renker *et al.* (1987, 1989), using inelastic neutron scattering. Such a correlation is a strong indication of the importance of the electron–phonon interaction.

The tunneling conductance of $Bi_2Sr_2CaCa_2O_8$ was measured by Shiina *et al.* (1995), Shimada *et al.* (1998), and Tsuda *et al.* (2007). They also observed a correspondence of the peaks in d^2I/dV^2 and the phonon density of states. Moreover, the McMillan–Rowell inversion was performed, and the result was clearly supportive of

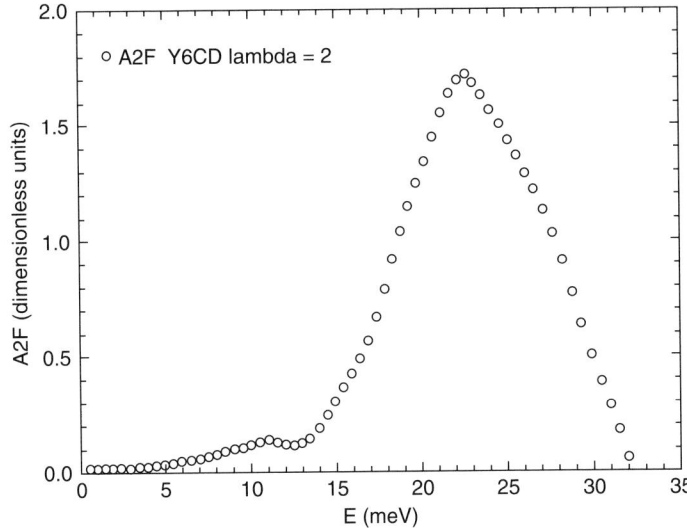

Fig. 10.7 Function $\alpha^2(\Omega)F(\Omega)$ for YBCO. (Figure reproduced from Dynes *et al.*, 1992.)

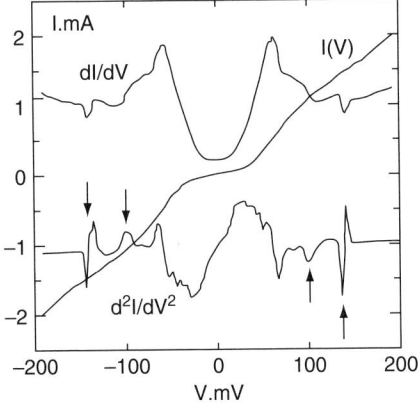

Fig. 10.8 $I(V)$, $\frac{dI}{dV}$, and $\frac{d^2I}{dV^2}$ characteristics for a BSCCO break junction. (Figure reprinted from Aminov *et al.*, 1994. © 1994, with permission from Elsevier.)

the electron–phonon scenario. The spectral function $\alpha^2(\Omega)F(\Omega)$ contains two groups of peaks at $\Omega \approx 15$–20 meV and $\Omega \approx 30$–40 meV. The positions of the peaks corresponds with a high degree of accuracy to the structure of the phonon density of states. The coupling constant λ appears to be equal to about 3.5; such strong coupling is sufficient (see eqn. 3.35) to provide the observed value of T_c ($T_c \approx 90$ K).

Another tunneling technique which appears to be a powerful tool in many studies is scanning tunneling microscopy (STM). This method is used widely in order to obtain information about the local structure of the order parameter, its inhomogeneity, and so on. This type of tunneling (STM) can also be used to perform a study that can probe the mechanism of high T_c. For example, Lee et al. (2006) carried out an STM analysis of the $Bi_2Sr_2CaCu_2O_{8+\delta}$ compound. As a part of the study, measurements of the tunneling current and its second derivative d^2I/dV^2 were performed. The locations of the peaks in the second derivative coincide with the position of specific phonon modes. This is a strong indication of the importance of electron–phonon coupling.

An interesting study was described by Shim et al. (2008). Bicrystal tunneling junctions were used to study the $La_{1.84}Sr_{0.16}CuO_4$ films. The second derivative $(\partial^2 I/\partial V^2)$ data contain twelve peaks, and their positions are in good agreement with Raman scattering data (Sigai et al., 1990). The tunneling data also agree with the picture observed with the use of neutron scattering. As a whole, tunneling spectroscopy continues to be a powerful tool.

Another set of interesting data was obtained with the use of the photoemission technique (see Section 6.5). After the discovery of high-T_c cuprates, the photoemission technique was developed as a powerful tool for obtaining information about the energy spectrum and electronic structure of these novel materials. Photoemission experiments indicating the presence of substantial electron–phonon coupling were published by Lanzara et al. (2001). They studied different families of hole-doped cuprates, Bi2212, LSCO, and Pb-doped Bi2212, and investigated the electronic quasiparticle dispersion relations. A kink in the dispersion around 50–80 meV was observed. This energy scale corresponds to the energy scale of some high-energy phonons; it is much higher than the energy scale for the pairing gap. Such a kink cannot be explained by the presence of a magnetic mode, because such a mode does not exist in LSCO, while the kink structure was also observed in this cuprate.

The structure observed by photoemission is consistent with the data on the phonon spectrum obtained by neutron spectroscopy. These measurements were also used in order to obtain a crude estimate of the electron–phonon coupling, since the quasiparticle velocity in the low-temperature region is renormalized by the electron–phonon coupling constant λ: $v = v_b(1+\lambda)^{-1}$, where v_b is the bare (unrenormalized) velocity, which corresponds to the high-temperature region. This estimate indicates substantial electron–phonon coupling.

A different type of spectroscopy—ultrafast electron crystallography—was employed by Gedik et al. (2007). The $La_2CuO_{4+\delta}$ compound was used, and doping by photoexcitation was performed. It is interesting to note that the number of photon-induced carriers per copper site was close to the density of chemically doped carriers in the superconducting compound. The study of time-resolved relaxation dynamics demonstrated the presence of transitions to transient states which are characterized by structural changes (noticeable expansion of the c axis). Such a large effect on the lattice caused by electronic excitations is a strong signature of electron–lattice interaction.

The study of thermodynamic properties can also provide information about the pairing mechanism, due to the fact that the effective mass and the electronic heat capacity are renormalized by the electron–phonon interaction (see Section 2.4 and eqn. (2.35)). The presence of the second term in the expression for $\gamma(T)$ (eqn. (2.34)) reflects the fact that the moving electron becomes "dressed" by the phonon cloud. As temperature increases, the "cloud" becomes weaker, so that $\gamma(T)$ decreases. As a result, measurements of electronic heat capacity at high temperatures and in the low-temperature region can be used to evaluate the value of the electron–phonon coupling constant λ, which determines T_c. Such measurements were performed by Reeves et al. (1993) for the YBCO compound. The main challenge was to evaluate the electronic contribution to the heat capacity at high temperatures, where the heat capacity is dominated by the lattice. The lattice contribution was calculated by using the phonon density of states obtained by neutron scattering. As a result, the value of $\lambda > 2.5$ was obtained, which means that there is strong electron–lattice coupling, sufficient to provide the observed high T_c.

The impact of the electron–lattice interaction does not exclude the additional contribution to the pairing, which arises from other, electronic or magnetic, excitations. The electronic mechanisms (exchange by excitons or acoustic plasmons) are discussed in Chapter 4. As we know, many novel superconducting systems (such as cuprates, MgB_2, princites, and so on; see Chapter 13) belong to the family of layered (quasi-two-dimensional) conductors and are characterized by strongly anisotropic transport properties. An interesting question can be raised. Why is layering a favorable factor for superconductivity? We can show (see Section 4.4.) that layered conductors have a plasmon spectrum that differs fundamentally from three-dimensional metals. In addition to a high-energy "optical" collective mode, the spectrum also contains an important low-frequency part ("electronic" sound; see Fig. 4.4). The screening of the Coulomb interaction is incomplete, and the *dynamic* nature of the Coulomb interaction becomes important. The contribution of the plasmons in conjunction with the phonon mechanism may lead to high values of T_c.

A detailed analysis based on eqn. (4.13) shows that the impact of dynamic screening is different for various layered systems. For example, for the metal-intercalated nitrides (see Section 13.7) the plasmon contribution dominates. As for the cuprates, the plasmon contribution is not so crucial but is noticeable: about 20% of the observed value of T_c is due to acoustic plasmons. The main role is played by phonons, and their impact leads to a high value of T_c.

The method based on precise infrared measurements (see Section 6.4) developed by Little et al. (1999, 2007) allows us to reconstruct the junction $\alpha^2(\Omega)F(\Omega)$ for an energy interval above the region of phonon energies. According to the analysis, the pairing in cuprates is provided by both phonon and electronic channels.

We should stress that the high-T_c superconducting state cannot be provided by weak electron–phonon coupling. Correspondingly, the analysis based on the BCS model, developed in weak coupling approximation, is not applicable. The electron–phonon interaction should be strong, and strong coupling effects should be incorporated in the theory. In connection with this, it is important that the cuprates display the polaronic effects. As described in Section 3.4, these effects lead to an effective increase of the electron–lattice interaction, and are very beneficial for the pairing.

10.7 Proposed experiment

The experiments described in the previous section provide strong support for the phonon mediated superconductivity in the cuprates. One more interesting experiment aiming at determination of the mechanism can also be proposed (a detailed description can be found in the review by Kresin and Wolf, 2009).

The method is based on the technique of using Josephson junctions for the generation of phonons (Eisenmenger and Dayem, 1967; Eisenmenger, 1969; Dynes et al., 1971; Dynes and Narayanamurti, 1973). This technique can be modified for any boson contributing to the pairing. A nonequilibrium superconducting state is formed by incoming radiation. The creation of excited quasiparticles is followed by a relaxation process. By the end of this process, a noticeable number of quasiparticles are concentrated at or very near the energy gap edge, $E \approx \varepsilon$, where ε is the pairing gap. The final stage of relaxation is the recombination of Cooper pairs. For conventional superconductors, this stage is accompanied by radiation of phonons.

In a classic experiment (Eisenmenger and Dayem, 1967; Eisenmenger, 1969), the generation and detection of phonons propagating through a sapphire substrate was demonstrated using two Josephson junctions located diametrically on opposite sides of a cylindrical sapphire block. The study was aimed at the generation of almost monochromatic phonons. We now look at such experiments from a different point of view. Indeed, these experiments were possible only because phonons were responsible for pairing in the electrodes of the emitting junction and are thus emitted when quasiparticle excitations relax to the gap edge and recombine to form pairs. In other words, one can observe the recombination of electrons with energies near the gap edge; these electrons can form Cooper pairs, and this process is accompanied by radiation of phonons with $\hbar\omega \approx 2\varepsilon$.

The following question can be raised. Why are other excitations not radiated, only phonons? The answer is obvious, and reflects directly the fact that phonons form the glue for pairing. In fact, radiation of phonons created by recombination is an additional support for the phonon mechanism of pairing in conventional superconductors. If pairing is provided, for example, by magnetic excitations, the recombination would be accompanied by radiation of magnons.

We can propose the following experiment. On one side of a high-quality sapphire (or other nearly defect-free single-crystal substrate) one can prepare a Nb or NbN tunnel junction as a detector of phonons. This detector will be most sensitive to phonons that are above the gap energy 2ε of the electrodes. On the other side of the substrate we prepare a cuprate junction. After the current or light has been removed, the generated quasiparticles will relax very rapidly to the gap edge, and as they recombine to form pairs they will emit phonons. The observation of a signal from the cuprate, similar in magnitude and temporal behavior to that of the control junction, would be extremely strong evidence that phonons were the primary excitation from the recombination of excited quasiparticles.

If we now assume that spin fluctuations are the primary pairing excitation, then we would replace the substrate that was a very good phonon propagator with a substrate that would not support the propagation of high-energy phonons but was magnetic and would be an excellent propagator of magnons. Perhaps single-crystal yttrium iron garnet (YIG) with appropriate impurities could be prepared into such a substrate. The same (for example, NbN) junction would be placed on one side of this substrate, and the same two experiments would be performed. The conventional emitter should give a very small signal, whereas the cuprate signal should be much larger—an indicator that magnons are the primary recombination excitation.

Note that the LaSrCuO compound has a relatively sharp gap edge (Murakami et al., 1994), so that it could be used for the first such experiment.

11

Inhomogeneous superconductivity and the "pseudogap" state of novel superconductors

The high-T_c oxides, as well as some other superconducting systems, display many properties in the normal state above T_c which are drastically different from those for conventional materials. This unusual normal (it is called "normal" because of finite resistance) state has been dubbed the "pseudogap" state. The first observation of this state was reported in 1989; that is, shortly after the discovery of high-T_c superconductivity. NMR measurements demonstrated the presence of an energy gap for spin excitations (Alloul et al., 1989; Warren et al., 1989).

The subsequent studies reveal a number of other features. It is clear that above T_c the sample is in a peculiar state which is intermediate between fully superconducting and normal (see Table 11.1). Indeed, as any normal metal, the sample displays a finite resistance. In addition, such a key feature as macroscopic phase coherence does not persists above T_c. However, one can observe some features typical for the superconducting state such as the energy gap, anomalous diamagnetism, and the isotope effect. The observation of the anomalous diamagnetism (the Meissner effect; see Section 11.1.1) is especially important. Note also that the manifestation of the "pseudogap" state depends on the doping level, being the strongest for the underdoped region.

Table 11.1 "Pseudogap" state versus conventional superconducting and normal states

	Superconducting state ($T < T_c$)	Normal state ($T > T_c$)	"Pseudogap" state ($T_c^* > T > T_c$)
Resistance	R = 0	R ≠ 0	R ≠ 0
Energy gap	$\Delta \neq 0$	$\Delta = 0$	$\Delta \neq 0$
Anomalous diamagnetism	Yes	No	Yes
Macroscopic phase coherence	Yes	No	No
Josephson effect	Yes	No	"Giant" effect
Isotope effect	Yes	No	Yes
Impedance Z	ReZ ≠ ImZ	ReZ = ImZ	ReZ ≠ ImZ

11.1 "Pseudogap" state: main properties

Let us, at first, compare the behavior of novel superconductors above T_c with that of the usual systems. Table 11.1 contrasts the two classes.

The most fundamental feature is the anomalous diamagnetism (the Meissner effect), usually observed below the critical temperature. As for the region above T_c, the magnetic response of conventional metals is relatively small and almost temperature independent.

As we know, superconductors have a finite resistance above T_c, whereas below the critical temperature they are in the dissipationless state ($R = 0$). In addition, a.c. transport behaves differently above and below T_c; namely, above T_c, with high accuracy (see, for example, Landau et al., 2004), ReZ = ImZ (Z is the surface impedance). Contrary to this, below T_c in usual superconductors one can observe a strong inequality ReZ $\neq$ ImZ.

The superconducting state is also characterized by macroscopic phase coherence. For example, such a remarkable phenomenon as the Josephson effect is related directly to this feature.

As can be seen in Table 11.1, the "pseudogap" state appears to be rather peculiar. Let us describe its properties in more detail.

11.1.1 Anomalous diamagnetism above T_c

Usual normal metals display relatively weak response to a small external magnetic field. Indeed, the electronic gas is characterized by small Pauli paramagnetism. The magnetic susceptibility of real metals consists of several contributions (see, for example, Ashchroft and Mermin, 1976), and the resultant response might be diamagnetic, but the total susceptibility is almost temperature independent.

The situation above T_c in the cuprates appears to be drastically different. Unusual magnetic properties of the "pseudogap" state have been observed by using various techniques.

Scanning SQUID microscopy was used by Iguchi et al. (2001) to study the underdoped $La_{2-x}Sr_xCuO_4$ compound. This technique allows one to create a local magnetic image of the surface: its "magnetic" map. The critical temperature of the underdoped LSCO films was $T_c \approx 18$ K. A peculiar inhomogeneous picture has been observed: the film contains diamagnetic domains, and their presence persists up to 80 K(!). The total size of the diamagnetic regions grows as the temperature is decreased (Fig. 11.1). As a result, the diamagnetic response appears to be strongly temperature dependent—a very unusual feature of the materials.

A very interesting experiment was described by Sonier et al. (2008): they measured the μSR relaxation. This method allows one to measure directly the local magnetic field. As we know, the μSR spectroscopy (μSR stands for muon spin relaxation; see Chapter 6) is the experimental technique based on the implantation of spin-polarized muons into the sample (see, for example, the reviews by Keller (1989) and Scheck (1985)). In many respects this technique is similar to the electronic paramagnetic resonance (EPR), and even more resembles the nuclear magnetic resonance (NMR) methods. The method allows one to obtain information about the local magnetic field,

"Pseudogap" state: main properties   149

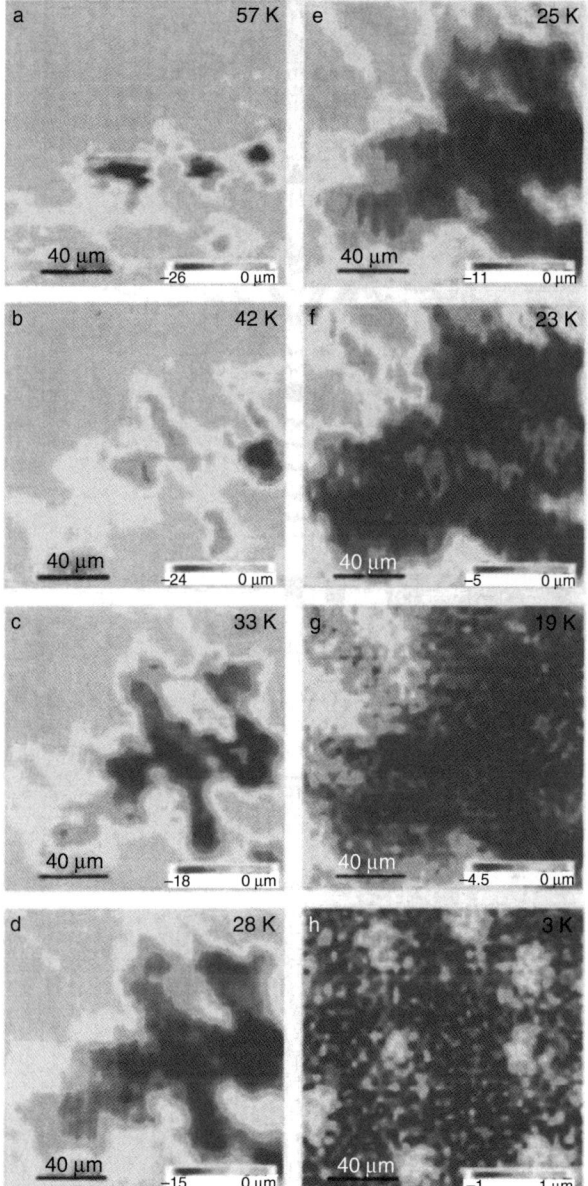

Fig. 11.1 Development of magnetic "islands" with temperature (magnetic imaging of LSCO films). (Figure reprinted from Iguchi *et al.*, 2001, by permission from Macmillan Publishers, © 2001.)

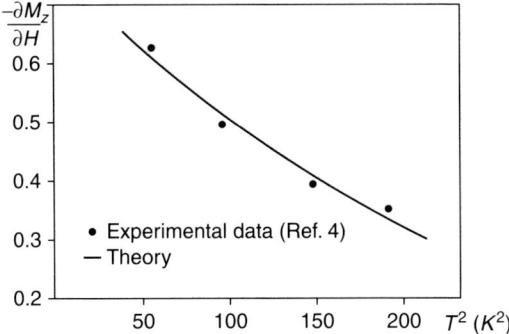

Fig. 11.2 Diamagnetic susceptibility for the $Tl_2Ba_2CuO_{6+\delta}$ ($T_c = 15$ K): experimental data; solid line, theory. (Figure reproduced from Ovchinnikov et al., 1999.)

since it affects the spin precession. The study by Sonier et al. (2008) is of special interest, because the μSR method, unlike the STM studies, is a bulk measurement. According to the data, a strong magnetic response is observed above T_c. More specifically, the relaxation process is described by the function $G(t) = \exp[-(\Lambda t)^\beta]$, where β = const. YBCO and LSCO samples have been studied. The relaxation was compared with that for Ag samples, Ag is not a superconducting metal. If the state at $T > T_c$ is a usual normal metal, one should expect behavior similar to that for Ag. However, according to Sonier et al. (2008) the behavior of Λ for LSCO and YBCO at $T > T_c$ is different; the quantity Λ–Λ_{Ag} depends strongly on temperature, and this corresponds to the presence of magnetic regions above T_c.

A strong temperature-dependent diamagnetic response has also been observed by Bergemann et al. (1998) by using the torque magnetometry technique for the overdoped $Tl_2Ba_2CaO_{+\delta}$ compound above $T_c \approx 15$ K. As in the case of LSCO, the diamagnetic moment was also strongly temperature dependent (Fig. 11.2). Torque magnetometry was also employed recently by Wang et al. (2005) to study the Bi2212 compound. Similarly, diamagnetic response was observed. It was essential that the analysis ruled out fluctuations as a key source of the observed diamagnetism.

11.1.2 Energy gap

The presence of the energy gap above T_c has been observed by using various experimental methods. Even the title "pseudogap state" reflects the existence of the gap structure. In connection with this it is worth noting that this title is confusing, since we are dealing not with a "pseudogap" but with a real gap; that is, with a real dip of the density states in the low-energy region.

The energy gap ε is a fundamental microscopic parameter. As we know, $\varepsilon = 0$ in the normal phase and opens up at T_c. One should note, however, that the energy gap is indeed an important parameter, but its presence, unlike the order parameters, is not a crucial factor for superconductivity. For example, one can observe "gapless" superconductivity (see Chapter 8) caused by the pair-breaking effect, for example, by the presence of localized magnetic moments.

Let us begin with tunneling spectroscopy, which allows one to perform the most detailed and reliable study of the gap spectrum. The data obtained by Renner et al. (1998) for an underdoped Bi2212 crystal are shown in Fig. 11.3. Scanning tunneling microscopy (STM) of the crystals cleaved in vacuum was employed. One can see directly the dip in the density of states (energy gap) which persists above $T_c \approx 83$ K and persists up to ≈ 200 K (!). It should be stressed that the gap structure changes *continuously* from the superconducting region $T < T_c$ to the "pseudogap" state ($T > T_c$); there is no noticeable change at T_c. This can be considered as an indication that the gap structure above T_c is related to superconducting pairing.

Tunneling spectroscopy was also employed by Tao et al. (1997). They concluded that the gap observed at $T > T_c$ reflects the presence of the paired electrons above T_c. Interlayer tunneling spectroscopy was employed by Suzuki et al. (1999), and the presence of the energy gap at $T > T_c$ has also been confirmed.

The presence of an energy gap for spin excitations has been established using NMR. Actually, as mentioned previously, the "pseudogap" state was observed initially by using this method. The gap has been observed in YBCO for different nuclei in both the Knight shift and spin-relaxation rate experiments: for ^{89}Y (Alloul et al., 1989), for ^{63}Cu (Warren et al., 1989; Walstedt and Warren, 1990), and for ^{17}O (Takigawa et al., 1989). It follows also from optical data (Homes et al., 1993; Fig. 11.4).

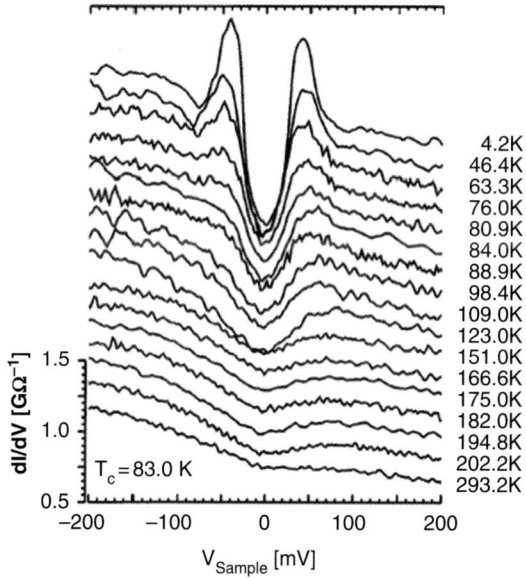

Fig. 11.3 Tunneling conductance of states for the underdoped Bi 2212 crystal. (Figure reproduced from Renner et al., 1998. © 1998 by the American Physical Society.)

Photoemission spectroscopy has also revealed the presence of an energy gap at $T > T_c$ (see the reviews by Shen and Dessau (1995) and by Randeria and Campuzano (1998)). For example, it has been demonstrated that the energy gap persists in an underdoped sample of Bi2212 at $T > T_c$ (Loeser et al., 1996). Ding et al. (1996) come to a similar conclusion.

Photoemission spectroscopy (Kanigel et al., 2008) has revealed the presence of disconnected Fermi arcs, some of which correspond to the normal metal. One can also observe gapped regions which correspond to the superconducting state. The picture is consistent with pairing above T_c.

We have described the data which present direct spectroscopic observation of the gap structure above T_c. A gap in the spectrum can also be inferred from heat capacity data. One should note that measurements carried out by Loram et al. (1994) and Wade et al. (1994) were some of the first observations of the "pseudogap" state. Measurements of the Sommerfeld constant $\gamma(T)$ display a loss of entropy caused by the gap structure. The data for the energy gap $\Delta(T)$ are derived from values of the electronic entropy $S(T, x)$ for $0.73 < x < 0.97$. Again, it has been observed that the energy gap persists for $T > T_c$, and the effect is especially strong for the underdoped samples of $YBa_2Cu_3O_{6+x}$.

11.1.3 Isotope effect

Another interesting property of the "pseudogap" state is the strong isotope effect. This effect has been observed by Lanzara et al. (1999) for $La_{2-x}Sr_xCuO_4$, using X-ray absorption near-edge spectroscopy (XANES). The effect has been also observed by Temprano et al. (2000) for the $HoBa_2Cu_4O_8$ compound. The slightly underdoped $HoBa_2Cu_4O_8$ sample was studied by using neutron spectroscopy. As we know (see, for example, the review by Mesot and Furrer, 1997), the opening of the gap, which could be associated with the "pseudogap," affects the relaxation rate of crystal field excitations. The isotopic substitution $^{16}O \rightarrow {}^{18}O$ leads to a drastic change in the value of the pseudogap temperature T_c^* ($T_c^* \approx 170\,\mathrm{K} \rightarrow T_c^* \approx 220\,\mathrm{K}$). Such a large isotope shift corresponds to a large value of the isotope coefficient. Note that, contrary to the ordinary superconductor, its value is negative.

11.1.4 "Giant" Josephson effect

The so-called "giant" Josephson proximity effect is another interesting phenomenon observed in the "pseudogap" region above T_c (Bozovic et al., 2004). Films of $La_{0.85}Sr_{0.15}CuO_4$ ($T_c \approx 45\,\mathrm{K}$) were used as electrodes, whereas the underdoped LaCuO compound ($T_c' \approx 25\,\mathrm{K}$) formed the barrier which was prepared in the c-geometry (the coherence length $\xi_c \approx 4\mathrm{A}$) The measurements were performed at $T_c' < T < 35$ K, so that the barrier was in the "pseudogap" state. Since $T > T_c'$, we are dealing with the SNS junction. As is known, for such a junction the thickness of the barrier should not exceed the coherence length, which is of order of ξ_c. However, the Josephson current was observed for thicknesses of the barrier up to 200A(!). Such

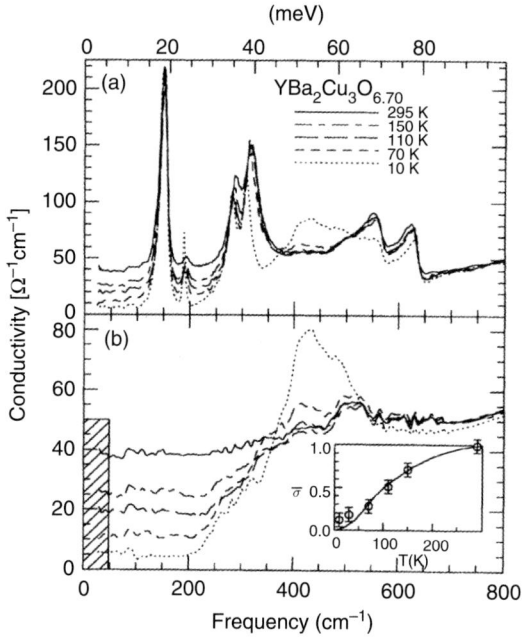

Fig. 11.4 The optical conductivity of $Yba_2Cu_3O_{6.7}$ along the c-axis. (Figure reproduced from Homes *et al.*, 1993. © 1993 by the American Physical Society.)

a "giant" effect cannot be explained by using conventional theory. We will discuss this effect in detail in Section 11.4.3.

11.1.5 Transport properties

The microwave properties and a.c. transport have been studied by Kusco *et al.*, (2002). It was shown that above T_c, that is, in the normal state, ReZ $\neq$ ImZ, where Z is the surface impedance. This is an unusual property, since in ordinary normal metals with a high degree of precision, the real and imaginary part of the impedance are equal (see, for example, Landau *et al.*, 2004), that is, $ReZ_n \cong ImZ_n$. The observed inequality ReZ $\neq$ ImZ is typically observed in superconducting materials.

The measurements of normal resistivity (Darmaoui and Jung, 1998; Yan *et al.*, 2000; Jung *et al.*, 2000) have revealed a strong inhomogeneity of YBCO and TBCCO samples. The presence of two different phases and inhomogeneous structure of the order parameter has been demonstrated.

Interesting data on thermal conductivity for various high T_c compounds were described by Sun *et al.* (2006). The measurements performed for different doping levels reveal the absence of universal dependence for the thermal flow and indicate the importance of strong inhomogeneity in the cuprates.

11.2 Inhomogeneous state

As described previously, intensive experimental studies reveal a number of unusual features of the cuprates above the resistive transition T_c. We described anomalous diamagnetism, which depends strongly on temperature, an energy gap structure, a strong inequality $\text{Re}Z \neq \text{Im}Z$ (Z is the surface impedance), a "giant" Josephson proximity effect, and an isotope effect on T_c^*.

The key issue, which is still controversial, is related to the nature of the "pseudogap" state. The main question is whether this state is a normal metal or a superconductor. At first sight, observation of finite resistance provides the answer to this question. Nevertheless, as described previously (see Table 11.1), in reality the situation is more complicated.

The most fundamental fact is the observation of the Meissner effect above $T_c = T_c^{res.}$. ($T_c^{res.}$ corresponds to the transition into the dissipationless macroscopic state with zero resistance).

It is essential that the contribution of fluctuations is not sufficient to explain the data for the underdoped region (Caretta et al., 2000; Lascialfari et al., 2002); indeed, the temperature scale for T_c^* is very large. Therefore, we are dealing with a serious challenge: we should explain the coexistence of finite resistance and anomalous diamagnetism.

The properties of the "pseudogap" state are caused by intrinsic inhomogeneity of the metallic phase (Ovchinnikov et al., 1999, 2001, 2002; and the reviews, Kresin et al., 2006; Kresin and Wolf, 2012).

11.2.1 Qualitative picture

Consider an inhomogeneous superconductor, so that $T_c = T_c(\mathbf{r})$. The system contains a set of superconducting regions—"islands" embedded in a normal metallic matrix (Fig. 11.5). Properties of such system correspond to the "pseudogap" state. Indeed,

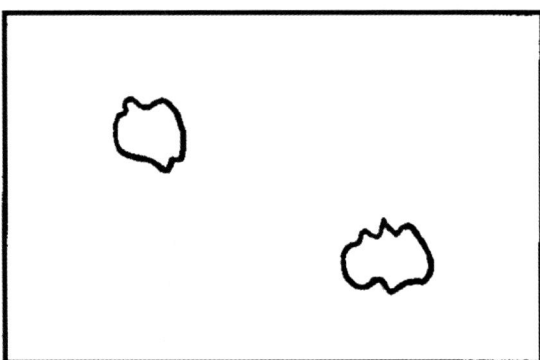

Fig. 11.5 Inhomogeneous structure. "Islands" are characterized by values of T_c higher than the matrix. (Reproduced from Ovchnnikov et al., 2001.)

the normal metallic matrix provides finite resistance, whereas the existence of the superconducting "islands" leads to the diamagnetic moment and energy-gap structure.

As mentioned previously (Section 11.1.1), the presence of diamagnetic "islands" has been observed directly by Iguchi et al. (2002); Fig. 11.1. These superconducting "islands" are embedded in the normal metallic matrix. As a result, we are dealing with the superconductor–normal-metal interface and the proximity effect (see Chapter 8) plays a crucial role. The proximity effect determines a minimum length scale of the superconducting regions which is of order of the coherence length ξ_0. Indeed, if a superconducting "island" has a size smaller than ξ_0, its superconducting state would be depressed totally by the proximity effect between the superconducting region and the normal metallic phase.

As temperature decreases toward T_c, the size of the superconducting regions increases, as does the number of "islands." The critical temperature T_c corresponds to the percolation transition; that is, to the formation of a macroscopic superconducting region ("infinite cluster" in terms of the percolation theory; see, for example, Shklovskii and Efros, 1984; Stuffer and Aharony, 1992), and to phase coherence and dissipationless superconducting phenomena.

In conventional superconductors the resistive and Meissner transitions occur at the same temperature, T_c. The picture in the "pseudogap" state is different. The resistive and Meissner transition are split. The Meissner transition (the appearance of diamagnetism) occurs at T_c^*, whereas the resistive transition—that is, the transition to the macroscopic dissipationless state—takes place at T_c, and $T_c < T_c^*$. Note that in the papers (Ovchinnikov et al., 1999, 2001) the notations $T_{c\text{res}}$ and $T_{c\text{Meis}}$ are used. Here we use only the notations T_c and T_c^*, so that $T_c \equiv T_{c\text{res}}$, $T_c^* \equiv T_{c\text{Meis}}$.

It is important to note also that the "pseudogap" state in the region $T_c < T < T_c^*$ is not a phase-coherent one; each superconducting "island" has its own phase. At $T = T_c$ the macroscopic superconducting region is formed, and below T_c we are dealing with macroscopically phase-coherent phenomena.

11.2.2 The origin of inhomogeneity

As noted previously, the presence of superconducting "islands" embedded in a normal metallic matrix implies an inhomogeneity of the compound. There are two possible scenarios for such an inhomogeneous structure:

1. Inhomogeneous distribution of pair-breakers.
2. Inhomogeneous distribution of carriers leading to spatial dependence of the coupling constant.

Both scenarios lead to inhomogeneous superconductivity. Pair-breaking can be caused by localized magnetic moments (Abrikosov and Gor'kov, 1961) (see Section 8.4). Qualitatively, the picture of pair-breaking can be visualized in the following way. A Cooper pair consists of two carriers with opposite spins (for singlet pairing; this is the case for both, s- or d-wave scenarios). A localized magnetic moment acts to align both spins in the same direction, and this leads to pair-breaking. It is known that for d-wave pairing, non-magnetic impurities are also pair-breakers.

As discussed in Section 8.4, a pair-breaking effect leads to a depression in T_c. Therefore, a non-uniform distribution of pair-breakers makes the critical temperature spatially dependent: $T_c \equiv T_c(\mathbf{r})$. Such a distribution is caused by the statistical nature of doping. The region which contains a larger number of pair-breakers is characterized by a smaller value of the local T_c.

Note that the pseudogap phenomenon is strongly manifested in the underdoped region. At optimum doping the distribution of pair-breakers becomes more uniform, and as a result, $T_c^* \simeq T_c$. The "pseudogap" state at $T > T_{c;opt}$ is characterized by the gap caused not by pairing, but by a CDW (see Section 11.3.1). In the underdoped state the spectrum is affected by both contributions.

As mentioned previously, an inhomogeneity can also be caused by an inhomogeneous distribution of carriers (see Ovchinnikov et al., 2001).

The picture described is directly related to the concept of phase separation. This concept was introduced by Gor'kov and Sokol (1987), shortly after the discovery of the high T_c oxides, and was then studied in many papers ((for example, Sigmund and Mueller, 1994). The concept implies the coexistence of metallic and insulating phases.

This picture of the "pseudogap" state is a next step in the inhomogeneous scenario. Namely, in addition to the mixture of metallic and insulating phases, the metallic phase is itself inhomogeneous; we are dealing with coexistence of normal and superconducting regions within the metallic phase.

11.2.3 Percolative transition

As mentioned previously, the transition at $T \equiv T^{res}$ corresponds to the formation of the macroscopic superconducting phase. This transition is of percolative nature (Ovchinnikov et al., 1999; Mihailovich et al., 2002; Alvarez et al., 2005). In terms of percolation theory the transition to the dissipationless macroscopic state corresponds to the formation of an "infinite" cluster.

The percolative nature of the transition at $T = T_c$ is due to the statistical nature of doping. The picture is similar to that introduced in manganites, which represents another family of doped oxides (see Chapter 12). Manganites (such as $La_{0.7}Sr_{0.3}MnO_3$) are characterized by the presence of ferromagnetic metallic regions embedded in the low conducting paramagnetic matrix above $T_c \equiv T_{Curie}$, where T_{Curie} is the Curie temperature. At $T_c = T_{Curie}$ one can observe a percolative transition to the macroscopic ferromagnetic metallic state. The transition from the "pseudogap" to the macroscopic dissipationless state is also of percolative nature.

11.2.4 Inhomogeneity: experimental data

In Section 11.1.1 we described an interesting study of the La-based compound (Iguchi et al., 2001) performed using the STM technique with magnetic imaging (Fig. 11.1), which has demonstrated directly the presence of diamagnetic "islands" embedded in a normal matrix, and which shows the percolative picture as $T \to T_c$. The same can be said about μSR spectroscopy (Sonier et al., 2008), also described in Section 11.1.1.

Nanoscale inhomogeneity was also described in an interesting paper by Zelikovic et al. (2012). In addition to observing the inhomogeneous picture in $Bi_{2-y}Sr_{2-y}CaCu_2O_{8+x}$, they were able to correlate the structure of the "pseudogap" state with different types of oxygen dopants. It turns out that the apical oxygen vacancies are especially effective. This is an interesting observation, because we know that the apical oxygen is very important for the charge transfer, and such a defect, indeed, can act as a strong pair-breaker.

One key study was performed, using scanning tunneling spectroscopy (STM), by Gomes et al. (2007) and Parker et al. (2010). The measurements were performed on the $Bi_2Sr_2CaCu_2O_{8+\delta}$ samples at finite temperature, so that the energy spectrum was measured above T_c, in the "pseudogap" region. These data have provided crucial information about local values of the gap and its evolution with temperature. The study of this evolution has led to the conclusion that the observed gap spectrum indeed corresponds to superconducting pairing. It is essential that the distribution of gaps turns out to be strongly inhomogeneous. As a whole, the presence of pairing persists for temperatures which greatly exceed those of T_c, especially for the underdoped samples. For example, for the sample with the doping level $x \approx 0.1$ ($T_c \approx 75$ K) the pairing persists up to $T_c^* \approx 180$ K. It is interesting also that from these measurements one can determine the local values of the gap Δ_{loc} and T_p^{loc}. One can conclude that the ratio appears $2\Delta_{\text{loc}}/T_c^{\text{loc}}$ to be equal to $2\Delta_{\text{loc}}/T_p^{\text{loc}} \approx 7.5$. Such a large value of this ratio greatly exceeds that for the usual superconductors (according to the BCS theory $2\Delta/T_c \approx 3.52$), and corresponds to very strong electron–phonon coupling (see Section 3.5). The inhomogeneous structure with superconducting regions ~ 3 nm in size was observed by Lang et al. (2002).

As stressed previously, one should distinguish the intrinsic critical temperature T_c^* and the resistive T_c ($T_c \equiv T_{c^{res}}$). According to the described approach, the pairing gap is related to T_c^*, since the Cooper pairing occurs at first at this temperature. That is why the changes in the value of T_c^* lead to corresponding changes in the gap value. As for $T_c \equiv T_{c^{res}}$, this temperature describes the percolation transition to the macroscopic superconducting state and is not related directly to the pairing interaction. Correspondingly, the changes in T_c should not affect the gap value. It is interesting that precisely this picture has been observed by Lubashevsky et al. (2011) The values of T_c^* and T_c were changed independently (for example, the value of T_c was modified by Zn substitution). Indeed, it has been observed that the value of the gap measured by using ARPES was sensitive to the changes in T_c^*, but was not affected by changes in the resistive T_c.

11.3 Energy scales

The real picture in the cuprates is complicated, as we are dealing with three different energy scales (Kresin et al., 2004, 2006) and, correspondingly, with three characteristic temperatures (we denote them T_c, T_c^*, and T^*).

11.3.1 Highest-energy scale (T^*)

The highest-energy scale, which we have labeled T^* ($\sim 5.10^2$ K) corresponds to the formation of the inhomogeneity and peculiar crystal structure of the compounds. For example, for YBCO the formation of the chains occurs at T^*.

An energy gap could open in the region below T^*. This gap is not related to the pairing, but, as mentioned previously, there are many other sources for the appearance of a gap. For example, the presence of a chain structure in YBCO is consistent with a charge density wave and, correspondingly, with a gap on part of the Fermi surface. Nesting of states might also lead to a CDW instability in other compounds.

Another important property of the compound below T^* is its intrinsic inhomogeneity; this is due to the statistical nature of doping, and is manifested in phase separation (Gor'kov and Sokol, 1987; see also Sigmund and Mueller, 1994). This property implies the coexistence of metallic and insulating phases. The periodic stripe structure (Bianconi, 1994a,b; Tranquada et al., 1995, 1997; Zaanen, 2000) also appears below T^*.

11.3.2 Diamagnetic transition (T_c^*)

If the compound is cooled to below T^*, then at some characteristic temperature we have labeled T_c^* ($T_c^* \approx 2.10^2$ K) one can observe a transition into the diamagnetic state.

The characteristic temperature T_c^* corresponds to the appearance of superconducting regions embedded in a normal metallic matrix (Fig. 11.5). The presence of such superconducting clusters ("islands") leads to a diamagnetic moment, whereas the resistance remains finite, because of the normal matrix. As for the energy gap, coexistence of pairing and a CDW determine its value below T_c^*. It is remarkable that the superconducting state appears at a temperature T_c^* which is much higher than the resistive T_c. This value of T_c^* corresponds to the real transition to the superconducting state (one can call it an "intrinsic critical temperature"; see Kresin et al., 1996).

Strictly speaking, the experimentally measured value of T_c^* lies below the intrinsic critical temperature, because of the impact by the proximity effect. Nevertheless, T_c^* is an important parameter measured experimentally. It corresponds to the appearance of diamagnetic "islands," and reflects the impact of pairing. The superconducting phase appears, at first, as a set of isolated "islands."

The picture of different energy scales T^* and T_c^* just described is in total agreement with interesting experimental data by Kudo et al. (2005a,b). The impact of an external magnetic field was studied by out-of-plane resistive measurements. According to the study, there are, indeed, two characteristic temperatures (Kudo et al. dubbed them as T^* and T^{**}; $T^* > T^{**}$). The behavior of the resistivity appears to be independent of magnetic field in the region $T^* > T > T^{**}$, but strongly affected by the field at $T < T^{**}$. According to Kudo et al., the state formed below T^{**} is related to superconductivity. The characteristic temperatures T^* and T^{**} correspond directly to the energy scales T^* and T_c^* (in our notation, $T^{**} \equiv T_c^*$) introduced above.

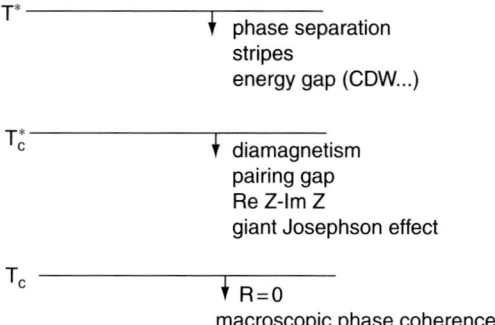

Fig. 11.6 Energy scales.

11.3.3 Resistive transition (T_c)

As the temperature is lowered below T_c^*, new superconducting clusters appear (Fig. 11.1) and existing clusters form larger "islands." This is a typical percolation scenario. At some characteristic temperature T_c the macroscopic superconducting phase is formed ("infinite" cluster in terms of the percolation theory; see, for example, Shklovskii and Efros, 1984). The formation of a macroscopic phase at T_c leads to the appearance of a dissipationless state ($R = 0$).

It is also important to stress that in the region $T_c^* > T > T_c$, each "island" has its own phase, so that there is no phase coherence for the whole sample. Macroscopic phase coherence appears only below T_c.

Therefore, there are three different energy scales and, correspondingly, three characteristic temperatures T^*, T_c, and T_c^* (Fig. 11.6).

The value of T_c is lower than T_c^* because of local depressions caused by the pair-breaking effect and an inhomogeneous distribution of pair-breakers (dopants). It is interesting to note that the value of T_c^* is close to an intrinsic value of the critical temperature. This value is noticeably higher than the resistive T_c.

To conclude this section, let us stress again that the inhomogeneous distribution of pair-breakers (dopants), along with local depressions in the value of critical temperature, leads to a spatial dependence of T_c; that is, $T_c(\mathbf{r})$. The value of T_c^* is close to an "intrinsic" critical temperature.

What are manifestations of the high-temperature superconducting state? Of course, the presence of the normal matrix at $T > T_c$ excludes the possibility of observing a state with zero dc resistance ($R = 0$). However, one can observe a number of various phenomena (see Section 11.1) which will be analyzed in the next section.

11.4 Theory

In this section we will present the theoretical analysis of the main features of the pseudogap state: diamagnetism, a.c. properties, and the "giant" Josephson proximity

160 Inhomogeneous superconductivity

effect. The analysis should take into account the inhomogeneity of the structure. There are two essential factors:

1. The proximity effect, since the superconducting regions are embedded into normal metallic matrix.
2. The pair-breaking effect.

11.4.1 General equations

Inhomogeneity of the system is a key ingredient of the theory. Because of it, it is convenient to use a formalism describing the compound in *real* space. That is why we employed the method of integrated Green's function developed by Eilenberger (1968) and independently by Larkin and Ovchinnikov (1969); see also the review by Larkin and Ovchinnikov (1986).

The main equations have the form:

$$\alpha\Delta - \beta\omega + \frac{D}{2}(\alpha\partial_-^2\beta - \beta\partial_r^2\alpha) = \alpha\beta\Gamma \tag{11.1}$$

$$|\alpha|^2 + |\beta|^2 = 1 \tag{11.1'}$$

$$\Delta = 2\pi T|\lambda|\sum_{\omega>0}\beta. \tag{11.1''}$$

Here α and β are the usual and pairing Green's functions averaged over energy, Δ is the order parameter, $\Gamma \equiv \tau_s^{-1}$ is the spin-flip relaxation time, and λ is the pairing coupling constant. Because of the inhomogeneity, all of these quantities are spatially dependent. In addition, $\partial_\pm = \partial_r \pm 2ie\boldsymbol{A}$, where $\boldsymbol{A}$ is the vector potential, $\partial_r = (\partial/\partial\boldsymbol{r})$. We consider the "dirty" case, so that D is the diffusion coefficient.

These equations contain the spatially dependent functions α, β, and Δ. The method is very effective for treatment of spatially dependent properties.

11.4.2 Diamagnetism

The Cu–O layers contain superconducting "islands," and their presence leads to an observed diamagnetic moment. Because of the dependence $T_c(\mathbf{r})$, the size of the superconducting region occupied by the "islands" decreases as temperature is increased. As a result, one can observe strongly temperature dependent diamagnetism.

Let us describe the evaluation of the diamagnetic moment (Ovchinnikov et al., 1999). Based on eqns. (11.1)–(11.1''), one can calculate the order parameter $\Delta(\mathbf{r})$ and then the current $j(\mathbf{r})$ Then, one can calculate the magnetic moment, since the magnetic moment for an isolated cluster is

$$M_Z = L \int d\boldsymbol{\rho}[\boldsymbol{\rho}\boldsymbol{j}]_z \tag{11.2}$$

Here, L is the effective thickness of the superconducting layer; axis z is chosen to be perpendicular to the layers, and $\boldsymbol{\rho}$ is perpendicular to OZ.

Assume that the sample contains a sufficient amount of magnetic impurities so that $\tau_s T_c^\circ \ll 1$; as a result, $T_c \ll T_c^\circ$, where T_c is the average value of the critical temperature, and T_c° corresponds to the transition temperature with no magnetic impurities. In this case, with the use of eqns. (11.1–11.1″), we obtain

$$\Delta = 2\pi T |\lambda| \sum_{\omega>0} \left(\Gamma + \omega - \frac{D}{2}\partial_-^2\right)^{-1} \left\{\Delta - \frac{\omega}{2}\beta_0|\beta_0|^2 + \frac{D}{4}\beta_0\partial_r^2|\beta_0|^2\right\} \quad (11.3)$$

The order parameter can be found in the form $\Delta = C\Delta_0$, where $C \equiv C(T)$, and Δ_0 is the solution of the equation:

$$\left[\Gamma - (D/2)\partial_-^2\right]\Delta_0 = (\Gamma_\infty + \gamma)\Delta_0 \quad (11.4)$$

Here Γ_∞ is the value of Γ outside of the "island," and γ is the minimum eigenvalue. As a result, one arrives at the following equation:

$$\ln(T_c^\circ/T) = \psi\left[0.5 + \frac{\Gamma_\infty + \gamma}{2\pi T}\right] - \psi(0.5) + \frac{C^2}{12\Gamma_\infty^2} \frac{(\Delta_0^{*2}, \Delta_0^2)}{(\Delta_0^*, \Delta_0)} \quad (11.5)$$

Here, ψ is the Euler function, and the notation (f, g) corresponds to the scalar product of the functions. The transition temperature T_c is determined by the equation which can be obtained from eqn. (11.5) if we insert $C = 0$ and $\gamma = 0$:

$$\ln(T_c^\circ/T_c) = \psi[1/2 + (\Gamma_\infty/2\pi T_c)] - \psi(1/2) \quad (11.6)$$

which is the well-known pair-breaking equation (Abrikosov and Gor'kov, 1961; see Section 8.12). Equation (11.5) is the generalization of eqn. (11.6) for the inhomogeneous case.

The current density is described by the expression (Larkin and Ovchinnikov, 1969):

$$j = -iev D\pi T \sum_\omega (\beta^*\partial_-\beta - \beta\partial_+\beta^*) \quad (11.7)$$

As a result, we can obtain the following expression for the current density of an isolated cluster:

$$j = -\frac{ievDC^2}{2\pi T}\psi'\left(1/2 + \frac{\Gamma_\infty + \gamma}{2\pi T}\right)(\Delta_0^*\partial_-\Delta_0 - \Delta_0\partial_+\Delta_0^*) \quad (11.8)$$

Here, Δ_0 is the solution of eqn. (11.4) and the vector-potential has been chosen as $\mathbf{A} = \frac{1}{2}[\mathbf{H}\mathbf{r}]$. Note also that because the cluster size is smaller than the penetration depth, one can neglect the spatial variation of the magnetic field.

Consider the most interesting case when the variation of the amplitude $\delta\Gamma = \Gamma_\infty - \Gamma$ has the form:

$$\delta\Gamma(\mathbf{r}) = \begin{cases} \delta\Gamma(\rho) & \rho < \rho_0 \\ 0 & \rho > \rho_0 \end{cases} \quad (11.9)$$

where ρ_0 is the "island" radius.

With use of eqns. (11.4–11.6) one can obtain the following expression for the magnetic moment:

$$M_z = -A\left(\tilde{B} - \tau^2\right) H \qquad (11.10)$$

Here

$$A = (8\pi^2 e^2 vDT_c^2/\Gamma_\infty)\rho_0^2 z_0^{-4} n\left(\tilde{x}_{3;2}\tilde{x}_{2;1}/\tilde{x}_{1;4}\right); \quad \tilde{B} = B + 1; c^2 \propto \alpha$$
$$\alpha = T_c^2\left(\Delta_0^*, \Delta_o\right)\left(\Delta_0^{*2}, \Delta_o^2\right)^{-1}; \quad \tau = T/T_c; \quad B = -6\Gamma_1\Gamma_\infty\Big/(\pi T_c^*)^2 \qquad (11.11)$$

and

$$\Gamma_1 = -\delta\Gamma + 0.5D(z_o/\rho_o)^2 \qquad (11.12)$$

where n_s is the concentration of superconducting clusters, and $\tilde{x}_{n;i} = \int_0^{z_0} dx \cdot x^n J_0^i(x)$. z_0 is the lowest zero of the Bessel function; $z_0 \cong 2.4$. If $\delta\Gamma \ll \Gamma_\infty$, then $T_c^* \gg T_c$.

The value of λ_1 (and therefore the value of $\tilde{\beta}$) depends on an interplay of two terms. The first term reflects the impact of pair-breaking, and the second term describes the proximity effect. It is natural that the impact of the proximity effect increases with a decrease in the size of the inhomogeneity ρ_0.

One can see directly from eqn. (11.10) that it is possible to observe a noticeable diamagnetic moment. Indeed, if we assume realistic values: $p_F = 10^{-20}$ gcm-sec^{-1}, $l = 40$ Å (l is a mean free path: $D = v_F l/3$), $T_c = 10$ K, $\Gamma_\infty = 10^2$ K, $\delta\Gamma = 50$ K we put $k_B = 1$, $\rho_0 = 80$ Å, and $n_s \cong 0.1$, we obtain the following values of the parameters: $A \cong 10^{-5}$, $B = 3$, $\gamma = 5$ K. Then, for example, at $T = 11$ K one can observe $\chi_D = M_z/H = -3 \times 10^{-5}$. This contribution greatly exceeds the usual paramagnetic response of a normal metal, $\chi_P \cong 10^{-6}$.

A diamagnetic response can be observed in the region $(T/T_c) < \tilde{B}$. This is natural, since the influence of the proximity effect (see, for example, Gilabert, 1977) to depress the superconductivity grows with a decrease in the size ρ_0 of the superconducting grain.

11.4.3 Transport properties; "giant" Josephson effect

The dc transport properties of the inhomogeneous system above T_c are determined by the normal phase, since only this phase can provide a continuous path. The situation with ac transport is entirely different, and the superconducting "islands" make a direct contribution to the ac conductivity and to surface impedance.

As we know, the real and imaginary parts of the surface impedance of a normal metal are almost equal (see, for example, Landau et al., 2004). Indeed, the surface impedance Z is determined by the relation:

$$Z = \left(\frac{\omega}{4\pi\sigma}\right)^{1/2} \exp(-i\pi/4) \qquad (11.13)$$

For normal metals the difference between ReZ and ImZ is negligibly small, and is connected with the dependence: $\sigma(\omega) = \sigma_0(1 - i\omega\tau_{tr})^{-1}$; in our case, $\omega\tau_{tr} \ll 1$. The situation in superconductors is entirely different (see, for example, Tinkham, 1996), and the same is true for the "pseudogap" state. Indeed, a metallic compound which contains superconducting "islands" is characterized by a strong inequality: $\mathrm{Re}(Z) \neq |\mathrm{Im}(z)|$. This can be shown theoretically (Ovchinnikov and Kresin, 2002) and measured experimentally (Kusco et al., 2002) for $HgBa_2Ca_2Cu_3O_{8-\delta}$ compound at $T > T_c$.

Let us discuss the situation with the Josephson current. We mentioned previously an interesting experimental study of S–N–S Josephson junctions (Bozovic et al., 2004). This phenomenon cannot be explained by the usual theory of S–N–S proximity junctions.

We focus on the especially interesting case of S–N'–S junctions where the electrodes are the high T_c superconducting films (such as $La_{0.85}Sr_{0.15}CuO_4$, or $YBa_2Cu_3O_7$), and the barrier N' is made of the underdoped cuprate; T_c' is the critical temperature of the underdoped barrier, and T_c is the critical temperature of the electrode. The generally accepted notation N' emphasizes a difference between S–N'–S and a typical S–N–S junction (then $T_c^N = 0$ K), so that $T_c' < T_c$. Here we consider temperatures when the barrier is in the normal resistive state because $T > T_c'$. The use of the underdoped cuprate as a barrier is beneficial for various device applications because the structural similarities between the electrodes S and the barrier N' eliminate many interface problems.

The "giant" phenomenon is manifested in a finite superconducting current through the S–N'–S Josephson junction with a thick barrier, so that $L \gg \xi_{N'}$, (where L is the thickness of the barrier, and $\xi_{N'}$ is the proximity coherence length). The configuration is such that the layers forming the barrier N' are parallel to the electrodes so that the Josephson current flows in the c-direction. Then the coherence length is very short $\xi_c \approx 4$ Å, so that we are dealing with the "clean" limit. This type of junction using the LaSrCuO material was studied by Bozovic et al. (2004). The films of $La_{0.85}Sr_{0.15}CuO_4$ ($T_c \approx 45$ K) were used as electrodes, whereas the underdoped LaCuO compound ($T_c' \approx 25$ K) formed the barrier. The atomic-layer-by-layer molecular beam epitaxy technique was used for these junctions and provides atomically smooth interfaces. The barrier was prepared in the c-axis geometry. As noted previously, the coherence length $\xi_c \approx 4$ Å. The measurements were performed at $T_c' < T < 35$ K. The Josephson current was observed for a thickness of L up to 200 Å(!). Such a "giant" effect cannot be explained with use of conventional theory. Indeed, as we know (see, for example, Barone and Paterno, 1982), the amplitude of the Josephson current for the "*clean*" *limit* is

$$j_m = j_0 \exp(-L/\xi_N) \tag{11.14}$$

The thickness of the barrier L should be comparable with the barrier coherence length ξ_N, and this condition is satisfied for conventional Josephson junctions. The picture described previously for the junctions with the cuprates is entirely different, since $L \gg \xi_{N'}$. The superconducting current in the c-direction occurs via an intrinsic Josephson effect between the neighboring layers (see Kleiner et al., 1992; Kleiner and Muller, 1994). If the barrier contains several homogeneous normal layers, then the Josephson current through such a barrier is practically absent.

To understand the nature of the "giant" Josephson proximity effect it is very important to stress that the barriers we are considering are formed by underdoped cuprates. As a result, the barriers are not in the usual normal state but in the "pseudogap" state; indeed, $T_c' < T < T_c^*$. For example, see the study (Iguchi et al., 2001; see Section 11.1.1) of the compound La$_{2-x}$Sr$_x$CuO$_4$ (x $\cong$ 0.1; $T_c \cong$ 18 K), in which the stoichiometry is close to that for the sample used by Bozovic et al. (2004) as the barrier. According to Iguchi et al. (2001), this compound has a value of $T_c^* \cong 80$ K whereas $T_c' = T_c^{res} \cong 18$ K describes the resistive transition to the dissipationless state. Since the diamagnetic moment measured by Iguchi et al. (2002) persists up to T_c^*, the question of the origin of the "giant" proximity effect is related directly to the general problem of the nature of the "pseudogap" state.

This effect can be explained (Kresin et al., 2003) by the approach described in this chapter and based on the intrinsically inhomogeneous structure of the compound. According to the model, the CuO layers forming the N'-barrier contain superconducting "islands," and these "islands" form the path for the Josephson tunneling current.

For typical S–N–S junctions the propagation of a Josephson current requires the overlap of the pairing functions F_L and F_R (see, for example, Kresin, 1986); F_R and F_L are pairing Gor'kov functions for left- and right-side electrodes. This overlap is caused by the penetration of F_L and F_R ("proximity") to the N-barrier. For the system of interest here the situation is quite different. Each "island" has its own pairing function with its own phase. As a result, the Josephson current is caused by the overlap of F_L and F_1, F_1 and F_2, and so on, where F_1 corresponds to the "island" located at the layer nearest to the left electrode, and so forth. The superconducting "islands" form the network with the path for the superconducting current.

The propagation of a Josephson current through the S–N'–S junction requires the formation of a channel between the electrodes. The transport of the charge in superconducting cuprates in the c-direction is provided by the interlayer Josephson tunneling (intrinsic Josephson effect). Therefore, the Josephson current through the barrier is measurable because of the superconducting state present in the layers.

The transfer of the Josephson current in the model described implies that the electrons tunnel inside the layers between the superconducting "islands" until one of them appears to be close to some "island" in the neighboring layer. Then the next step—the interlayer charge transfer via the intrinsic Josephson effect—occurs, and so on. As a result, the chain formed by the superconducting "islands" provides the Josephson tunneling between the electrodes, and the path represents a sequence of superconducting links. It is important to note that the amplitude of the total current is determined by the "weakest" link in the chain.

The density of the critical current is determined by the equation:

$$j = A \int \exp(-r/\xi) dP \qquad (11.15)$$

or

$$j = (A/\xi) \int_0^\infty dRP \exp(-R/\xi) \qquad (11.15')$$

where $\xi \equiv \xi_{11}$ is the in-plane coherence length, R is the distance between the "islands" on the same layer, $A \propto n_s j_{C\perp}$, $n_s \equiv n_s(T)$ is the concentration of the superconducting region, so that $S_{\text{sup.}} = n_s S$ is the area occupied by the superconducting phase, S is the total area of the layer, $j_{C\perp}$ is the amplitude of the Josephson interlayer transition, and P is a probability of formation of chain with length R for the links, so that $P = p^{m-1}$ (see, for example, Stuffer and Aharony, 1992), m is the number of layers forming the barrier, and p is the probability for two neighboring in-plane "islands" to be separated by distance $r < R$.

Assume that p is described by a Gaussian distribution; that is:

$$p = (c/\pi\delta)^{1/2}\xi^{-2} \int_0^R dr\, r \exp\left\{-\delta^{-1}\left[(r/\xi) - n_s^{-1/2}\right]^2\right\} \quad (11.16)$$

where δ is the width of the distribution, and $c = \text{const}$. Then the integral (11.16) can be calculated by the method of steepest descent, and we obtain

$$j_m = \tilde{A} f(m) \exp\left(-1/n_s^{1/2}\right) \quad (11.17)$$

where $\tilde{A} = \text{const}$. Equation (11.17) can be written in the form:

$$j_{\max} = j_0 \left(-1/\eta(T)\right) \quad (11.18)$$

Here

$$j_0 = j_{\max}(T_c),\, \eta(T) = n_s^{1/2}\left(1 - n_s^{1/2}\right)^{-1}$$
$$f(m) \approx \exp\left[-\delta \ln^{1/2}\left(m\tilde{l}\right)\right],\, \tilde{l} = (\pi\delta)^{-1/2}$$

We have assumed that $\delta \ll 1$, $m \gg 1$.

It can be seen directly from eqns. (11.17) and (11.18) that the current amplitude depends strongly on temperature and is determined mainly by the dependence of the area occupied by the superconducting "islands" on T: $n_s \equiv n_s(T)$. In addition, there is a weak dependence of $j_{\max}$ on the barrier thickness.

The dependence $n_s(T)$ is different for various systems and is determined by the function $T_c(\mathbf{r})$; that is, by the nature of the doping. Note that near T_c^* the value of $n_s(T)$ is very small and the current amplitude is negligibly small. However, the situation is different in the intermediate temperature region and in the region $T \ll T_c^*$, which is not far from T_c. This is true for the data by Bozovic et al. (2004): $T_c' = 25\,\text{K}$ and $T_c' < T < 35\,\text{K}$. For example, the value $T = 30\,\text{K}$ is relatively close to T_c' but is much below $T^* = 80\text{--}100\,\text{K}$. At $T = 30\,\text{K}$ there are many superconducting "islands", so that the value of $n_s(T)$ is relatively large. At temperatures close to T_c one can use eqn. (11.18) with $\eta = a(t-1)^v$; $t = T/T_c$, and $a = \text{const}$ (we have chosen $a = 10$). One can see (Fig. 11.7) that such a dependence with $v = 1.3$ is in good agreement with the experimental data. Note that this value for v is close to the value of the critical index for the correlation radius in the percolation theory (see, for example, Shklovskii and Efros, 1984).

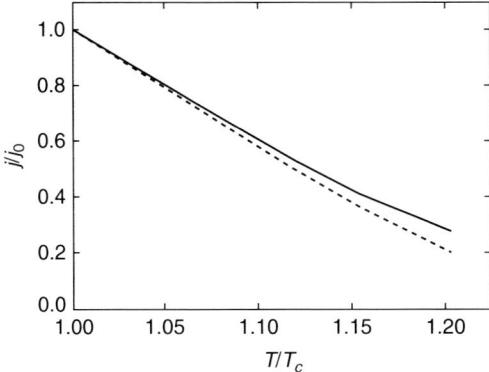

Fig. 11.7 Dependence of the Josephson current on temperature: dotted line, experimental data; solid line, theory. (Reproduced from Kresin et al., 2003.)

In principle, one can use a junction with a barrier grown in the ab direction, so that the c-axis is parallel to the S electrodes. Then the path contains S–N–S junctions formed by the "islands" with metallic N' barriers. Since $\xi_{ab} \gg \xi_c$ ($\xi_{ab} \cong 20\text{–}30\text{Å}$), one should expect even a larger scale for the "giant" Josephson proximity effect with thickness L up to $10^3\,\text{Å}$. The calculation performed by Covaci and Marsiglio (2006) also supports the conclusion that the giant proximity effect benefits from the presence of superconducting regions.

Therefore, the "giant" Josephson proximity effect is also caused by intrinsic inhomogeneity of the cuprates. The "giant" scale of the phenomenon is provided by the presence of "superconducting" islands embedded in the metallic matrix and forming the chain transferring the current. The use of superconductors in the "pseudogap" state represents an interesting opportunity for "tuning" the Josephson junction on a "giant" scale.

11.4.4 Isotope effect

As mentioned previously, according to an interesting experiment (Temprano et al., 2000), the "pseudogap" state is characterized by a strong isotope effect; that is, by a large shift in T_c^* caused by isotope substitution. They studied the $HoBa_2Cu_4O_8$ compound with a value of $T_c^* \approx 170$ K. The isotopic substitution $^{16}O \rightarrow ^{18}O$ leads to a drastic change in the value of T_c^*. One can observe the following change: $T_c^* \approx 170\,\text{K} \rightarrow T_c^* \approx 220\,\text{K}(!)$. We are dealing with a giant isotope effect; in addition, its value is negative.

Such an isotopic dependence of T_c^* is also a consequence of the presence of the superconducting regions, and reflects the fact that the superconducting pairing persists above the resistive transition. It is interesting that the isotope coefficient has a negative sign. This unusual feature is consistent with the dynamic polaron model of the isotope effect (see Section 9.7 and eqn. (9.12)). Indeed, a strong non-adiabaticity (axial oxygen in YBCO is in such state) results in a peculiar polaronic isotope effect:

$$\alpha = \gamma \frac{n}{T_c^*} \frac{\partial T_c^*}{\partial n} \tag{11.19}$$

Based on eqn. (11.19) one can explain why the isotope coefficient has a negative value. Indeed, $\alpha \propto \partial T_c^*/\partial n$ and $(\partial T_c^*/\partial n) < 0$: an increase in doping in the underdoped region leads to a decrease in the value of T_c^*. (At optimum doping $T_c^* \cong T_c$, and this is due to an increase in a number of dopants—that is, the pair-breakers—so that the distribution of pair-breakers becomes more uniform.)

11.5 Other systems

Intrinsic inhomogeneity is an essential feature of the high-T_c oxides, and this feature is manifested in a peculiar "pseudogap" behavior. However, the scenario is a more general one, and the "pseudogap" state can also be observed in other inhomogeneous systems. Let us describe some of these systems.

11.5.1 Borocarbides

Borocarbides represent an interesting family of novel superconductors, because they allow us to study an interesting interplay between superconductivity and magnetism (see, for example, Canfield et al., 1998; Schmiedeshoff et al., 2001). According to data by Lascialfari et al. (2003), the borocarbides YNi_2B_2C display precursor diamagnetism above T_c ($T_c = 15.25$ K). Analysis of magnetization data taken with a high-resolution SQUID magnetometer led to the conclusion that the unusual results were caused by an inhomogeneity of the compound and by the presence of superconducting droplets with a local value of the critical temperature higher than the usual T_c corresponding to macroscopic dissipationless current. The presence of the superconducting isolated droplets is due to an inhomogeneity similar to that observed in the cuprates and described in this chapter. Indeed, an unconventional temperature dependence of the critical fields H_{c1} and H_{c2} was observed (Suh et al., 1996; Schmiedeshoff et al., 2001). This dependence can be explained by the presence of pair-breakers (Ovchinnikov and Kresin, 2000; see Section 10.4.2). A Statistical distribution of pair-breakers leads to spatial dependence of the critical temperature: $T_c \equiv T_c(\mathbf{r})$ and to the inhomogeneous "pseudogap" picture (Fig. 11.5).

11.5.2 Granular superconductors; Pb+Ag system

Granular superconducting films were studied intensively before the discovery of high-T_c superconductivity (see, for example, Dynes and Garno, 1981; Simon et al., 1987). These films also represent inhomogeneous superconducting systems. Such inhomogeneous films could display diamagnetic moment above T_c. It would be interesting to carry out a study of their magnetic properties.

An interesting study of the Pb+Ag system was described by Merchant et al. (2001). An electrically discontinuous (insulating) Pb film was covered with an increasing thickness of Ag (Fig. 11.8). The Ag acts to couple the superconducting Pb grains via the

Fig. 11.8 Pb/Ag proximity system. (Figure reproduced from Merchant *et al.*, 2001. © 2001 by the American Physical Society.)

proximity effect. The resistive transition, as well as tunneling spectra, has been taken on a series of these films. The most insulating film has no resistive transition but a full Pb gap as revealed by the tunneling spectra. This gap is reduced as silver is added, reflecting the decrease in the mean-field T_c of the Pb grains. At some point, the composite film becomes continuous and superconducting with a low resistive transition temperature. The evolution of the mean-field transition temperature and the resistive transition temperature with increasing Ag thickness mimics the phase diagram of the cuprates with doping. The mean-field transition temperature resembles the pseudogap onset temperature, and the resistive T_c resembles the superconducting transition temperature, with the mean-field transition temperature lying above the resistive transition.

The results of the Pb/Ag artificial inhomogeneous superconductor model the behavior of the cuprates, which are doped substitutionally and inhomogeneously. At some concentration of doping there are regions with a sufficiently high concentration of carriers to superconduct locally and therefore reduce the low-energy density of states. The evolution of these islands into a percolating dissipationless state would resemble the percolating proximity coupling described previously. Therefore, it is not surprising that the phase diagrams would be nearly identical.

11.6 Ordering of dopants and potential for room-temperature superconductivity

The "pseudogap" scenario and its manifestation, which is especially strong in the underdoped region of the cuprates, implies that the transition into a superconducting state is a two-step process.

The superconducting regions appear below some characteristic temperature T_c^* (we call it an "intrinsic" critical temperature); its value greatly exceeds the usual resistive $T_c \equiv T_{c;res}$. For example, for the underdoped LaSrCuO compound ($T_c \approx 25$ K) the value of $T_c^* \approx 80$ K. For the YBCO compound the value of $T_c^* \approx 250$ K; that is, it is close to room temperature. Decreasing the temperature toward $T_c \equiv T_{c;res}$ leads to an increase in the number of "islands" and to an increase in their size. The second step is the transition at T_c which is of percolative nature and corresponds to the formation of a macroscopic superconducting phase ("infinite cluster" in terminology of the percolation theory) capable of carrying a macroscopic supercurrent. It is obvious that the set of superconducting clusters embedded in a normal matrix structure is unable to carry a superconducting current. Such a structure reflects the intrinsic inhomogeneity of the sample caused by the statistical nature of doping.

It would be attractive to "order" the superconductive regions, and instead of statistically distributed superconducting "islands" to have a continuous superconducting phase; then one can expect the supercurrent to flow at high temperatures. In other words, it would mean an effective increase in the value of resistive T_c. In the following we describe the concept of "ordering of dopants," which appears promising in achieving this goal.

The delocalized carriers whose presence is responsible for the metallic and, correspondingly, the superconducting state are created by doping. At the same time, the superconducting pairing could be depressed by the pair-breaking effect (see Sections 8.4 and 11.2.2). The dopants play a double role: they provide delocalized carriers, and are also responsible for the pair-breaking. The statistical nature of doping leads to a random distribution of dopants, and at relatively low doping the spatial distribution is rather broad. As a result, the regions with a smaller number of dopants have a larger value of the critical temperature. These regions form the superconducting "islands" inside the normal matrix.

An increase in T_c can be achieved by a procedure that allows for a special ordering of defects (Wolf and Kresin, 2012). To clarify the concept, let us consider a specific example: the YBCO compound. The doping of the parent $YBa_2Cu_3O_6$ sample is provided by adding oxygen to the chain layer. The mixed-valence state of the in-plane Cu leads to the plane-chain charge transfer and the appearance of a hole, initially on the Cu site. Because of diffusion, the hole enters the system of delocalized carriers responsible for the metallic and, correspondingly, superconducting behavior: superconductivity is caused by pairing of such holes. The added oxygen ion and corresponding in-plane Cu form the defect mentioned previously, with its pair-breaking impact.

The picture described could be affected strongly by specific ordering of the oxygen ions. More specifically, let us consider the thin film formed by a layer-by-layer deposition technique; the film is growing in the c-direction. The film should be built as a set of columns, the top layer could be imagined as a set of strips, and an additional oxygen should be placed inside specific columns. As a result, these columns would contain the chains with a composition close to $YBa_2Cu_3O_7$. The chains are parallel to the inter-strip boundaries. Such columns could be called the reservoir columns, since holes are created in these areas. The holes are delocalized and diffuse into the neighboring columns, which are free of defects and represent the high-T_c regions. Such a structure could be built, for example, using a nano-implantation technique (see, for example, Schenkel et al., 2009; Toyli et al., 2010) with appropriate annealing or other methods for providing an ordered defect structure similar to what has been described here.

As a whole, the film would then contain alternating reservoir and high-T_c columns. Because of the presence of the oxygen defects, the reservoir columns have value of T_c lower than in the neighboring columns which are free of defects. It is expected that the high-T_c columns would have values of T_c close to the intrinsic critical temperature, T_c^*, which in fact could be close to room temperature. Moreover, such a structure would provide a continuous path for the supercurrent, which means an effective increase in the resistive T_c.

Note that the value of T_c in these high-T_c columns could be depressed by the proximity effect with the reservoir columns, and this leads to some limitation on the width of the high-T_c columns, W_H. Indeed, we have the S_L–S_H proximity system, where L and H correspond to the reservoir and high-T_c columns ($T_{cH} > T_{cL}$). We know that the scale of the proximity effect is of order of the coherence length, ξ_H, of the high-T_c region. To minimize the impact of the proximity effect, the width of the high-T_c column, W_H, should be larger than ξ_H. However, the value of ξ_H is rather small. Indeed, $\xi_H \approx v_F / 2\pi T_{c;H}$. If we take the values $v_F \approx 10^7$ sm/sec, and $T_c \approx (2\text{--}2.5)\, 10^2$ K, we obtain $\xi_H \approx 5\text{--}10$ Å. Because of such a small value of the coherence length, the condition $W_H \gg \xi_H$ is perfectly realistic. Note that similar ordering could be effective for an increase in T_c for other cuprates: for example, LaSrCuO or Bi-2212 compounds.

The proposed method conceptually is analogous to the observed increase in T_c for the cuprates caused by pressure (see, for example, Schilling and Klotz, 1992; Hochheimer and Etters, 1991; and the analysis by Kresin et al., 1996), by an applied field (Mannhart et al., 1991; Ann et al., 2006), or by the photoinduced effect (Yu et al., 1992; Pena et al., 2006). Indeed, the external pressure, radiation, or an applied field affects the doping without creating defects; that is, pair-breakers. The specifics of the structure proposed here is that the carriers appear in a region (high-T_c column) which is free of defects; they are produced in a different spatially separated column. As stressed previously, we observe the impact of such separation in the existing cuprates (in the "pseudogap state"), especially in the underdoped region, where the superconducting clusters (high-T_c regions) are embedded in a normal matrix. Due to such a separation one can observe the anomalous diamagnetism (Meissner effect) at temperatures higher than the resistive T_c. The proposed ordering leads to a noticeable increase of the critical temperature for the resistive transition and thus a much higher effective transition temperature.

The ordering leads to an increase in the size of superconducting clusters, and as a result, the percolative transition occurs at a higher temperature. This effective increase could be enhanced in a material with a larger dielectric constant (Mueller and Shengelaya, 2013). More specifically, they proposed to create a multilayer structure, so that the underdoped high-T_c oxide thin film (in the range of ∼1–10 nm) is sandwiched between the high-dielectric-constant insulator layers; their presence reduced the Coulomb repulsion between the superconducting "islands." Examples of such materials are $SrTiO_3$ (Mueller and Burkard, 1979, with $\varepsilon \simeq 10^4$), or other ferroelectrics, such as $Sr_{1-x}Ba_xTiO_3$ or $Pb(Zr_xTi_{1-x})O_3$ ($\varepsilon \simeq 3\times 10^2$).

An additional increase in T_c can be achieved with the use of isotope substitution (Mueller, 2012). Indeed, as noted in Section 9.2.3, one can observe a large negative isotope effect, and as a result, the $O_{16} \rightarrow O_{18}$ substitution leads to an increase of order of 30 K(!).

The ordered doping manifested itself in experiments by Liu et al. (2006). The ordering of apical oxygen has been observed for $Sr_2CuO_{3+\delta}$ superconductor ($T_c \simeq 75$ K). This superconductor is characterized by a peculiar feature: the apical oxides sides are not fully occupied, and the hole doping occurs through the charge transfer between the apical oxygen and the Cu–O plane. The additional oxygen can be deposited on these sides in an ordered way, and as a result, an increase in T_c up to 95 K was observed.

Correlation between structure and its ordering and superconducting properties— that is, a noticeable rise in T_c—was described by Fratini et al. (2010) and

Pokkia et al. (2011). X-ray radiation was used to create ordered superconducting regions in the La_2CuO_{4+y} samples, and it was accompanied by an increase in T_c ($T_c \simeq 33.5$ K $\to T_c \simeq 41.5$ K). The onset of high-quality superconductivity was associated with 2D defect ordering.

11.7 Remarks

The approach to the "pseudogap" state described in this chapter is rather general in the sense that it is valid for any pairing force (phonons, magnons, plasmons, excitons, and so on), but, nevertheless, we are dealing (in the region $T_c < T < T^*$) with real Cooper pairs, so that T_c^* is the "intrinsic" critical temperature. Moreover, each superconducting "island" has its own phase. It is an essential feature of this picture that the superconducting regions are embedded in a normal metallic matrix which provides normal dc transport. As a result, the proximity effect between the superconducting regions and the normal metal plays an important role.

Until recently, the presence of inhomogeneities was considered as a signature of a poor-quality sample (except for the "pinning" problem). However, we think that the situation is similar to that in the history of semiconductors. Indeed, initially the presence of impurities in these materials was considered as a negative factor (they were called "dirty" semiconductors). But later, when scientists developed tools allowing the precise control of the impact of various impurities (donors and acceptors), it became clear that the presence of impurity atoms is a critical ingredient; even the language has changed and sounds more "respectful" ("doped" semiconductors). The analogy between inhomogeneous novel superconductors and semiconductors is even stronger, because we are dealing with doping for both classes of materials.

12
Manganites

12.1 Introduction

This chapter is concerned with the properties of the so-called manganites, which are not superconductors. They are magnetic materials, and therefore the question arises of why a book on superconductivity contains such a chapter. It can be justified because both families—cuprates and manganites—are the mixed-valence compounds, and there is a large similarity between them.

Manganites are named after the manganese ion which is a key ingredient of the compounds. Their chemical composition is $A_{1-x}R_xMnO_3$; usually $A \equiv$ La, Pr, Nd, and $R \equiv$ Sr, Ca, Ba. These materials were first introduced by Jonner and van Santen in 1950. Unlike in the usual ferromagnetics, the transition of manganites to ferromagnetic state (at $T = Tc$, Tc is a Curie temperature) takes place at finite "doping", $x \neq 0$, and is accompanied by a drastic increase in conductivity. This simultaneous transition from an insulating to a metallic and magnetic state is one of the most remarkable fundamental features of these materials.

One year later, Zener (1951) explained this unusual correlation between magnetism and transport properties by introducing a novel concept called the "double exchange" mechanism (DE). Zener's pioneering work was followed by more detailed theoretical studies by Anderson and Hasegawa (1955) and de Gennes (1960).

A revival of interest in the manganites and their properties came about after the remarkable discovery of the colossal magnetoresistance effect (CMR) by Jin *et al.* (1994). The very name of the phenomenon originates from the observation of a thousand-fold(!) change in the resistivity of the La–Ca–Mn–O films near $T = 77$ K in the presence of applied magnetic field, $H \approx 5\ T$.

It is important to mention that the discovery of CMR in the magnetic oxides (manganites) was preceded by the discovery of high-temperature superconductivity in the copper oxides (cuprates) by Bednorz and Mueller (1986). As noted previously, despite the obvious difference in the two phenomena (superconductivity versus ferromagnetism), there is a deep analogy between the two classes of material. Both classes are doped oxides. The parent (undoped) compounds (for example, the cuprate $LaCuO_4$ or the manganite $LaMnO_3$) are antiferromagnetic insulators. It is "doping" that leads to the insulator–metal transition for both systems. Of course, there are profound differences between these compounds, but discovery and the following intensive study of the high-T_c cuprates was a factor, very beneficial for the progress in understanding of manganites. It is worth noting that the discovery of the CMR effect was made possible with the use of high-quality thin films. The preparation of such films (Chahara et al., 1993) was based on a method developed for high-temperature superconducting oxides.

A study of manganites remains very active (see, for example, the reviews by Coey et al., 1999; Tokura, 2003; Gor'kov and Kresin, 2004). We focus here on the fundamentals of these materials and on similarities between their properties and those of the high-T_c superconducting oxides. Here we follow the analysis by Gor'kov and Kresin (2004).

12.2 Electronic structure and doping

12.2.1 Structure

Let us start with an undoped (parent) compound, LaMnO$_3$. Consider its cubic perovskite structure (Fig. 12.1). The Mn^{3+} ions are located at the center, and the La ion at the corner of the unit cell. Each Mn^{3+} ion is caged by the O^{2-} octahedron (Fig. 12.2); locally, this forms an MnO$_6$ complex with the Mn ion in the symmetric central position surrounded by six light oxygen ions.

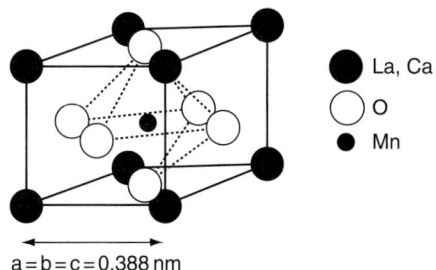

Fig. 12.1 Parent compound LaMnO$_3$; unit cell.

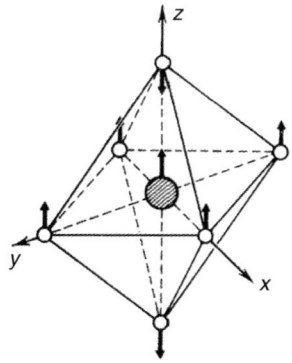

Fig. 12.2 MnO$_6$ octahedron.

As noted previously, the LaMnO$_3$ crystal contains Mn^{3+} ions. Such valence state of Mn is determined by the simple neutrality count, since La ion has "+3" valence state, and each oxygen ion is in the O^{2-} valence state.

Electrons of a free Mn atom form the incomplete d-shell (...) 3d^5 4s^2; (...) $\equiv$ 1s^{2}2s^{2}2p^{6}3s^{2}3p^6 (see, for example, Landau and Lifshitz, 1977). Therefore, the Mn^{3+} ion, (...) 3d^4, contains four d-electrons.

The five-fold orbital degeneracy is split by the cubic environment into two terms, t$_{2g}$ and e$_{2g}$. The t$_{2g}$-level contains three electrons that form the so-called "t-core". The last d-electron (e$_{2g}$ electron) is well separated in energy and forms a loosely bound state (see Fig. 12.3). This e$_{2g}$ electron plays a key role in conducting and other properties of manganites, as well as in determining its magnetic order. The analysis of its behavior in the lattice is a major subject of the microscopic theory.

Hund's rule demands that the three d-electrons forming the "t-core" have the same spin-orientation; and as a result, the localized "t–core" has the total spin $S = 3/2$. The e$_{2g}$ electron is also affected by the same strong Hund's interaction. Therefore, its spin must be polarized along the same direction as for the t-core.

It is very essential also that the e$_{2g}$ term is a doubly degenerate one. As a result, we meet with the situation in which the Jahn–Teller effect becomes an important factor that may lead to a lattice instability.

The parent compound LaMnO$_3$ is an insulator and its transition to the conducting state is provided by doping; the scenario is similar to that in the cuprates. The doping is realized through a chemical substitution: for example, La^{3+} $\rightarrow$ Sr^{2+}; that is, by placing a divalent ion into the local La^{3+} position. The substitution La^{3+} $\rightarrow$ Sr^{2+} leads to the change in manganese-ion valence. The four-valent Mn ion loses its e$_{2g}$ electron (Fig. 12.3). The missing electron can be described as a creation of a hole. The Sr^{2+} ion is located at the center of the cubic cell. As to the hole itself, it is spread over the unit cell, being shared by eight Mn ions.

As a result, one obtains a crystal La$_{1-x}$Sr$_x$MnO$_3$, where a number of La ions are substituted randomly by the Sr ions. Even in the presence of some holes, the

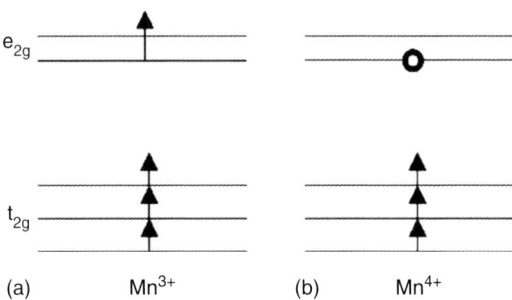

Fig. 12.3 (a) d-shell of the Mn^{3+}; total spin of the t-core is $S = \frac{3}{2}$; for clarity, the e$_{2g}$ degenerate level is split; (b) Mn^{4+} ion; o-hole; the La^{3+} $\rightarrow$ Sr^{2+} substitution leads to the Mn^{3+} $\rightarrow$ Mn^{4+} transition.

crystal at first continues to behave as an insulator. In other words, each hole remains being localized on the scale of one unit cell. In this concentration range, localization corresponds to the formation of local polarons. Such an insulating state is preserved with an increase in doping up to some critical value $x = x_c \approx 0.16$–0.17.

At $x = x_c$ the material makes a transition into the conducting (metallic) state. This can be seen directly from the data by Urushibara et al. (1995), Fig. 12.4. Note that the conductivity of the best samples of the Sr-doped films at low T is of order $\sigma = 10^4$–$10^5 \Omega^{-1}$ cm^{-1}; that is, we are now dealing with a typical metallic regime.

It is remarkable that the transition at $x = x_c$ is also accompanied by the appearance of the ferromagnetic state. The correlation between conductivity and magnetism is the fundamental feature of manganites, and we address this problem in what follows.

So far, we have concentrated on manganites in the low-temperature region and the evolution of their properties with doping. Take now the sample in the metallic ferromagnetic state (FM) with a fixed carrier concentration—for example, $x = 0.3$—and then increase the temperature. Such an FM state persists up to the Curie temperature $T_c \approx 170$ K. Above this temperature the compound makes the transition into the paramagnetic state with much higher resistivity. Once again one sees that there is a correlation in electronic properties that manifests itself in an almost simultaneous change (at $T = T_c$) in both conductivity and magnetization.

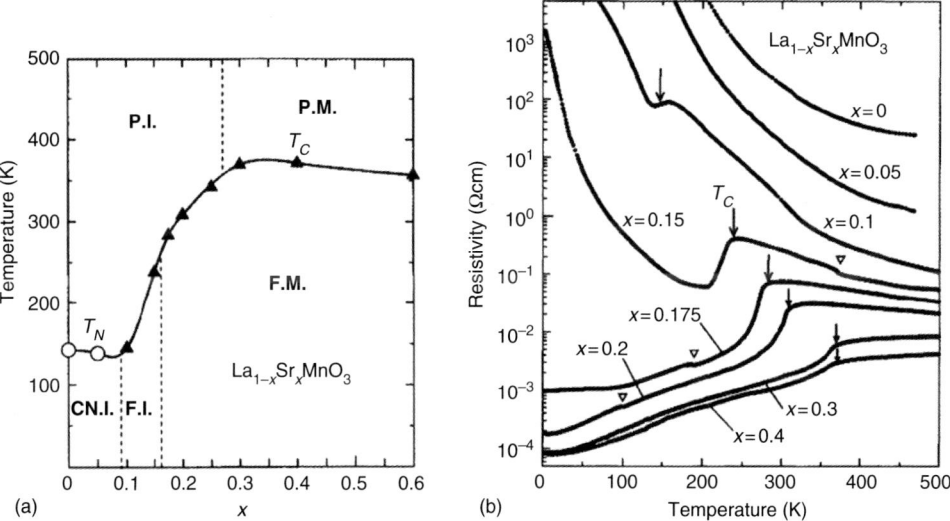

Fig. 12.4 Phases of manganites (La$_{1-x}$Sr$_x$MnO$_3$). (a) Phase diagram: T_N, Néel temperature; FM, ferromagnetic metal, AI, antiferromagnetic insulator (A-phase); FI, ferromagnetic insulator; PI, paramagnetic insulator; PM, disordered paramagnetic with large resistance; (b) evolution of the dependence $R(T)$ with doping. (Figure reproduced from Urushibara et al., 1995. © 1995 by the American Physical Society.)

12.2.2 Magnetic order

The type of magnetic structure is determined by the doping level. The parent compound, LaMnO$_3$, belongs to the so-called antiferromagnetic insulating phase. The ferromagnetic ordering in layers is combined with an antiferromagnetic order in the direction perpendicular to the layers (the A structure; Fig. 12.5). The value of the Néel temperature (see Fig. 12.4a) for this antiferromagnetic phase is relatively low: $T_N \approx 150$ K.

The doping leads eventually to an appearance of the 3D ferromagnetic (F) state (for $x \gtrsim 0.16$). This state persists up to $x \approx 0.5$ for (LaSr)MnO$_3$, and then the crystal, while continuing to be in the metallic state, can change its magnetic structure, which becomes the *metallic* A state. The magnetic order in this state is similar to that of the parent (underdoped) compound (Fig. 12.5), but it has metallic conductivity. Such a compound is a natural spin valve system. Indeed, as is known, the "giant" magnetoresistance effect (GMR) has been observed by using a special artificial multilayer structure. Contrary to it, the metallic manganites with magnetic A structure are natural 3D systems which display the GMR phenomenon. In addition, this material can be used for making the new type of Josephson junction (SAS); see Section 12.7.2.

12.2.3 Double-exchange mechanism

Here we discuss qualitatively the nature of the observed ferromagnetic spin alignment. As noted previously, the concept—the so-called "double exchange" (DE) mechanism—was introduced by Zener (1951) almost immediately after the discovery of manganites.

Here the situation is opposite to that of the cuprates. The mechanism of high-T_c superconductivity is still a controversial issue, though the phenomenon was discovered in 1986. As for manganites, the nature of their unusual magnetism and simultaneous transition into metallic and ferromagnetic state were explained rather shortly.

Let us discuss the DE mechanism. If one of the Mn^{3+} ions becomes four-valent (Mn^{3+} → Mn^{4+}, this is the result of the doping; that is, for example, the La^{3+} → Sr^{2+} substitution), a hole appears on this site. It allows for another e_{2g} electron localized initially at the neighboring Mn^{3+} ions to jump on the new vacant place (such a hopping corresponds to the hole moving in the opposite direction). But, as noted previously, the e_{2g} electron is spin-polarized (Fig. 12.3) because of the Hund's interaction with its

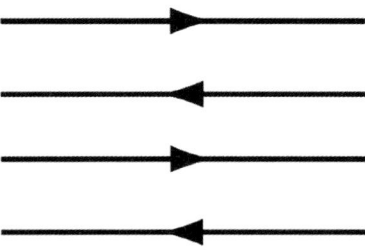

Fig. 12.5 A-structure.

t-core. The total spin of each t-core is equal to $S = 3/2$, but their mutual orientations on different sites were independent. Relative orientation of spins of the e_{2g} electron and the "vacant" t-core is the crucial factor for the hopping because of the strong Hund's interaction. Indeed, imagine that the direction of the spin of the core for the Mn^{4+} ion is opposite to that for the e_{2g} electron of the neighboring Mg^{3+} ion. Then the hopping is forbidden. At the same time, such hopping, as any increase in a degree of delocalization, is energetically favorable. In other words, it produces a gain in the kinetic energy, and the ground state of the ferromagnetically ordered system (all spins are polarized along one direction) lies below the paramagnetic state. As a result, the t-cores become ferromagnetically coupled, and this, in its turn, favors hopping of the e_{2g} electrons.

This simple picture describes qualitatively the origin of the ferromagnetism in the manganites, and demonstrates the direct interdependence between hopping and ferromagnetic ordering. This correlation leads to the observed interplay between conductivity and magnetism. A quantitative analysis requires a rigorous treatment of hopping along with the Hund's interaction, and will be described in what follows.

It is important to emphasize that the charge transfer in the conducting ferromagnetic manganites is provided by spin-polarized electrons. Such a conductor is different from a usual metal where spins of conduction electrons have both directions. Because of such spin specifics the conducting state in manganites is called "half-metallic."

12.2.4 Colossal magnetoresistance (CMR)

The huge magnetoresistance effect has been observed in the ferromagnetic metallic films $La_{0.67}Ca_{0.33}MnO_3$ (Jin et al., 1994). The magnetoresistance is defined as

$$\Delta R/R_H = [R(T, H) - R(T, 0)]/R(T, H) \tag{12.1}$$

The magnetoresistance has a sharp peak near $T \approx 190$ K; that is, near the Curie temperature for the Ca-doped manganites, and the change in resistivity caused by an applied magnetic field ($H \approx 5\,T$) is very large: $\Delta R/R_H \approx -1.3 \times 10^3 (!)$

Such a drastic change in resistivity is caused by the aforementioned correlation between magnetic ordering and conductivity, and therefore is directly related to the double-exchange mechanism realized in manganites. Indeed, below the Curie temperature, T_c, conductivity has a metallic band mechanism, whereas above T_c mechanism of conductivity bears a polaronic hopping character. The presence of an external magnetic field is a favorable factor allowing us to establish ferromagnetic ordering at temperatures higher than $T(H = 0)$. The magnetic order triggers the conductivity increase (through the double-exchange mechanism), and this leads to a large resistivity change $R_H = R(T,H) - R(T,0)$. The shift is negative; indeed, it corresponds to the transition into ferromagnetic metallic state.

The "colossal" magnetoresistance effect (CMR) greatly exceeds in its value the so-called "giant" magnetoresistance effect (GMR); see, for example, the reviews by Fert (2007), von Helmholt et al. (2008), and Parkin (1995). The GMR effect is observed mainly in artificial multilayer systems with alternating magnetic structure, and is used in many applications. It reaches a value $R_H/R_H \approx 50\%$. The scale of the CMR phenomenon is much larger.

The nature of the CMR and GMR phenomena are entirely different. It is interesting, nevertheless, that, as mentioned previously, the metallic manganite with the A structure (say, in $La_{2-x}Sr_xMnO_3$ with $x > 0.5$) forms a natural GMR system, and this might lead to the GMR effect in manganites.

12.3 Percolation phenomena

12.3.1 Low doping: transition to the ferromagnetic state at low temperatures

Substitution creates a hole located inside one unit cell. The Coulomb attraction prevents the hole from larger delocalization; namely, from a spreading through the whole lattice. An increase in doping leads to an increase in the number of such unit cells containing holes. In addition, if the Sr substitution occurs for two or more neighboring units, a larger cluster forms covering two and more unit cells with holes delocalized along such bigger clusters. From Zener's double-exchange mechanism we expect that spins of the Mn ions have the ferromagnetic alignment inside each cluster.

The random character of the Sr substitution leads to statistical (chaotic) distribution of these clusters. As a result, growth of the clusters can be treated by means of percolation theory (see, for example, Sklovskii and Efros, 1984; Deutscher, 1987; Stuffer and Aharony, 1992).

As noted previously, an increase in the doping level of the $La_{1-x}R_xMnO_3$ crystal (R $\equiv$ Sr, Ba, ...) leads to an increase in the clusters' size and their overlap. Finally, at some critical value of $x = x_c$ (percolation threshold) the system forms the so-called "infinite" cluster piercing the whole sample. In other words, at $x = x_c$ one first sees the appearance of connected islands of a *macroscopic* metallic ferromagnetic phase. As other phase transitions, the percolative insulator-metal transition in manganites can be characterized by some critical indices. This theoretical approach to the transition in manganites was first introduced and developed by Gor'kov and Kresin, (1998, 1999; and review, 2004; see also Dzero et al., 1999, 2000). Later, this approach was employed in many papers (for example, Jaime et al., 1999; Dagotto et al., 2001).

The approach based on the percolation concept implies that the system is intrinsically inhomogeneous. This inhomogeneity may manifest itself through "phase separation." The phenomenon of "phase separation" corresponds to the simultaneous existence of the mutually penetrating sub-phases. As noted previously, the progress in theory of manganites was strongly influenced by the preceding discovery and studies of high-temperature superconductivity in the copper oxides. For instance, the concept of phase separation (coexistence of insulating and metallic sub-phases) was first suggested by Gor'kov and Sokol (1987) in relation to the high-T_c superconducting oxides. The idea received further development and experimental attention in many papers (for a review, see, for example, Sigmund and Mueller, 1994). The phase separation in manganites was analyzed by Nagaev (1996). *Ab initio* study (Pineiro et al., 2012) shows that the nanoscale doping inhomogeneity appears to be stable.

The percolation can be interpreted as a more subtle case of phase separation where "phases" coexist on a mesoscopic scale or even as interweaving clusters. Macroscopic properties are determined mainly by one sub-phase, where there are also inclusions of another phase. At $x > x_c$, metallic paths are formed and gradually become more and more dominant, though the insulating regions are also present.

12.3.2 Percolation threshold

Experimentally, the metallic behavior of a compound at low T sets in at the critical value $x = x_c \approx 0.16$ (Urushibara et al., 1995). At cooling, the temperature at which conductivity sharply increases almost coincides with the onset of the ferromagnetism, especially in good-quality samples. For example, for the $La_{0.67}Ca_{0.33}MnO_3$ sample ($T_c = 274$ K) studied by Heffner et al. (1996), these temperatures coincide to within 1 K (see also Schiffer et al., 1995, and the review by Coey et al., 1999). Such closeness of the onset temperatures is a strong indicator in favor of the double-exchange mechanism.

Let us discuss the threshold value $x_c \approx 0.16$. Recall that material may be prepared by various methods, but always at high temperatures. As a result, positions of atom R, which substitute for a parent atom, are completely random. Divalent atom R locally creates a "hole" localized on adjacent Mn sites. The Coulomb forces in the dielectric phase keep the "hole" close to the negative charge at R^-. When the concentration is small, average distances between R ions are large, and the holes remain isolated, forming trapped states (polarons).

Consider the concentration at which the nearest-neighbor R atoms start, in accordance with the percolation picture, to form infinite clusters piercing the whole crystal. Usually one attacks percolation in one of two discrete mathematical models on the cubic lattice: the "site" and "bonds" problems. The picture of a hole localized at a single center would correspond to the so-called "site" problem (contrary to the "bond" problem) of the percolation theory. The critical concentration for the "site" problem depends on the type of lattice, and for a simple cubic structure is equal $x_c = 0.31$ (see, for example, Shklovskii and Efros, 1984). However, this is not our case. While ionic substitution takes place at the center of the cubic unit, the formed hole is spread over several Mn sites around the R^- ion. At the same time, the charge transfer due to forming the larger cluster occurs only along Mn–O–Mn bonds. Hence, the picture of a critical cluster, constructed from the R^-, ions is not correct; such an initial (nucleating) cluster is not a point-like formation, and already has a finite size ("thickness"). The size is even bigger at a large enough dielectric constant that would weaken the Coulomb attraction to the R^- ion. According to numerical calculations (Scher and Zallen, 1970), this circumstance (the involvement of a scale of a few lattice constants into the percolation problem) strongly decreases the value of the critical concentration for percolation. It is interesting that now the critical value x_c depends only on the spatial dimension and appears to be invariant for all lattices. For the 3D case (see, for example, Shklovskii and Efros, 1984):

$$x^{3D} \approx 0.16 \qquad (12.2)$$

The "site" problem corrected by the finite hole radius becomes similar to the "continuum" percolation. As to the continuous limit, it is only natural that independent numerical studies (Shklovskii and Efros, 1984) lead to close values for x_c. Therefore, the invariant of the percolation theory $x_c \approx 0.16$ describes the threshold, above which the formation of an infinite cluster first takes place. In its manifestation it corresponds to the appearance of a new macroscopic state; that is, in our case, to the transition into a metallic ferromagnetic state.

It is remarkable that the experimentally observed value of the critical concentration $x_c \approx 0.16$ corresponds, indeed, with a good accuracy, to the value obtained in the framework of the percolation theory.

The measurements by Urushibara et al. (1995) were performed for $La_{1-x}Sr_xMnO_3$. Analysis of the phase diagram for $La_{1-x}Ca_xMnO_3$ (Schiffer et al., 1995; see also the review by Coey et al., 1999) shows that the values of the critical concentration for different manganites are close to the value (12.2).

The percolative description means the situation when one phase manifests itself as tiny inclusions ("islands") embedded into another, the dominant macroscopic phase. Here lies the difference between the percolative picture and that of the macroscopic electronic phase separation.

12.3.3 Increase in temperature and percolative transition

We have described the evolution of the phase diagram, including the metal–insulator transitions as driven by doping at low temperatures. Now let us consider the compound with a fixed level of doping. For example, the manganite with $x = 0.3$ has a well-established metallic ferromagnetic state in the low-temperature region. An increase in temperature leads to the transition (at $T = T_c$) to the paramagnetic and low-conducting state. This transition can also be treated by means of percolation theory, and one can apply the ideas similar to those described previously. Indeed, the high-temperature resistivity $(T > T_c)$ is very large, and one may take approximately the conductivity here: $\sigma(T > T_c) = \rho^{-1}(T > T_c) \approx 0$. The fast increase in $\sigma(T)$ at $T < T_c$ is then expected to correspond:

$$\sigma(T < T_c) \propto (T_c - T)^\gamma \tag{12.3}$$

where γ is a critical index.

The statement that the transition at T_c is also of a percolative nature implies the intrinsic inhomogeneity of the sample—that is, the phase separation—quite similar to that discussed earlier.

It is essential to stress the similarity between the manganites and the cuprates. The statistical nature of doping leads to the percolative transitions observed for both systems. As described in Chapter 11, the transition of the cuprates at $T_c = T_c^{res}$ into the macroscopic dissipationless state is of percolative nature. It is similar to that in manganites for their transition at T_c into a ferromagnetic metallic state.

12.3.4 Experimental data

The theoretical approach based on the percolation theory has strong experimental support. To start with, the experimentally measured critical concentration $x_c \cong 0.16$ for the La_{1-x}-$Sr_x MnO_3$ compound and a close value for other manganites is in an excellent agreement with the value predicted by the 3D percolation theory (see eqn. (12.2)). This agreement is a direct quantitative indication in favor of the percolative nature of the metal–insulator transition.

The percolation always implies some inhomogeneity of the system. Let us consider first the low-temperature region and trace in more detail the doping dependence. The neutron pulsed experiments (Louca and Egami, 1997, 1999) directly indicate the presence of such local inhomogeneites. These experiments probe the local arrangements of the oxygen octahedra. The presence of such insulating inclusions in $La_{1-x}Sr_x MnO_3$ is seen up to $x \approx 0.35$—well above x_c—in the metallic region. The inhomogeneous structure has also been observed with the use of tunneling spectroscopy (Becker et al., 2002).

Note also that the value of magnetization M depends on the doping level: $M \equiv M(x)$. The observed dependence is close to the simple relation $M(x) = (4-x)\mu_B$, but it is *less* than this value (see Coey et al., 1999), indicating that the admixture of the "non-metallic" (non-ferromagnetic) phase still persists at these concentrations. This fact, and the concentration range, agree well with the value $x \approx 0.35$ estimated from neutron experiments (Louca and Egami, 1999).

At the same time, there are data which indicate the existence of the metallic ferromagnetic regions below x_c, inside the insulating phase. The presence of such metallic islands should manifest itself in the linear (electronic) term in the heat capacity. Indeed, according to Okuda et al. (1998), the finite value of Sommerfeld's constant at $x < 0.16$ was observed in $La_{1-x}Sr_x MnO_3$.

Interesting results on the $La_{1-x}Zn_x MnO_3$ compound were reported by Felner et al. (2000). The authors measured the magnetization and magnetic susceptibility, and observed an additional ferromagnetic signal at $T = 38$ K for $x = 0.05$ and $x = 0.1$. The samples with $x = 0.05$ and $x = 0.1$ are in the insulating state, but the interesting results described previously were explained by assuming that, in accordance with the percolation picture, they contain the ferromagnetic metallic (FM) clusters. The presence of such clusters leads to an additional signal.

As discussed previously, the transition at $T = T_c$ from the low-conducting high-temperature state to the metallic ferromagnetic state, in turn, can be treated in percolation terms. Therefore, one should expect to observe insulating paramagnetic inclusions below T_c. The phenomenon, indeed, has been directly demonstrated in the STM experiments at the surface of $La_{1-x}Ca_x MnO_3$ ($x \approx 0.3$) compound (Fath et al., 1999). The STM images were taken in magnetic fields between 0 and $9\,T$. The insulating regions were observed at temperatures below bulk $T_c(!)$ With the magnetic field increase, insulating regions shrink and convert into metallic ones (as they should for the double-exchange mechanism), though some insulating regions survive even at fields as high as $9\,T$.

Detailed analysis of the local structure can be performed with use of X-ray adsorption fine-structure spectroscopy (see Li et al., 1995).

According to Booth et al. (1998), above T_c one can observe not only localized states, but also free carriers. These results strongly support the percolation picture.

Other experimental support comes from Mössbauer spectroscopy measurements (Chechersky et al., 1999). The strong paramagnetic signal has been observed at $T > T_c$. Decrease in temperature leads at $T < T_c$ to the appearance of the strong signature of the ferromagnetic state, seen as the "six-peaks" structure in the Mössbauer signal. However, even at $T < T_c$ the paramagnetic signal persists down to $T \cong 20$ K, which is much below T_c.

For some manganites—for example, $Nd_{1-x}Sr_xMnO_3$—one can observe (at $x \gtrsim 0.5$–0.6) transition to the charge-ordered state (see, for example, Mahendiran et al., 1999). It is interesting that the transition between this and the ferromagnetic metallic states is also described by the percolation theory. Here we focus on the lower doping regions because of their strong similarities with the cuprates.

According to the data obtained by Billinge et al. (2000), using the X-ray diffraction technique, the sample of $La_{1-x}Ca_xMnO_3$ crystal above T_c, in the paramagnetic phase, is characterized by three different bond lengths: b_1 for Mn–O, b_2 for Mn^{4+}–O, and b_3 for Mn^{3+}–O, so that $b_2 < b_1 < b_3$. Below T_c, close to $T = 0$ K, there is one bond length only, which corresponds to Mn–O bond. It is interesting that at higher temperature T which is still below T_c one can observe an appearance of two other bonds b_2 and b_3, in accord with the percolation scenario.

All these data (see also the review by Gor'kov and Kresin, 2004) strongly support the approach based on the percolative picture.

12.3.5 Large doping

The 3D metallic ferromagnetic phase of $La_{1-x}Sr_xMnO_3$ persists up to $x \approx 0.5$. Further increase in doping leads to a rather sharp transition (at sufficiently low temperature) to the so-called metallic A-phase. This phase is also metallic as far as low-temperature conductivity is concerned (similar to the FM phase at smaller x), but it has a different magnetic structure: it consists of metallic ferromagnetic layers with the magnetization orientation alternating in the direction perpendicular to the layers.

This metallic A-phase persists up to $x \approx 0.55$. For the compound with such a large concentration it might be more convenient to consider first the opposite end of the phase diagram, $x = 1$. The limit $x = 1$ described the compound $RMnO_3$ (such as $SrMnO_3$). This material is an insulator and contains only Mn^{4+} ions. Such a compound does not contain e_{2g} electrons at all. Starting from this end, one can describe the phase diagram as the result of substitutions A $\rightarrow$ R (such as Sr $\rightarrow$ La substitution); that is, as the electron doped compound. The composition of the sample can be written as $La_ySr_{1-y}MnO_3$. At $y = 0.45$ (it corresponds to $x = 0.55$ in the "hole-doping" picture) there is the transition to the metallic state. Because of the antiferromagnetic ordering is along the c-direction (the c-axis is chosen to be perpendicular to the layers), we start dealing with almost 2D transport (hopping in the c-direction is spin-forbidden due to the Zener double-exchange).

It is rather temping to interpret the metal–insulator transition at $y = 0.45$ as a percolative transition in 2D model. Indeed, as noted in the previous section, the percolation threshold depends on the dimensionality of the system; for the hole-doping (3D case), $x_c = 0.16$. According to the percolation theory, for the 2D case (see Shklovskii and Efros, 1984), $y_c = 0.45$. This value of the invariant is in remarkable agreement with the experimentally observed value $y_c \approx 0.45$ ($x_c^{\text{large}} \approx 0.55$) for the $\text{La}_{1-x}\text{Sr}_x\text{MnO}_3$ compound.

12.4 Main interactions: Hamiltonian

As noted previously, the e$_{2g}$ electron is a key player in the physics of manganites. Its hopping provides both the mechanism of conductivity as well as ferromagnetism (DE mechanism).

One should add also that because two neighboring octahedra share a common oxygen ion (along the Mn–O–Mn bond), the deformations are not independent, and, referring to the Jahn–Teller effect, we always mean a *cooperative* Jahn–Teller effect (see, for example, Kaplan and Vekhter, 1995); that is, the corresponding structural change of a whole lattice.

The total Hamiltonian for the e$_{2g}$ electron has the form:

$$\hat{H} = \sum_i \left(\sum_{i,\delta} \hat{t}_{i,i+\delta,} - J_H \hat{\sigma} \cdot S_i + g \hat{\tau}_i Q_i + J_{\text{el}} Q_i^2 \right) \quad (12.4)$$

The first term describes the hopping process from a site i to its nearest neighbor $i + \delta$. Note that because of double degeneracy of the e$_{2g}$ level $\hat{t}_{i,i+\delta}$ becomes a two-by-two matrix.

The second term represents the Hund interaction. The Hund's coupling is rather strong ($J_H S \cong 1 - 1.5$ eV), and this is the largest energy scale in the theory.

The third term is the Jahn–Teller interaction, manifested in the spontaneous lattice distortion (Fig. 12.6). Q_i are the vibrational modes (Kanamori, 1961), and $\hat{\tau}$ is the pseudospin matrix (Kugel and Khomskii, 1982). The importance of the JT term has been stressed by Millis et al. (1995). The last term accounts for the "elastic" energy part of the local JT mode.

As mentioned previously, in the approach described the largest energy scale in Hamiltonian (12.4) corresponds to the Hund's interaction ($\sim$1–1.5 eV). As for the hopping and the Jahn–Teller terms, they have a competitive strength—the same order of magnitude—so that $t \sim g \sim 0.1$ eV, with $t, g \ll J_H$.

The hopping term in eqn. (12.4) leads to delocalization of the e$_{2g}$ electron and, correspondingly, results in the band picture. A large value of Hund's coupling constant J_H is an important feature of manganites. The double degeneracy of the e$_{2g}$ orbitals plays an important role for manganites' properties. We refer to it as the two-band scenario.

The fact that the band approach can capture the main physics of manganites is far from being obvious. Another approach often uses explicitly the generalized local Hubbard model to account for strong on-site electron–electron interactions. Actually,

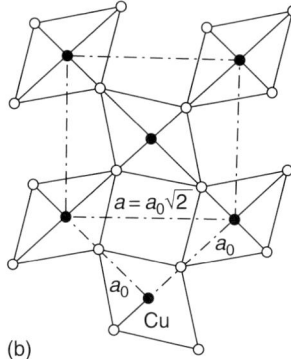

Fig. 12.6 Jahn–Teller deformation of the octahedron: (a) antiferrodistortion, 3D picture; (b) top view.

instead of a direct Hubbard on-site interaction $U > 0$ which hinders the double occupancy of a single Mn site, one may consider the local Jahn–Teller effect as an alternative way to describe the same physics. Indeed, a single electron positioned on the degenerate e_g orbital on the Mn site will cause a local lattice distortion, reducing the energy of the system. On the other hand, *two* electrons on the same site *do not* lead to Jahn–Teller instability, and the Jahn–Teller energy gain does not realize itself. Therefore, locally it is always more favorable energetically to have a single electron on a given site. We should also stress the large values of the Hund's coupling ($J_H \sim 1$–1.5 eV) responsible for the spin alignment of all electrons on a given Mn site.

Based on the Hamiltonian (12.4) one can find the electron energy spectrum, and afterward study numerous electronic properties of the system.

12.5 Ferromagnetic metallic state

12.5.1 Two-band spectrum

Let us determine the single-electron spectrum of the 3D metallic ferromagnetic manganite as it follows from the Hamiltonian (12.4); the results apply at $0.16 < x < 0.5$ (Dzero et al., 2000). We can use the tight binding approximation, assuming that the overlap of electronic wavefunctions for neighboring Mn ions is rather small.

The Bloch electronic wavefunction in the tight-binding approximation may, as usual, be chosen in a form

$$\Psi = \sum_n e^{i\vec{p}\vec{a}_n} \psi(\mathbf{r} - \bar{a}_n) \qquad (12.5)$$

The summation is over all lattice centers, $a_{\bar{n}}$, $\bar{p}$ is the quasimomentum, and ψ denotes the column formed by localized (atomic) wavefunction. This is the only difference that derives from the fact that the state of the e_{2g} electron is the double-degenerate one.

In order to calculate the one-electron spectrum one should choose the local basic set. It is convenient to use the normalized basic set of the form (Gor'kov and Kresin, 1998, 2004)

$$\begin{aligned} \psi_1 &\propto z^2 + \varepsilon x^2 + \varepsilon^2 y^2 \\ \psi_2 &\equiv \psi_1^* \end{aligned} \tag{12.6}$$

where $\varepsilon = \exp(2\pi i/3)$. This choice allows us to account for the cubic symmetry of the initial lattice.

In principle, one can pick up another basis which is often used in the literature: the set of real functions:

$$\begin{aligned} \varphi_1 &\propto d_{z^2} \equiv 3z^2 - (x^2 + y^2) \\ \varphi_2 &\propto d_{z^2-y^2} \equiv x^2 - y^2 \end{aligned} \tag{12.7}$$

There is a simple connection between these two sets:

$$\begin{aligned} \psi_1 &= (\varphi_1 + i\varphi_2)/\sqrt{2} \\ \psi_2 &= \psi_1^* \end{aligned} \tag{12.8}$$

Of course, the expression for the energy spectrum does not depend on the choice of the local basic set.

The equation of motion for an electron determining its energy spectrum is (for the metallic phase the spectrum is determined by the hopping term only; see Anselm, 1982):

$$E \begin{Bmatrix} \psi_1(\boldsymbol{r}_i) \\ \psi_2(\boldsymbol{r}_i) \end{Bmatrix} = \sum_{\boldsymbol{a}_n} \hat{T}_{i,i+m} \begin{Bmatrix} \psi_1(\boldsymbol{r}_i + \boldsymbol{a}_n) \\ \psi_2(\boldsymbol{r}_i + \boldsymbol{a}_n) \end{Bmatrix} \tag{12.9}$$

$$\hat{T}_{ij} = \begin{Bmatrix} T^{11} & T^{12} \\ T^{21} & T^{22} \end{Bmatrix} \tag{12.9'}$$

As a result, one can obtain the following expression for the energy spectrum:

$$E_{1,2}(\boldsymbol{k}) = -|A|(c_x + c_y + c_z \pm R) \tag{12.10}$$

where

$$c_i = \cos(k_i a), i = x, y, z, |A| \cong 0.16\,\text{eV} \tag{12.11}$$

The constant $|A|$ is determined from the experimentally measured parameters (this is the single constant which enters the theory); a is the lattice period, and

$$R = \left(c_x^2 + c_y^2 + c_z^2 - c_x c_y - c_y c_z - c_z c_x\right)^{1/2} \tag{12.12}$$

One can see that the spectrum consists of the two branches. The two-band structure of the electron spectrum is an important feature of metallic manganites. It explains naturally the observed asymmetry of the phase diagram (hole versus electron doping).

Moreover, it is essential for a detailed description of various features of the compounds, and especially, their optical properties (see Section 12.5.4).

As assumed previously, the Hund term is the largest one, so that in what follows we use the strong inequality $J_H \gg |A|$. Therefore, all branches of the electronic spectrum are shifted up or down by the energy $\pm J_H S$, depending on the e$_{2g}$ and t$_{2g}$ mutual spin orientation. Itinerant spins for each of the branches would merely split into two by adding the $\pm J_H \langle S \rangle$ energy term.

Therefore, the lowest two bands correspond to the parallel orientation of the e$_{2g}$ and local t$_{2g}$ spins:

$$E^L_{1,2} = -J_H \langle S \rangle - |A|(c_x + c_y + c_z \pm R) \tag{12.13}$$

c_i and R are defined by eqns. (12.11) and (12.12). There are also two upper bands:

$$E^U_{1,2} = J_H \langle S \rangle - |A|(c_z + c_y + c_z \pm R) \tag{12.14}$$

However, the ground state is always formed by making use of the two lowest bands, $\varepsilon^L_{1,2}$.

With use of eqns. (12.12) and (12.14) one can estimate the total width of the spectrum ΔE:

$$W = 6|A| \tag{12.15}$$

In accordance with eqn. (12.11), ΔE is of the order of 1 eV. In practice, at all concentrations, the Fermi level lies at lower energies. Therefore, in metallic manganites we are dealing with relatively narrow energy bands.

With use as the spectrum (12.10)–(12.12) we can calculate the concentration dependence of the main parameters of manganites (Fig. 12.7).

12.5.2 Heat capacity

As is known, there are several contributions to the low-temperature heat capacity. In ferromagnetic manganites, in addition to the common electronic ($\propto T$) and phonon ($\propto T^3$) terms, there is also a contribution from spin waves (magnons). Their dispersion law ($\omega \propto k^2$) leads to the contribution $C_{\text{mag}} \propto T^{3/2}$. In addition, there is a kink in the density of states at $x \approx 0.3$ (Fig. 12.7). The kink takes its origin from the fact that this concentration is the point of the "Lifshitz" 2.5 singularity (Lifshitz, 1960; see Blanter et al., 1994).

As mentioned, the total magnon contribution into specific heat is proportional to $T^{3/2}$. The proximity to the Lifshitz "2.5" transition results in the appearance of a term in the electronic specific heat term with the same T-dependence. This observation makes the procedure of extracting the "pure" magnon $\propto T^{3/2}$ terms less transparent.

The spectrum (12.13) can be used in order to calculate the density of states per single spin ν and then the usual linear electronic term $C = \gamma T$ with the Sommerfeld constant $\gamma = \nu/3$. With use of the value at $x = 0.3$ (see Fig. 12.6) and the value $|A| \approx 0.16$, we obtain $\gamma \approx 6$ mJ/mole K^2. This value is in a reasonable agreement with the experimental data (see, for example, Viret et al., 1997).

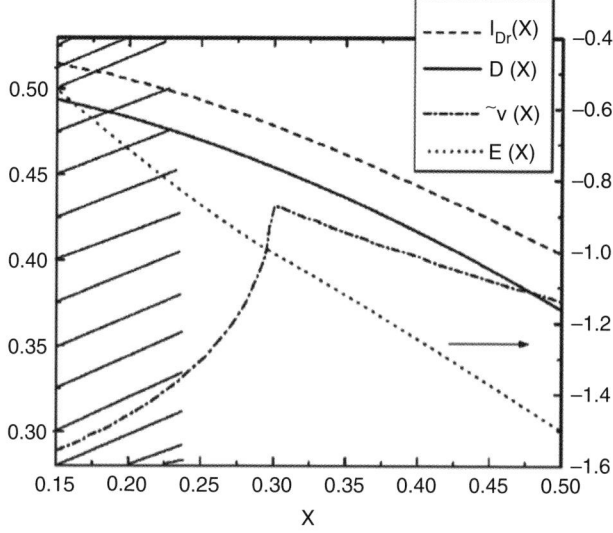

Fig. 12.7 The dimensionless Fermi level (in units of $|A|$), $E(x)$, density of states, $\bar{\nu}(x)$, the spin coefficient, $D(x)$, and the Drude conductivity, $I_{Dr.}(x)$, plotted as a function of concentration, x. The meaning of different lines is indicated in the onset. The shaded area shows the percolative regime. (Figure reproduced from Dzero et al., 2000.)

12.5.3 Isotope substitution

As we know, molecules as well as various complex systems can display the dynamic Jahn–Teller effect. In Section 9.7 this phenomenon was discussed as the origin of a peculiar isotope effect in superconductors. It turns out that manganites also display an isotopic dependence on the critical temperature for the transition into a 3D ferromagnetic metallic state.

It turns out that the oxygen isotope substitution ($O^{16} \to O^{18}$) leads to a large shift in the Curie temperature (Zhao et al., 1996; Franck et al., 1998; Gordon et al., 2001; see also the review by Belova, 2000), $T_c \propto M^{-\alpha}$; for instance, for $La_{0.8}Ca_{0.2}MnO_3$ the isotope coefficient is quite large, and equals $\alpha \approx 0.85$. Similarly to high-T_c oxides, this shift is also caused by the dynamic polaronic effect (Kresin and Wolf, 1998).

The oxygen ion is located between the Mn ions, and its dynamics is directly involved in the process of a charge transfer. Seated just between two Mn ions, the oxygen ion, in addition, is characterized by *two* minima (Fig. 12.8). This picture is similar to that in the high-T_c superconducting oxides (see Chapter 10). The presence of two minima leads to the appearance of symmetric and asymmetric energy terms and corresponding energy levels splitting.

The energy splitting corresponds to inverse time for the charge hopping between the two Mn ions ($\Delta E \propto \tau^{-1}$). Since the ferromagnetic ordering is caused by the electron hopping, the splitting provides for the estimate the value of the critical temperature

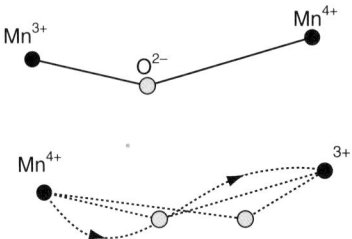

Fig. 12.8 Mn–O–Mn unit; two close minima for the O ion. (Figure reproduced from Sharma et al., 1996. © 1996 by the American Physical Society.)

for the ferromagnetic ordering; in other words, it is natural to assume that $T_c \propto \Delta E$ (Millis et al., 1996).

The expression for the splitting can be written in the form (see eqn. (2.41):

$$\Delta E \propto L_0 F \qquad (12.16)$$

where L_0 is the electronic factor (at $R = R_0$), and F is the Franck–Condon (FC) factor, determined by eqns. (2.42) and (2.43).

The electronic factor L_0 determines the hopping parameter, A (and, therefore, T_c in terms of the usual double-exchange model) with the *frozen* lattice. It can be seen from eqn. (12.16) that the account of the O ion dynamics leads to the appearance of the additional FC factor. The obvious inequality, $F < 1$, leads to a decrease in the energy-splitting. Moreover, the presence of the FC factor leads to the oxygen isotope effect.

Let us add some qualitative remarks. The charge transfer for the extra electron can be visualized as a multistep process; first the electron makes a transition from the Mn^{4+} site to the oxygen, and then the oxygen goes over to another minimum, from where the electron jumps into the other Mn ion. Therefore, actually, the charge transfer includes such an important additional step as the oxygen movement between two minima. This additional step leads to a drastic increase, described by the FC factor, in the characteristic time for the jump between two Mn ions. Naturally, this factor also decreases the strength of the ferromagnetic coupling, and consequently, T_c. In the traditional language, charge transfer is accompanied by the transition to another electronic term. This process is similar to the Landau–Zener effect (see, for example, Landau and Lifshitz, 1977).

As noted previously, the dependence on the mass of the oxygen isotope is brought into the problem through the appearance of the FC factor. Indeed, this factor contains overlap of the nuclear wavefunctions, and therefore depends on the nuclear mass M (in our case the mass of the O ion). The FC factor was estimated in Section 8.7. It can be written in the form $F \sim \exp(-\kappa^2)$, $\kappa = (\Delta \rho / 2\rho)$, where $\rho = (1/M\Omega)^{1/2}$ is the vibration amplitude. If we take the value $\kappa \approx 1.25$–1.5 (consistent with the data by Sharma et al., 1996), we obtain $F \approx 0.18$–0.2. The value of the isotope coefficient appears to be in good agreement with the experimental data. Indeed, defining the isotope coefficient by the relation $T_c \propto M^{-\alpha}$, one rewrites it as $\alpha = -(M/\Delta M)(\Delta T_c/T_c)$, and with $T_c \propto \Delta E$ we obtain

$$\alpha = (M/\Delta M)(\Delta T_c/T_c) = -(M/\Delta M)(\Delta F/F) \quad (12.17)$$

where $M = M - M^*$. We used the relation $T_c \propto \Delta E$ and eqn. (12.16). Based on the expression for the FC factor, we obtain $\alpha = 0.5\kappa^2$. Using the value $\kappa \approx 1.25$–1.5, we obtain the values in the range of 0.8–1, in good agreement with the value obtained by Zhao et al. (1996).

The charge transfer in the metallic manganites is accompanied by the motion of the oxygen ion between two minima: the electron becomes "dressed" by this motion. This dynamic polaronic effect affects the value of the critical temperature.

12.5.4 Optical properties

Study of the optical properties of metallic manganites is of special interest, because it is necessary to take into consideration the two-band structure of the spectrum. For manganites, the interband transitions appear to be the key factor in the infrared region.

The a.c. conductivity is described by the well-known Kubo–Greenwood expression:

$$\begin{aligned}\sigma_{ij} = &-\frac{e^2}{V\omega}\sum_{k,k'} f_0(k)[1 - f_o(k')]\\ &\times <\psi(k)|\hat{v}_i|\psi(k)><\psi(k')|\hat{v}_j|\psi(k')>\\ &\times [\delta(\varepsilon(k') - \varepsilon(k) - \omega) - \delta(\varepsilon(k') - \varepsilon(k) + \omega)]\end{aligned} \quad (12.18)$$

where $f_0(k)$ is the Fermi distribution function, k is a quasi-momentum, $\hat{v}$ is a velocity operator, and $\psi(k)$ is the two-column electron wavefunctions. For the cubic crystal, $\sigma_{ij} = \sigma_{ji}$.

To start with, we determine the velocity operator, $\hat{v} = \dot{r}$ (see Lifshitz and Pitaevsky, 1999):

$$v(k) = \frac{\partial \varepsilon_l(k)}{\partial k} + i[\varepsilon_l(k) - \varepsilon_{l'}(k)] <lk|\hat{\Omega}|l'k> \quad (12.19)$$

The off-diagonal operator $\hat{\Omega}$ is defined by the relation:

$$<lk|\hat{\Omega}|l'k> = i\int u_k^{*l'}(r)\frac{\partial u_k^l}{\partial k}d^3r \quad (12.20)$$

and $u_k^l(\mathbf{r})$, the periodic Bloch functions.

Since l,l' are the band indexes, the second term in eqn. (12.19) includes explicitly the interband transitions. The light absorption comes about due to these contributions. These are the direct transitions, and the importance of these transitions for the light absorption is simply due to a negligibly small value of the photon's momentum. As a result, conservation of energy and momentum can be satisfied simultaneously (in the clean limit) only for the interband transitions.

The matrix element for the interband transition is

$$<l\boldsymbol{k}|\hat{\Omega}|l'\boldsymbol{k}> = ia\frac{\sqrt{3}}{4}\frac{(-\sin k_x)(c_y - c_z)}{|t_{12}|^2} \tag{12.21}$$

As a result, one arrives at the following expression for the "optical" (interband) contribution

$$\sigma_{opt}(\omega, x) = \frac{3\pi e^2}{a\hbar}\frac{1}{\tilde{\omega}^3}\int \frac{d^3p}{(2\pi)^3}\sin^2 p_x (c_y - c_x)^2 \\ \times n(\varepsilon_+(\boldsymbol{p}))[1 - n(\varepsilon_-(\boldsymbol{p}))]\delta(\tilde{\omega} - 2R(\boldsymbol{p})) \tag{12.22}$$

The interband transitions affect the optical conductivity (see, for example, Okimoto and Tokura, 2000) which is also very sensitive to the sample quality.

12.6 Insulating phase

12.6.1 Parent compound

The parent compound AMnO$_3$—such as LaMnO$_3$—is an insulator and antiferromagnet. In addition, its magnetic ordering represents the so-called A-structure with relatively low Néel temperature ($T_N \approx 140$ K). Then the ferromagnetic ordering in layers is combined with an antiferromagnetic order in the direction perpendicular to the layers. It turns out that even the properties of parent manganite can be understood in the framework of the two-band picture (Gor'kov and Kresin, 1998; Dzero et al., 2000). Therefore, the band approach represents a unified description of the low-temperature properties of manganites, applicable to all phases (metallic and insulating).

Let us first make several preliminary remarks. The "right" stoichiometric end ($x = 1$) of the phase diagram for the A$_{1-x}$R$_x$MnO$_3$ compound—for example, for SrMnO$_3$—is also an insulator, which is not surprising. Indeed, in this case all manganese ions are Mn^{4+}, the itinerant e$_{2g}$ electrons are absent, and the ions contain only the strongly bound t$_{2g}$ groups.

The situation with the parent compound, AMnO$_3$ (the "left" side of the phase diagram, $x = 0$) is entirely different, because of presence of loosely bound e$_{2g}$ electrons. These electrons are responsible for metallic conductivity in the doped manganites. Nevertheless, despite the presence of such electrons the parent material behaves as an insulator. One can show that the two-band picture introduced in Section 12.5, but extended to incorporate the complete Hamiltonian—including the Jahn–Teller effect—allows us to understand fully the insulating ground state.

There is one more remark to be made. One might think that this limiting case ($x = 0$) should also be treated, like $x = 1$, in the framework of the localized picture; this would mean that the Mn^{3+} e$_{2g}$ shells should behave as localized e$_{2g}$ orbitals. And, indeed, the pertinent properties of manganites are often interpreted in terms of generalized microscopic Hubbard model (see, for example, Kugel and Khomskii, 1973). The key feature of the Hubbard model is the assertion that for two electrons to be placed on the same site the energy cost is very high (the famous Hubbard $U > 0$ due to the on-site Coulomb repulsion!). The Hubbard Hamiltonian approach has been challenged by Gor'kov and Kresin (1998) (see also Dzero et al., 2000). First, there are experimental motivations for such a challenge. For instance, it was shown experimentally that in doped manganites, La$_{1-x}$Sr$_x$MnO$_3$, at rather low concentrations, say, $x = 0.2$, the system may display excellent metallic behavior at low temperatures (Lofland et al., 1997). Meanwhile, the nominal number of e$_{2g}$ electrons per Mn site, $N = 1 - x < 1$, was changed only by one fifth in this study. In addition, from the theoretical point of view it is, of course, clear that dealing with the strictly atomic d-orbitals would be a strong oversimplification. If an Mn ion is placed in the oxygen octahedron environment, the e$_{2g}$ terms are formed by the whole ligand, so that the "pure" d-functions become considerably hybridized with the surrounding oxygen states (see, for example, the discussion by Anderson, 1959b, and by Pickett and Singh, 1996). Hence the electronic polarization would undoubtedly reduce the magnitude of the "Hubbard"-like (on-site) interactions. Recall also (as was stated in Section 12.4) that the Jahn–Teller instability also makes unfavorable the situation when two electrons occupy the same site. A study of the optical properties of La$_{1-x}$Ca$_x$MnO$_3$ at various x (Lung et al., 1998) demonstrated the presence of the JT splitting. Finally, for the Jahn–Teller effect (which itself is nothing but another form of the Coulomb interaction) to arise there is no need to use the Mn^{3+} localized states picture.

Let us describe the band insulator qualitatively (for a more detailed analysis, see Gor'kov and Kresin, 2004). The parent compound, AMnO$_3$, has magnetically the A-structure; that is, the neighboring layers have an opposite magnetization. As a result, the transport would have a 2D nature, since the hopping in the c-direction is spin-forbidden. Each 2D unit cell contains one delocalized electron. In the usual one-band picture such a system would stay metallic, since at the Pauli band filling the electrons will occupy half of the band. However, the double-exchange mechanism, in cooperation with a large value of J_H, makes all electrons have only one spin direction. Therefore, a single band in a "half-metallic" picture would be fully occupied. However, in manganites, e$_{2g}$ electrons occupy not one but two bands. If we neglect the JT term, the spectrum will be described by eqn. (12.13), and this should lead again to a metallic, not insulating, state. Therefore, the analysis, based on the Hund's (DE) and hopping terms only, is not sufficient, since it would produce the metallic state. Matters become different if one takes into account static deformations caused by the JT interaction. Indeed, as the result, the "superstructure" imposed by the JT deformations, as shown in Fig. 12.16, makes the 2D Brillouin zone now double-folded; that is, reduced by a factor of 2. After a superlattice is imposed at the JT collective transition, the same number of electrons may completely fill up the reduced Brillouin zone, producing an insulator.

12.6.2 Low doping: polarons

The previous section was concerned with the parent compound; that is, with the undoped manganites. The chemical substitution (such as La → Sr) of La^{+3} by the divalent ion leads to formation of a "hole" in the unit cell (see Section 12.2.1). Experimentally, at the low doping level the material remains an insulator. The fact that at small concentration doped manganites preserve the insulating state raises an interesting problem.

The insulating behavior at light doping means that the introduced "holes" remain localized. One obvious reason for this is certainly the Coulomb attraction to the doping ion (Sr^{2+}) that prevents the hole from immediately joining the top of one of the conduction bands. However, even if the Coulomb forces were screened on large distances in the presence of a finite hole concentration, strong electron–phonon interaction originated from the local JT term may change significantly the band characteristic. Local distortions may result in forming self-consistent trapping centers for hole/electron; that is, creating a new type of "carrier"—the so-called "polaron." In the 3D case for a band carrier to become trapped into the polaronic state (usually with a much heavier effective mass), energy is needed to overcome the energy barrier separating the band and the trapped polaronic states. With the polaron density increase, the latter should merge gradually into the band spectrum. The situation is more interesting and less trivial in the case of the 2D electronic spectrum. It is worth saying a few words in this regard, since, as we have just discussed, for the parent $LaMnO_3$ with its A-type magnetic structure, the conducting network of the MnO-planes bears two-dimensional features.

In the 2D, case the energy barrier for polaron formation may be equal to zero. Correspondingly, carriers may be either itinerant or localized (having heavy masses), depending on the numerical value of some parameter, C. This parameter characterizes competition for the energy gain between the gain in elastic energy (see the linear and quadratic terms in eqn. (12.4)), due to the JT distortion g^2/J_{el}, and the kinetic energy gain due to the finite band width which is proportional to t. If the value

$$C \approx g^2/J_{el}t \tag{12.23}$$

exceeds a threshold, usually of order of unity, the doped hole would inevitably go into a trapped state (Rashba, 1982; Toyozawa and Shinozuka, 1983).

In simple terms, eqn. (12.23) tells us whether, due to lattice deformation, the hole energy goes below the bottom of the band, and thus remains "localized." Recall again that for the A-structure the transport has mainly a 2D nature. Since, experimentally, at low doping manganites first remain in insulating state, we conclude that criterion (12.23) favors absence of the potential barrier for localization of introduced holes—or in other words, holes are trapped into "heavy" polarons.

The polaronic picture and the criterion (12.23) makes sense in the limit of low enough carrier concentration only. Increase in doping leads, eventually, to percolation and the phase separation picture described in Section 12.3. Here we focus on the low-temperature region. It is worth nothing, however, that the study performed by combining X-ray and photoemission spectroscopies (Mannella *et al.*, 2004) demonstrated that the state of disorder above T_c is consistent with the presence of the JT distortion and the polaronic picture.

At the percolation threshold, $x_c \cong 0.16$, an itinerant conduction network develops, leading to the transition into the metallic (and ferromagnetic) ground state. As we have shown previously, somewhat above the percolative concentration threshold the formed macroscopic metallic phase can be again described in terms of the band theory (see Section 12.5).

12.7 Metallic A-phase: S–N–S Josephson effect

12.7.1 Magnetic structure

We have discussed the properties of the ferromagnetic metallic phase of manganites $A_{1-x}R_xMnO_3$. Such a phase occupies the doping region $0.16 < x < 0.4$–0.5.

The study of $La_{1-x}Sr_xMnO_3$ (Akimoto et al., 1998; Tokura and Tomioko, 1999; Izumi et al., 2000; Moritomo et al., 2001) has led to a remarkable observation that the larger doping ($0.5 < x < 0.55$) gives rise to the appearance of a metallic antiferromagnetic phase: the A-phase. For the A-phase, the core (t_{2g}) spins are aligned ferromagnetically in each MnO plane (for example, the ab plane), and antiferromagnetically along the axis perpendicular to the planes (c-axis; Fig. 12.5). In the ferromagnetic phase, all electrons, including the e_{2g} electrons, are fully polarized ("half-metallic" state). Now the "half-metallic" state is realized only inside each of the ab planes of the A-phase.

Therefore, the magnetic A-structure combines the ferromagnetic order in the layers with antiferromagnetism in the c-direction. Such magnetic order leads to highly anisotropic transport, because the charge transfer in the c-direction is spin forbidden.

12.7.2 Josephson contact with the A-phase barrier

As is known, the superconducting state is characterized by macroscopic phase coherence. The Josephson effect (Josephson, 1962; see, for example, Barone and Paterno, 1982; Tinkham, 1996) represents the most remarkable manifestation of this coherence. The Josephson contact is comprised of superconducting electrodes separated by a tunneling barrier. The d.c. current flows through the contact without any external voltage, and is described by the equation:

$$J = j_0 \cos \varphi \tag{12.24}$$

where $\varphi = \varphi_1 - \varphi_2$ is the phase difference of the superconducting order parameters on each side of the barriers. The most common case is the S–I–S Josephson junction, where I denotes a thin insulating barrier. Another well-known case is the S–N–S junction; here, N is a normal metal. The amplitude of the superconducting current for such a junction (see, for example, Barone and Paterno, 1982; Kresin, 1986) is proportional to

$$j_0 \propto \exp(-L_N/\xi_N) \tag{12.25}$$

if $L_N > \xi_N$. Here, ξ_N is the normal coherence length, defined as $\xi_N = v_F/2\pi T$ ("clean" case; Clarke, 1969; see Section 8.1). One can see from eqn. (12.25) that for the effect to be observable the thickness of the barrier L_N should not noticeably exceed the coherence length ξ_N.

Let us consider the barrier formed by a magnetic metal. It is well understood that for two superconductors with singlet (s-wave or d-wave) pairing a ferromagnetic barrier (S–F–S junction) would present a strong obstacle. Indeed, the Josephson current is a transfer of a Cooper pair with its spins of the two electrons in opposite directions.

The exchange field in the ferromagnetic is trying to align the spins in the same direction, and this leads to the pair-breaking effect. The situation is entirely different for the tunneling through antiferromagnetic metallic (S–AFM–S) system (Gor'kov and Kresin, 2001, 2002). Contrary to the pair-breaking F case, superconducting currents might penetrate through an AFM barrier much easier.

We consider the junction in such geometry, that the Josephson current would flow along ferromagnetic layers. The layers are weakly coupled electronically. It is important, because the Josephson current is a transfer of *correlated* electrons. The manganite which is in the metallic A-phase form such a barrier. The A-phase manganite is a natural spin-valve system (see, for example, Kawano et al., 1997).

To demonstrate the effect we assume that the barrier is thick enough to neglect phenomena taking place in the immediate proximity of the boundary. Then one can use the interface Hamiltonian in the form:

$$H = V\Delta^{(i)}\psi^+(i)\psi(i) \qquad (12.26)$$

Here, $\Delta^{(i)}$ are superconducting order parameters on each side, i, ψ^+, ψ are the field operators for the carriers inside the barrier, and V is a tunneling matrix element at the boundary of the barrier. Integration along the contact surface is assumed.

As a next step it is practical to evaluate not the current itself directly, but to find, instead, the "surface" contribution to the thermodynamic potential, $\delta\Omega$, caused by the presence of the barrier separating two bulk superconductors. The current is then determined as a derivative $\delta\Omega/\delta\varphi$, where $\varphi = \varphi_1 - \varphi_2$ is the phase difference between the two superconductors, so that $\Delta^{(1)} = |\Delta^{(1)}|\exp(i\varphi_1)$, $\Delta^{(2)} = |\Delta^{(2)}|\exp(i\varphi_2)$. The amplitude of the Josephson current, j_m, turns out to be proportional to the matrix element of the Cooper diagram:

$$K(1,2) = |V|^2 \sum_{\omega_n} \Delta^{(1)}_{\sigma\sigma'} G_{\sigma\sigma''}(1,2;\omega_n) G_{\sigma'\sigma'''}(1,2;-\omega_n) \Delta^{*(2)}_{\sigma''\sigma'''} \qquad (12.27)$$

For a singlet superconductor, $\hat{\Delta}$ is the matrix of the form: $\Delta^{(i)}_{\sigma\sigma'} = \Delta^{(i)}(i\sigma_y)_{\sigma\sigma'}$. A method of thermodynamic Green's function is employed, and a summation over repeating spin indices is assumed. To properly evaluate the current through the magnetic barrier, one should pay special attention to the spin structure of eqn. (12.27), because the Green functions, $G_{\sigma\sigma''}$ and $G_{\sigma'\sigma'''}$, inside the barrier are not diagonal in spin indices. The calculation leads to the following general expression for the amplitude of the Josephson current:

$$j_m = r\gamma\pi T \sum_{\omega_n > 0} \int dp_z \exp(-\omega_n L/v_F) \exp(iLt_{||}M/S v_\perp)$$
$$\gamma = (1 - M/S)^2; \quad r \propto |V|^2 |\Delta_1|^2 |\Delta_2|^2 \qquad (12.28)$$

Assume that the width of the barrier $L \gg \xi_N$, where ξ_N is the coherence length for the normal layer, eqn. (12.25). Then one can keep only the first term of sum in eqn. (12.28). In the tight-binding approximation, $t_{(||)}(p_z) = t_0 \cos(p_z d)$ (d is the interlayer distance). Integrating over p_z, we arrive at the following expression:

$$j_m = j_m^0 \, e^{-\frac{L}{\xi_N(T)}} J_0(\beta M/S) \qquad (12.29)$$

Here, $\xi_N(T)$ is the coherence length inside the barrier, $j_m^0 = [1 - (M/S)]^2 r$, $\beta = (t_0/T_c)(L/\xi_0)$, $\xi_0 = \hbar v_F/2\pi T_c$, and $v_F = v_F^0$ is the maximum value of the component of the Fermi velocity along L. It is essential that $\beta \gg 1$; indeed, for manganites, $t_0 \gg T_c$ and $L \gg \xi_0$. If the canting M/S is not negligibly small, we can use an asymptotic form of the Bessel function, $J_0(x)$, and we obtain

$$j_m \approx (\pi \beta M/2S)^{-1/2} j_m^0 e^{-\frac{L}{\xi_N(T)}} \cos\left(\beta M/S - \pi/4\right) \qquad (12.30)$$

Equation (12.30) is valid if $L \gg \xi_N(T)$.

Therefore, the antiferromagnetic barrier does transfer the Josephson current, and in this case the exchange field does not break the Cooper pairs.

It is also very interesting that the mutual magnetic orientations of layers can be controlled by an external magnetic field. Thus, the AFM structure can be transformed into the ferromagnetic configuration (AFM -> FM). The complete AFM -> FM reorientation would result in a drastic impact, forbidding the Josephson current for the FM case ($j_m = 0$). This follows from eqn. (12.29); indeed, then $M = S$ and $j_m = 0$. Therefore, one can change the current by an external magnetic field.

It is very interesting that according to eqn. (12.30) one observes the oscillatory dependence of the amplitude of the Josephson current on an external magnetic field: $j_0 \propto \cos(\beta M/S - \pi/4)$. Canted moments may be induced by even a weak field, and would result at $\beta \gg 1$ in rapid and large ("giant") changes in the Josephson current ("giant" magneto-oscillations for the Josephson contact with the presence of A-structure as a weak link). Note that the oscillating dependence was theoretically obtained by Buzdin et al. (1982) for the ferromagnetic barrier; he considered the oscillations to be caused by *change in the barrier* thickness. As for the antiferromagnetic barrier, one can observe this effect as due to the slight variation in the external magnetic field.

12.8 Discussion: manganites versus cuprates

Manganites, similarly to the cuprates, are complex oxides—mixed-valence compounds. Both systems are characterized by rich phase diagrams, and the state is affected greatly by doping.

It has been demonstrated that in manganites an insulator–metal transition to the ferromagnetic and metallic state at finite doping represents a peculiar type of the transition that must be described in percolation terms (Gor'kov and Kresin, 1998; review, 2004). The percolative approach is also applicable for the transition at the Curie temperature. A similar picture can be observed for the cuprates (see Chapter 10). The systems are intrinsically inhomogeneous, and we are dealing with phase separation.

As demonstrated previously, many results could be obtained analytically. It is clear that such an approach has a serious advantage, not only because it allows one to gain an additional insight, but also because the calculations are tractable, whereas the numerical results are sometimes contradictory.

A sensible quantitative theory of manganites that explains self-consistently the main peculiarities in manganites can be built by using a relatively simple Hamiltonian (12.4) which contains, as the key ingredients, the hopping term, and the Hund's and

Jahn–Teller interactions. The band approach (the generalized two-band model), utilizing all these interactions, provides an adequate and unifying approach. Note that the corresponding two-gap scenario is essential for such a cuprate as YBCO compound (see Chapter 10).

Both systems (cuprates and manganites) are greatly affected by the Jahn–Teller instability and polaronic effects. The oxygen ions display dynamic Jahn–Teller effects, and this leads to peculiar isotopic dependences.

Manganites and cuprates share many similarities. Thus, for example, it was not by accident that de Gennes, who had made important contributions to both fields—superconductivity and manganites (de Gennes, 1960, 1966)—made an effort to explain the origin of high-T_c superconductivity in the cuprates by utilizing their similarity with manganites (Plevert, 2011). As noted in Section 12.2.3, the situation is opposite to the manganite case. There, the unusual transition into the ferromagnetic and metallic state in manganites was explained shortly after the its discovery (Jonner and van Santen, 1950; Zener, 1951). With the cuprates, on the other hand, the mechanism of high T_c is still a controversial, issue and a general consensus has not yet been achieved.

13
Novel superconducting systems

In this chapter, several novel superconducting compounds and superconducting systems are discussed. These range from the newly discovered Fe-based pnictides and chalcogenide superconductors to magnesium diboride, the A-15 superconductors, granular superconductors, ruthenate superconductors, and nitride superconductors.

13.1   Fe-based pnictide and chalcogenide superconductors

In 2006, superconductivity was discovered in an Fe-based layered compound: LaFePO (Kamihara et al., 2006). Although the T_c was only 4 K, the fact that an iron-based compound was superconducting at all was surprising, since conventional wisdom suggested that superconductivity in compounds with ferromagnetic transition metals are extremely rare, and that long-range magnetism and superconductivity are typically incompatible. Hosono's group, who made this discovery, were not searching for new superconductors, but in fact were looking for new magnetic semiconductors! Even more surprising was the fact that about a year later the T_c of this compound was raised to 26 K by replacing some of the oxygen with fluorine (Kamihara et al., 2008). These results are shown in Fig. 13.1 (Hosono, 2012). Over the next few years a host of new Fe-based superconductors were discovered (Hosono, 2012), with various crystal structures ranging from the simple two-element FeCh (where Ch = Se, Te) compound that has been labeled 11 (T_c max 37 K under pressure) to the $Sr_2VFeAsO_3$ compound (21113) with an ambient pressure T_c also of 37 K. Currently, the record T_c in this class of Fe-based materials is 55 K for both the fluorine-doped $SmFeAsO_{1-x}F_x$ compound (1111) or the oxygen-deficient $Sm(Nd)FeAsO_{1-x}$ (also 1111).

There are many similarities between the Fe-based superconductors and the cuprates and also a number of significant differences. The Fe-based compounds are layered structures consisting of Fe planes surrounded by a near tetrahedron of either arsenic, phosphorus, Se, or Te ions (As-based compounds have significantly higher T_c at ambient pressure). The Fe layers in all the compounds form a square array of just Fe. The cuprates are also highly two-dimensional structures, but the planes consist of copper and oxygen ions with a layer of oxygen between the Cu–O planes and the doping layer(s). Whereas oxygen is a crucial element in the cuprates, oxygen is not required at all for superconductivity in the Fe compounds, and there are many without oxygen—some of them (such as $Ba_{1-x}K_xFe_2As_2$, 122 structure) with T_c up to 38 K. Similar to the cuprates, these compounds achieve superconductivity by doping, and the T_c versus doping profile is quite similar to the cuprates, as it resembles an inverted dome, as shown in Fig. 13.2 for the 1111 and 211 compounds. (Hosono,

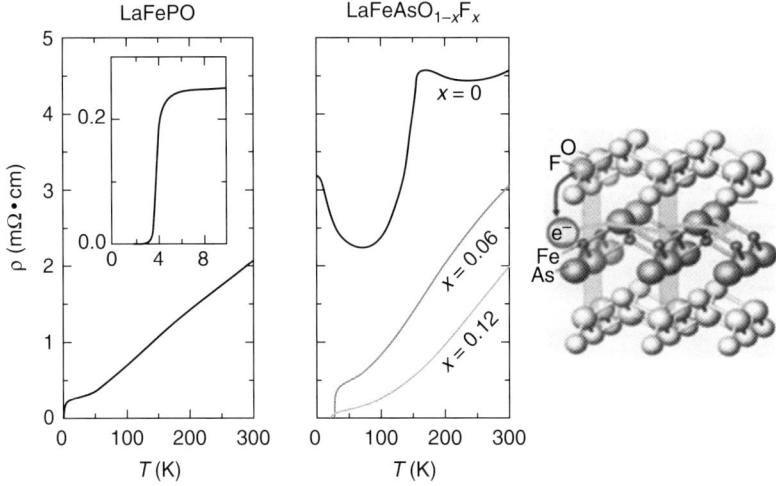

Fig. 13.1 Temperature dependence of resistivity in LaFePO and LaFeAsO$_{1-x}$F$_x$. Right: electron doping to FeAs layers by replacement of oxygen with fluorine. (Figure reproduced from Hosono, 2012.)

2012). On the other hand, the parent compound in the Fe-based system is an orthorhombic structure antiferromagnetic metal, and the parent compound in the cuprates is an antiferromagnetic insulator. However, there is an important and very distinct feature of the Fe compounds related to the crystal structure. The superconducting state is always tetragonal, whereas the normal state at temperatures above the

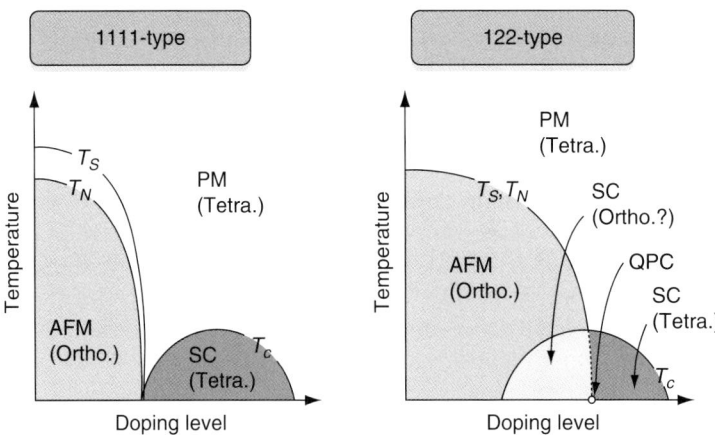

Fig. 13.2 Phase diagrams for 1111 and 122 iron pnictide superconductors. T_s: crystallographic transition temperature; T_n: magnetic transition (Néel) temperature; QCP: quantum critical point (phase transition at 0 K). (Figure reproduced from Hosono, 2012.)

transition temperature can be either tetragonal (for the 1111 compounds, for example) or doping-level dependent (for the 122 compounds, for example), either orthorhombic below the peak in T_c or tetragonal above the peak. For the underdoped samples that are orthorhombic in the normal state, the transition to the superconducting state is accompanied by the structural orthorhombic to tetragonal transition and the loss of the SDW at T_c (see Fig. 13.2). As regards the symmetry of the order parameter, the cuprates have exhibited d-wave symmetry for at least optimally doped YBCO, but the symmetry of the order parameter in the Fe-based materials seems to be s-wave. One additional feature of the Fe compounds is their transport anisotropies (parallel and perpendicular to the Fe planes). It is around 2–4 for the 122 compounds, compared to at least 5 for YBCO that is the least asymmetric of the cuprates.

A common feature of these compounds is the contribution of two bands to the transport and also to the superconductivity. For example, for the 1111 LaFeAsO compound, both the Fe 3d band and the As 4d band contribute to the carrier density at the Fermi level. A strong correlation exists between the height of the pnictigen (chalcogen) from the Fe plane and electronic structure and T_c. The larger this distance, then the higher the T_c. This correlation can be understood by ARPES measurements that correlate the density of states at E_F with this distance. Also, ARPES measurements have observed superconducting gaps that are non-nodal for both bands, indicating that the symmetry of the order parameter is s-wave or extended s-wave for both bands. Two-band or two-gap superconductivity is treated in significant detail in Chapter 7. In another similarity with the cuprates, a pseudogap state has been observed above the superconducting transition temperature for these compounds, and is most likely a result of the inhomogeneous nature of these compounds which require doping for their superconductivity. (See Chapter 11, which describes the pseudogap state in the cuprates in great detail.)

The question about the mechanism of superconductivity in this class of compounds is still open. A large isotope exponent of around 0.4 for the substitution of Fe^{56} for Fe^{54} implies that the lattice plays a significant role in the superconductivity. However, in the 122 compounds (see Fig. 13.2) there is a significant part of the phase diagram where the compound has a spin density wave still observable at the superconducting transition (but not below), and the presence of this spin density wave (SDW) has been observed by neutron scattering experiments (Ewings et al., 2008). The SDW transition temperature for these compounds is also affected by Fe isotope substitution. Various theoretical models have been proposed, including coupling to spin fluctuations (Cao et al., 2008; Dai et al., 2008; Ma et al., 2008) or electronic fluctuations (Onari et al., 2010) as the mechanism for the high T_c.

13.2 Magnesium diboride: MgB$_2$

In 2001, superconductivity was discovered in MgB$_2$ by Nagamatsu's group (Nagamatsu et al., 2001), at 39 K—the highest T_c for an intermetallic compound. This was indeed a remarkable discovery, since this was a well-known material whose transport and magnetization had not been investigated earlier at low temperatures. Also interesting was the fact that because of incorrect conventional wisdom surrounding the use of the McMillan expression for T_c valid only in the weak to intermediate coupling regime, many believed that the maximum T_c for a BCS superconductor was capped at 30 K, so finding a simple binary metallic superconductor with such a high T_c created significant interest in the scientific community, though part of the motivation was not correct! In fact, as clearly shown in Chapter 3, the phonon mechanism can provide transition temperatures above room temperature. The sufficient electron–phonon coupling is provided by the boron in-plane stretching (e_{2g} phonon mode). The structure is quite simple, as shown in Fig. 13.3, with alternating layers of boron and magnesium.

A key feature of this compound is the presence of two bands (σ and π bands; Liu et al., 2001; Choi et al., 2002) and, consequently, two rather different gaps in the excitation spectrum. Although two gaps are not unique to this compound (YBCO also has two gaps—one for the planes and one for the chains), it is still rather unique for an intermetallic compound. ARPES measurements clearly show these two bands (p and s), as reported by (Uchiyama et al., 2002). The symmetry of the bands was shown to be s-wave (Tsuda et al., 2003). The temperature dependence (Fig. 13.4; Tsuda et al., 2003) shows that both gaps close at the same temperature, as expected from the two-gap theory (see Chapter 7). The heat capacity measurements also presented clear evidence for the two-gap superconductivity (Bouquet et al., 2001).

Although these materials have not yet become significant technologically, there are expectations that because of the relatively high T_c, MgB_2 will be able to compete with the A-15s as materials for large and very-high-field magnets.

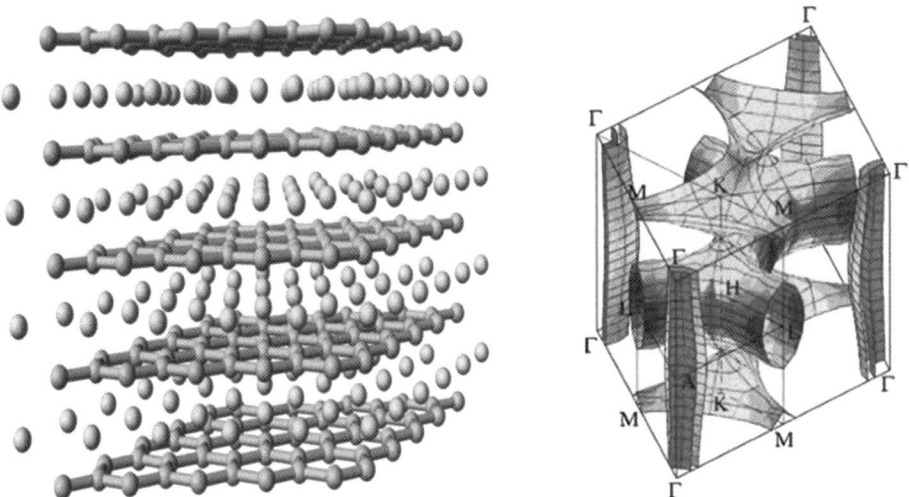

Fig. 13.3 At left is the crystal structure of magnesium diboride: isolated spheres, magnesium; connected spheres, boron atoms. At right is the Fermi surface, clearly showing the two bands.

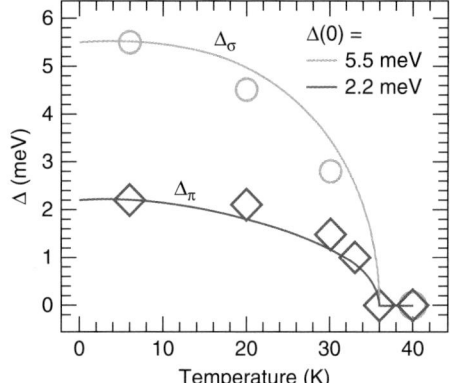

Fig. 13.4 Temperature dependence of superconducting gaps. Open circles and diamonds show the gap values on the σ and π bands. (Figure reproduced from Tsuda et al., 2003. © 2003 by the American Physical Society.)

13.3 A-15 structure superconductors

The A-15 structure materials, also known as the beta-tungsten structure, are intermetallic compounds with the formula A_3B where A is a transition metal and B can be almost any other element. The crystal structure is shown is Fig. 13.5. These materials became of significant interest when superconductivity at 17 K was discovered in V_3Si in 1953 (Hardy and Hulm, 1953). Until the discovery of the cuprates by Bednorz and

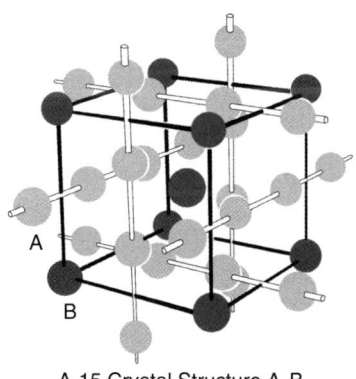

A-15 Crystal Structure A_3B

Fig. 13.5 The crystal structure of the A-15 materials; the grey sphere is the transition metal (Nb, V, and so on), and the black sphere is the other element, such as Ga, Al, Sn, Si, Ge, and so on.

Mueller (1986), this class of compounds held the record for the highest T_c: 23 K for Nb_3Ge. Despite their relatively low superconducting transition temperature compared to the cuprates, A-15 materials, and in particular Nb_3Sn, are of significant technological interest, because they can be made into robust conductors with very large critical current densities and very high critical magnetic fields. They have widespread use for the windings of superconducting magnets for magnetic resonance imaging applications, as well as accelerator magnets for the Large Hadron Collider (LHC)—the largest high-energy physics experiment in the world.

According to Labbe, Barisic, and Friedel (1967), the high T_c in these compounds is due to strong electron–phonon coupling to a very narrow d-band that has a large density of states at the Fermi level. The calculated width of this band is small compared to the width of the phonon spectrum. Thus, all these states should contribute to the superconductivity, and the frequency and width of the phonon spectrum should not be as important for the T_c as it is for conventional BCS superconductors in which the width of the electronic spectrum contributing to the pairing is set by Ω_D. This also explains the very small isotope coefficient in these high-T_c compounds.

Some of these compounds are also rather unique in that there is a structural transition from cubic to tetragonal called the martensitic transition, which occurs a few degrees above the superconducting transition. In the cubic high-temperature phase there is a stiff one-dimensional chain of transition-metal ions that creates the very large density of states at the Fermi surface due to a van Hove singularity created by this low-dimensional structure. The free energy of this structure can be lowered by a decrease in the symmetry via a cubic-to-tetragonal transition and a dimerization of the chains. The density of states at the Fermi surface is lowered, but the lattice is softened. The overall effect on T_c, however, has been shown to be relatively small. This scenario was proposed by Labbe, Barisic, and Friedel and published in a series of articles from the late 1960s through the 1970s, all of which are cited in a review by Bok and Bouvier (2012). They have applied a very similar methodology for explaining the properties of the cuprates.

13.4 Granular superconductors

Granular superconductors are by their very nature inhomogeneous materials that in many respects behave in unexpected ways. Typically granular superconductors consist of small grains with sizes comparable or smaller than the coherence length. These grains are often in a matrix that is normal or insulating, but the grains themselves are close enough that they couple by either S–N–S or S–I–S coupling. Depending on the coupling strength and the temperature, these superconducting islands can behave like zero-dimensional superconducting particles above T_c with strong fluctuations (Wolf and Lowrey, 1977; Deutscher et al., 1973), or as arrays of junctions whose superconducting properties mimic a single very-high-quality junction despite there being many junctions in the granular array (Wolf et al., 1979).

The inhomogeneous nature of granular "conventional" superconductors allows them to serve as a model system for the behavior of superconductors that become superconducting as a consequence of doping. This includes the cuprates as well as the

Fe-based superconductors and the superconducting semiconductors (such as Nb-doped SrTiO$_3$). For a random granular superconductor that contains grains of different sizes and varying coupling strengths, the transition from the normal state to the superconducting state is not sharp but rather broad, reflecting the inhomogeneities inherent in the system. As a granular sample (particularly one that contains S–I–S coupling) is cooled from high temperature, the highest T_c grains (they may not be the largest) begin to superconduct and a superconducting gap opens up on these grains, though only locally. At this point the resistance is still dominated by the intergranular resistance, and the overall resistivity does not show any sharp significant feature. As the temperature is lowered further, more and more grains become superconducting, and neighboring superconducting grains can couple and form small networks. At this point a small Meissner effect can be observed, since there can be a shielding current over finite regions of the sample, though long-range superconductivity is still not present. As the temperature is lowered further, superconductivity percolates and the resistance drops to zero, well below the superconducting transition of the highest T_c grains. This scenario has been observed clearly in NbN grains embedded in a BN matrix (Simon et al., 1987). Similar results were obtained by Merchant et al. (2001), using grains of Pb in an Ag matrix where, for a range of grain-to-grain coupling strengths, they clearly observed a pseudogap region and then a transition to a fully superconducting state. They pointed out the very distinct analogy with the pseudogap region above the resistive transition in the cuprates.

Another very important aspect of granular materials is the enhancement of T_c that occurs for some films. A good example of this effect is for Al films; it was observed that the critical temperature was $T_c \approx 2.1\,\text{K}$ (Shukhareva, 1963; Strongin et al., 1965). An increase was also observed for granular systems. Although bulk aluminum has a transition temperature of 1.2 K, granular Al that consists of Al grains embedded in an Al oxide matrix can have a transition temperature as high as 3.1 K (Deutscher et al.,1973). The origin of this enhancement is not fully understood, but the leading candidate for it is due to size quantization effects that lead to a discreet energy spectrum rather than a continuous spectrum (Kresin and Tavger, 1966; Parmenter, 1968; Shanenko et al., 2006). This explanation is supported by spectroscopic and T_c measurements (Alekseevskii and Vedeneev, 1967; Guo et al., 2004). Another possible explanation has been proposed (Deutscher, 2005, 2012) which relates the enhancement to the interplay of correlation effects and Kondo physics which occurs in the vicinity of the metal–insulator transition; that is, as the films become more resistive. Note that in principle such an increase could be explained by the change in the phonon spectrum, but this explanation has been ruled out by heat-capacity measurements of Al nanoparticles (Hock et al., 2011). It has been demonstrated that the value of the Debye frequency for these nanoparticles is similar to that for bulk Al.

13.5 Sr$_2$RuO$_4$: a very novel superconductor

The ruthenates form an extremely interesting class of layered compounds that exhibit very unusual behaviors, including ferromagnetism in SrRuO$_3$ and quantum metamagnetism in Sr$_3$Ru$_2$O$_7$. However, the most interesting may be the Sr$_2$RuO$_4$ compound

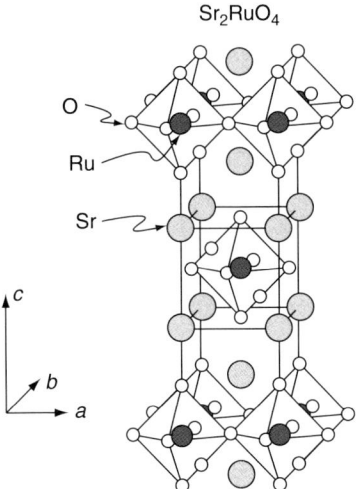

Fig. 13.6 The crystal structure of Sr_2RuO_4, showing the ruthenium–oxygen planes. (Reproduced from Maeno et al., 1994, by permission from Macmillan Publishers, © 1994.)

that is isostructural with LaSr CuO_4—meaning it is a layered perovskite, and is superconducting without the presence of Cu. The structure of this compound is shown in Fig. 13.6, illustrating the layering of the ruthenium–oxygen planes. Its superconductivity was discovered in 1994 (Maeno et al., (1994), with transition under 1 K, and to date the highest transition temperature has been observed in high-quality single crystals (Mackenzie et al., 1998). The most unusual property of this compound is the symmetry of the order parameter. Most conventional superconductors have s-wave symmetry, which means that the Cooper pairs have oppositely directed spins. Some of the cuprates, especially near optimum doping, have d-wave symmetry, which means that the sign of the order parameter changes in different regions of k space, but still the Cooper pairs have oppositely directed spins. Sr_2RuO_4 has odd-parity p-wave symmetry confirmed by muon spin relaxation (μSR) measurements (Luke et.al., 1998) on very high quality single crystals. The weak magnetism and the uniformity of the magnetism and the breaking of time reversal symmetry are all aspects of triplet pairing and all of these effects were confirmed by the μSR measurements.

13.6 Ruthenium cuprates

There are a few compounds that have mixed layers reminiscent of both the cuprates (see Chapter 10) and the ruthenates (above) that exhibit the remarkable phenomenon of coexistence of superconductivity and magnetism. This was first demonstrated by Felner et al. (1997). It turns out that the compound $Gd_{1.4}Ce_{0.6}RuSr_2Cu_2O_{10-\delta}$ is superconducting ($T_c \approx 42$ K), and also displays magnetic ordering at $T_N \approx 180$ K. This effect can also be observed for Eu substituting for Gd.

Magnetism is caused by the presence of the Ru sublattice, whereas superconductivity is confined to the CuO$_2$ planes, as in the cuprates. It is important to note that we are not dealing with phase separation, but with a single phase combining superconductivity and magnetism. The effect of such coexistence has been also observed for the RuSr2GdCu2O8 compound, by Chu et al. (2000).

13.7 Intercalated nitrides: self-supported superconductivity

Transition-metal intercalated nitrides form an interesting family of superconducting materials. The value of the critical temperature could be relatively high. For example, the compound Li$_x$(THF)$_y$HfNCl, discovered by Yamanaka et al. (1998), has $T_c \approx$ 26 K. Here, THF is the C$_4$H$_8$O molecule: tetrahydrofuran. The compound has a layer structure, and the intercalation of the Li and THF into the mother compound β–HfNCl (Fig. 13.6) leads to an increase in the distance between the Cl layers up to 18.7 Å. The intercalated compound shows superconductivity with the above-mentioned high value of T_c.

According to experimental (Tou et al., 2001, 2003) and theoretical (Weht et al., 1999; Heid and Bohnen, 2005) studies, the electron–phonon interaction is not sufficient to provide the observed value of T_c. Indeed, the heat-capacity experiments allow us to measure the renormalized value of the Sommerfeld constant $\gamma \propto (1 + \lambda)$; see Section 2.4. According to the analysis, $\lambda \simeq 0.25$, and this leads to the value of $T_c < 1$ K, which is much below the observed value ($\sim$25 K). Note also that there is no evidence for the presence of strong correlations, and the material can be described by the Fermi-liquid theory. Moreover, the absence of magnetic ions excludes magnetic interactions as the mechanism of pairing.

On can show (Bill et al., 2003) that the superconducting state in the layered metallonitrides is caused by an electronic mechanism, and more specifically, by exchange of low-frequency ("acoustic") plasmons (see Section 4.4.2). For the Li$_{0.45}$(THF)$_{0.3}$HfNCl compound one can use the following set of parameters obtained from experimental data and band-structure calculations: the interlayer distance $L = 18.7$ Å, $\left(\frac{m^*}{m_e}\right) \simeq 0.6$, $\mu^* \simeq 0.1$, $\varepsilon_M = 1.8$. (The large value of ε_M leads to an applicability of the RPA approximation; see eqn. (4.6)). As a result, one can obtain the value of $T_c \simeq 24.5$, which is close to the value observed experimentally.

It is remarkable that the superconducting state in these layered materials is provided by collective modes ("acoustic" plasmons) of the same electrons as those undergoing the superconducting transitions. This is the first system observed with self-supported superconductivity.

14
Organic superconductivity

14.1 History

Organic superconductors form a relatively young family of superconducting materials. The phenomenon of organic superconductivity was observed for the first time in 1980 (Jerome et al.), and the complex material $(TMTSF)_2PF_6$ (bis-tetramethyl-tetraselenafulvalene-hexfluorophosphate) displayed this feature ($T_c \simeq 0.9\,\text{K}$). Since the first observation, the field of organic superconductivity has grown very noticeably (see Fig. 14.1, and the reviews by Ishiguro, 1998; Jerome, 2012; Jerome and Greuzet, 1987; see also Kresin and Little, 1990).

The concept of organic superconductivity was introduced by Little (1964), who developed the model of organic polymer (see Chapter 4); this prediction was confirmed by the latest discoveries. The organic superconductors, similarly to oxides, including

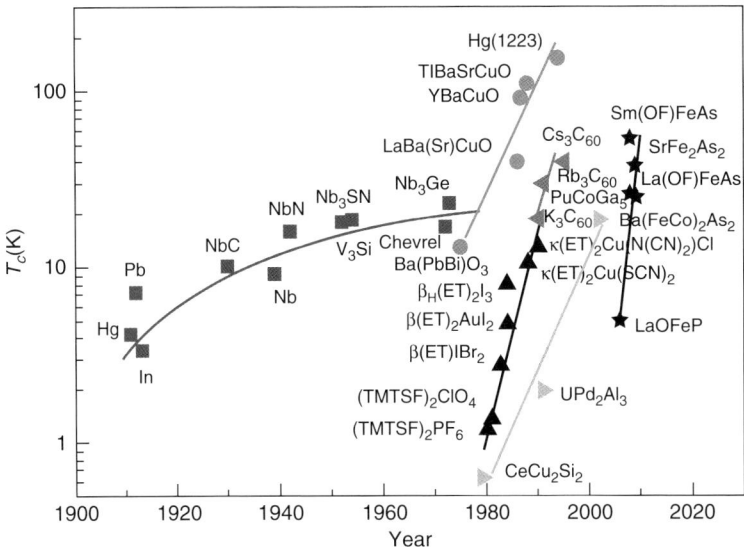

Fig. 14.1 Evolution of T_c in various series of superconductors over the years. (Figure reproduced from Jerome, 2012, with kind permission from Springer Science and Business Media.)

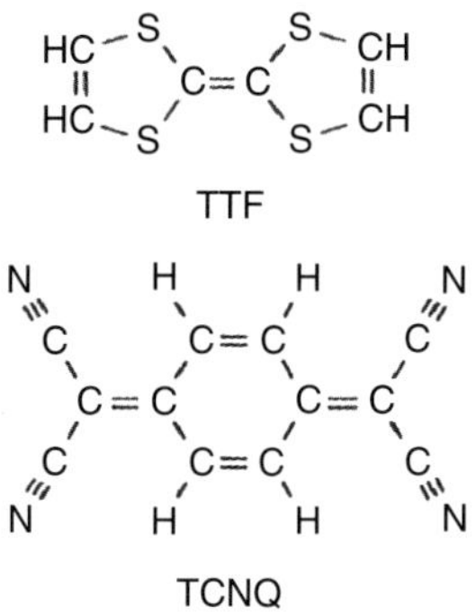

Fig. 14.2 TTF and TCNQ molecules. (Figure reproduced from Ishiguro et al., 1998.)

the high-T_c oxides, and pnictides, are complex systems. The synthesis is playing a key role, and as a result, this field is on the overlap of solid-state physics, materials science, and organic chemistry.

Progress in the area of organic superconductivity was based on previous developments, leading to the synthesis of organic conductors. The point is that the usual bulk organic materials are insulators. The possibility of organic metallicity was studied by London (1937), who analyzed aromatic organic molecules. These molecules (see Section 14.6) contain delocalized electrons (so-called π-electrons), and this feature (delocalization) makes these electrons similar to conduction electrons in metals. It is worth noticing that an analysis of these molecules had resulted in a formulation of the famous London equations, which are a key ingredient in the physics of superconductivity.

The observation of high conductivity in the perylene bromide complex (Akamutsu et al., 1954) has led to the development of the field of organic metals. The next important step was the creation of a stable organic metal formed by the TTF-TCNQ units (Ferraris et al., 1973; Coleman et al., 1973). The unit is a combination of two flat molecules (Fig. 14.2).

All this development was a precursor for the discovery of organic superconductivity. As mentioned previously, the first such superconductor was created in 1980 (Jerome et al.) with $T_c \simeq 0.9\,\text{K}$ under the pressure of around 9 kbar. Since then, many new compounds with higher values of T_c have been synthesized.

An important addition to the family of organic superconductors has been formed by doped fullerines (fullerides) (Haddon et al., 1991; Hebard et al., 1991; Rosseinsky et al., 1991; Holczer et al., 1991). The fullerides belong to this family, because the conductivity is also provided by the presence of carbon atoms, and, correspondingly, by π-electrons.

14.2 Organic superconductors: structure, properties

Let us discuss various types of organic superconductors. At first, one should describe the $(\text{TMTSF})_2\text{X}$ systems (so-called Bechgaard salts; Bechgaard et al., 1979), where X denotes an electron acceptor molecule, such as PF_6, AsF_6, ClO_4, ReO_4, and so on.

208 Organic superconductivity

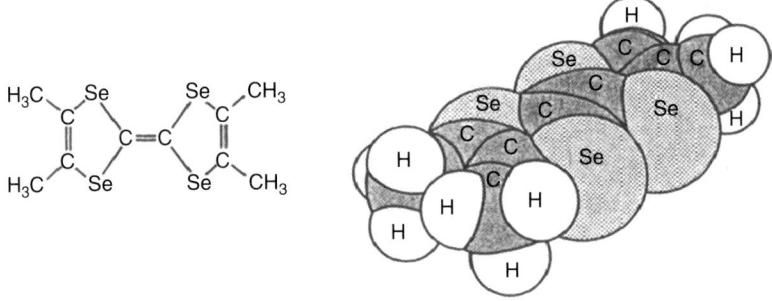

Fig. 14.3 Structure of TMTSF molecule. (Figure reproduced from Ishiguro et al., 1998.)

As mentioned in Chapter 1, superconductivity of the $(TMTSF)_2X$ system with $X \equiv PF_6$ was the first observation of organic superconductivity (Jerome et al., 1980). The structure of the TMTSF molecule can be seen in Fig. 14.3.

The crystal formed by $(TMTSF)_2X$ units is highly anisotropic, so that we are dealing with a quasi-1D crystal. For example, if $X \equiv ClO_4$, the ratio of the conductivities at room temperature in the ab plane is $\sigma_a/\sigma_b \simeq 25$, and $\sigma_a/\sigma_b \simeq 200$ for $X \equiv PbO_6$. Correspondingly, the electron transfer energies are: $t_a = 0.25\,eV, t_b = 0.0025\,eV, t_c = 0.0015\,eV$. The molecules form columns, and the carriers move in an easy way along these columns (along the a-axis).

The quasi-1D structure leads to nesting at the Fermi surface. The typical phase diagram ($X \equiv AsF_6$) is plotted in Fig. 14.4. In the low-temperature region at ambient pressure the compound is an SDW insulator. The SDW instability is consistent with the nesting feature and with the quasi-1D nature of the structure.

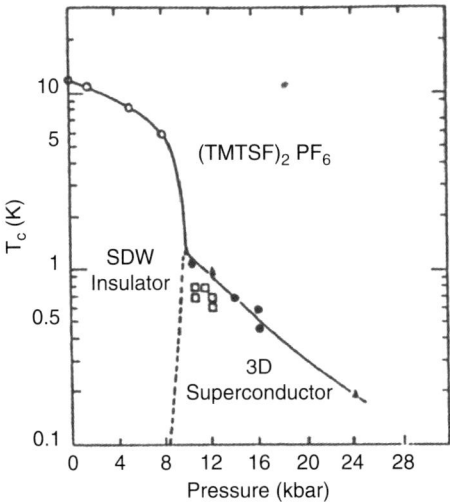

Fig. 14.4 Phase diagram of $(TMTSF)_2PF_6$. (Figure reproduced from Ishiguro et al., 1998.)

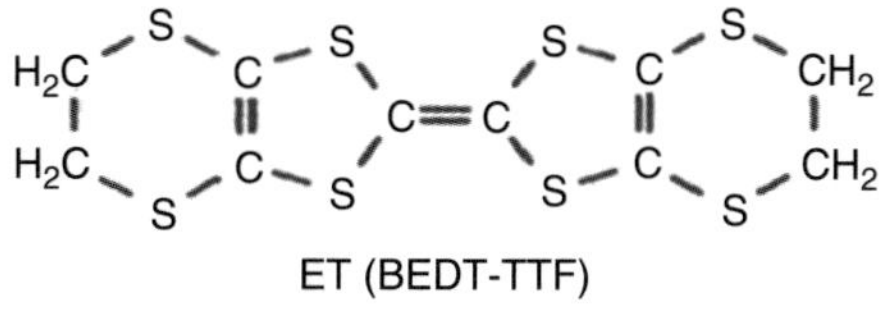

ET (BEDT-TTF)

Fig. 14.5 ET molecule. (Figure reproduced from Ishiguro et al., 1998.)

One should stress that an external pressure greatly affects the properties of organic crystals. This is not surprising, because we are dealing with molecular crystals and the intermolecular coupling is rather weak, especially in the c-direction. One can see from Fig. 14.4 that the pressure strongly decreases the temperature of the metal–insulator transition. As a result, the metallic phase persists to a lower temperature and allows observation of the transition into a superconducting state. Only the compound with $X \equiv ClO_4$ becomes superconducting at the ambient pressure.

As for some major superconducting parameters, they have values: $2\epsilon(0)/T_c \simeq 3.7, c_s/\gamma T_c \simeq 1.67$. These values are close to those for the conventional superconductors (see Section 3.5).

Another group of organic superconductors is also characterized by low-dimensional structures. Here we are referring to the so-called ET salts. The basic molecule denoted as BEDT-TTF (or shorter, as ET) is depicted in Fig. 14.5.

Contrary to the $(TMTSF)_2X$ family, the materials have quasi-2D structure (see Fig. 14.6). Correspondingly, the Fermi surface contains a cylindrical part. Note also that the in-plane coherence length is of the order $\xi_\| \simeq 6 \cdot 10^2 \text{Å}$, whereas $\xi_\perp \simeq 30 \text{Å}$.

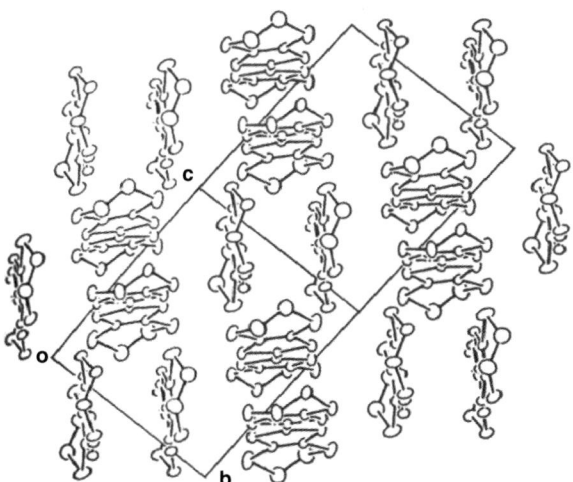

Fig. 14.6 Arrangement of ET molecules of κ-$(ET)_2I_3$. (Figure reproduced from Ishiguro et al., 1998.)

This anisotropy reflects the fact that the pairing occurs mainly in the layers, and the interlayer coupling is rather weak. This is confirmed by the observation of the internal Josephson coupling; that is, the transfer in the c-direction below T_c is dominated by the Josephson interlayer tunneling (Muller, 1994; Ito et al., 1994).

One should mention also an unusual dependence of the critical temperature on the non-magnetic scattering. For example, the superconductivity was suppressed in $(TMTSF)_2X$, $X \equiv ClO_4$ with its alloying (Tomic et al., 1983). This observation excludes the usual s-wave pairing state (see Section 8.4), which means that we are dealing with an unconventional pairing (for example, with d-wave symmetry; see Ichimura et al., 2008). A similar conclusion follows from the measurements of the penetration depth (Carrington et al., 1999). Another possibility is an appearance of the inhomogeneous FFLO state (Ferrell and Fulde, 1964; Larkin and Ovchinnikov, 1964). The manifestation of such a state is discussed by Varelogiannis (2002), Singleton and Mielne (2002), Cho et al. (2009), and Croutoru and Buzdin (2013).

There has been considerable progress in understanding the mechanism of superconductivity in organic materials, though controversy still remains. There are strong arguments that for the $(TMTSF)_2X$ compound the magnetic mechanism (spin fluctuations) is the dominant one (see Jerome, 2012). Indeed, there is a strong proximity between the SDW and superconducting phases (see Fig. 14.4). The superconducting state arises below some critical pressure near the SDW ground state. It is natural to expect the magnetic fluctuations to play a dominant role in the pairing. A detailed study performed by using various methods (see the review by Jerome, 2012), along with theoretical analysis, for example, by Duprat and Bourbonnais (2001), support this conclusion.

On the other hand, the low-frequency phonon modes can also contribute to the pairing. The experimental studies supporting such a conclusion were performed (Hawley et al., 1986; Gray et al., 1987; Nowack et al., 1986, 1987) on some $(ET)_2X$ superconductors. The point-contact tunneling spectroscopy proposed by Yanson (1974) was employed (see the review by Ishiguro et al., 1998). This point of view has also been supported by heat-capacity measurements (Elsinger et al., 2000), and the picture is well described by strong coupling theory (see Chapter 3) with $\lambda \simeq 2.5$, so that $2\epsilon(0)/T_c \simeq 5.4$, and s-wave pairing is present. The electronic heat capacity decreases exponentially, in accordance with the picture of a fully gapped order parameter.

On the whole, study of this fundamental problem is not complete. It might be that the pairing in some superconducting materials, such as $(TMTSF)_2X$, is dominated by the contribution of spin fluctuations, whereas, for example, in $(ET)_2X$ it is caused by the electron–phonon interaction. Perhaps we are dealing with some combined mechanism. One can expect that this question will be resolved in the near future.

14.3 Intercalated materials

There are a number of materials which are intrinsically insulators or semiconductors, but they become superconducting upon intercalating specific metallic ions. The most remarkable example of such systems is the family of fullerenes, and we will discuss their properties in the next section.

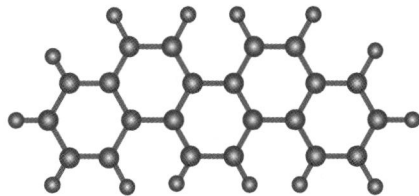

Fig. 14.7 Molecular structure of picene. (Reproduced from Kubozono et al., 2011, with permission of the Royal Society of Chemistry.)

As a first example of such organic systems, let us mention the intercalated graphite. The first observation of the superconducting transition in the alkali-metal-doped graphite K_8C ($T_c \simeq 140\,\text{mK}$) was described in 1965 (Hannay et al.). The highest value of T_c for such systems ($T_c \simeq 11.5$ K) was observed recently by Emery et al. (2005) for Ca-doped graphite.

An interesting organic system, which displays a local superconducting state with quite a high value of T_c, was described by Felner et al. (2013). They analyzed amorphous carbon (aC) films containing tungsten inclusions. It turned out that such aC-W films, doped with sulfur, contain superconducting regions with $T_c = 34.4\,\text{K}(!)$. The presence of such regions was confirmed by magnetic measurements. This is an example of an inhomogeneous superconducting system (see Chapter 11) based on an organic component. One can expect that further study will lead to the creation of a high-T_c percolative network capable of transporting a macroscopic supercurrent.

A drastic increase in T_c up to $T_c \simeq 18$ K has been observed for K-doped picene compound (Mitsuhashi et al., 2010); see also a review by Kubozono (2011). Picene is an aromatic hydrocarbon, and its molecule can be seen in Fig. 14.7.

The crystal structure of K_x picene can be seen in Fig. 14.8. The herringbone structure contains low-dimensional units. It is interesting that the superconducting

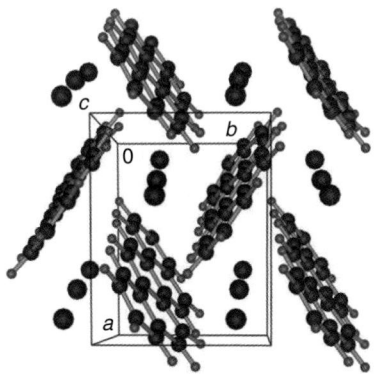

Fig. 14.8 Crystal structure of K_3 picene. The large spheres represent K. (Reproduced from Kubozono et al., 2011, with permission of the Royal Society of Chemistry.)

phase is formed for $X \simeq 3$, and the value of T_c is very sensitive to the doping level X. Indeed, there is no superconductivity for $X \lesssim 2$. The phase with $T_c \simeq 7$ K corresponds to $X \simeq 2.9$, whereas the much higher value, $T_c \simeq 18$ K(!), has been observed for the compound K_x picene with $X = 3.3$. The intercalated potassium atoms donate electrons, but also, according to calculations (Kosugi et al., 2009), provide strong hybridization of electronic orbitals and modify the electronic structure of the undoped crystal.

The aromatic molecules such as picene have a strongly developed π-electron system of delocalized electrons (see Section 14.5). The search for intercalated compounds containing other aromatic molecules seems very promising.

14.4 Fullerides

The building block of the fullerides is the C_{60} cluster (Fig. 14.9). Therefore, this organic material is an example of the cluster-based compound (see Section 15.5). The C_{60} cluster (Buckminster fullerene) has an almost spherical shape.

The C_{60} clusters aggregate and form a cubic crystal with a lattice constant $\alpha \simeq 14$Å. Such a crystal is a semiconductor with a direct band gap which is of order 1.9 eV.

The situation changes drastically if alkali metals are intercalated into the space between the clusters. These metallic ions act as donors, so that the clusters' lowest unoccupied orbital (HUMO orbital) becomes filled. As a result, the crystal becomes a metal (Haddon et al., 1991), and below some critical temperature, T_c, makes a transition into a superconducting state (see the review by Hebard, 1992).

It is interesting that the normal resistance above T_c increases with temperature, but there is no usual saturation (Ioffe–Regel limit) corresponding to the situation when the mean free path is of the order of the lattice period. This means that the electron–lattice scattering is affected greatly by the intracluster vibrational modes.

The superconducting state was observed for the K_3C_{60} compound (Hebard et al., 1991) with $T_c \simeq 18$ K. It was followed by the observation (Rosseinsky et al., 1991; Holczer et al., 1991) for Rb_3C_{60} ($T_c \simeq 28$ K) and for Cs_2RbC_{60} ($T_c \simeq 33$ K). The highest value of the critical temperature for the fulleredes was observed for Cs_3C_{60} (Tanigaki et al., 1991). Under pressure ($\simeq 15$ kbar), the superconducting state persists

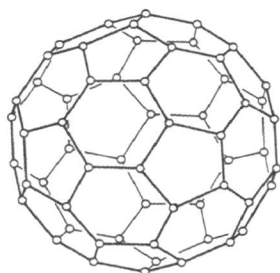

Fig. 14.9 Buckminster fullerene, C_{60}. (Figure reproduced from Ishiguro et al., 1998.)

up to $T_c \simeq 40$ K(Palstra et al. 1995). At present, this is the highest value of T_c for organic superconductors.

The pairing in the C_{60}-based materials is provided by electron–phonon interaction; it follows from isotope effect measurements. Indeed, the isotope substitution (Ramirez et al., 1992) leads to the value of the isotope coefficient, close to $\alpha \simeq 0.4$, which is close to an ideal BCS value (see Chapter 9). Probably the main contribution comes from the intracluster vibrations. The spectrum of the intracluster vibrations spread up to $\Omega_{max} \simeq 0.2$ eV. If the main contribution comes from the intermediate frequency range (Varma et al., 1991; Schluter et al., 1992), it would correspond to the intermediate strength of the interaction.

It is also interesting to note that the fullerides are characterized by a small value of the Fermi energy. For example, $E_F \simeq 0.2$ eV for Rb_3C_{60}, and $E_F \simeq 0.3$ eV for K_3C_{60}. As a result, the condition $\tilde{\Omega} \ll E_F$ (the Migdal theorem; Section 2.5) does not hold in this case. Such a strongly non-adiabatic situation deserves a special study.

In principle, there could be a peculiar scenario of superconductivity for the cluster-based materials; namely, the π-electrons can form the Cooper pairs on an isolated cluster. For example, for the C_{60} cluster such a pairing was described by Bergomi and Jollicoeur (1994) and Jollicoeur (1998). Then the transport is provided by the Josephson tunneling between such clusters. Such a mechanism, which is entirely different from the usual band picture and realistic for cluster-based materials (see Section 15.5), was introduced by Friedel (1992) after the discovery of superconductivity in fullerides. However, the impact of external pressure on T_c studied by Sparn et al. (1992) allows us to conclude that the superconducting state in the C_{60} materials is provided by pairing of the delocalized band states formed by the overlap of the electronic wavefunctions corresponding to the neighboring clusters. Indeed, the value of the critical temperature decreases under the pressure. This decrease is caused by decreasing the intercluster distance. In turn, it results in the larger overlap of the electronic wavefunctions and corresponding broadening of the energy bands. As a result, the density of states v_F is becoming smaller, which leads to a decrease in the value of the coupling constant λ (see eqn. (3.23)) and, correspondingly, to the observed smaller value of T_c.

14.5 Small-scale organic superconductivity

Let us start with an organic superconductor: $(BETS)_2GaCl_4$ (Kobayashi et al., 1995). The bulk sample of this superconductor has $T_c \simeq 8$ K, and it has a layered structure. The building blocks are the BETS molecules (Fig. 14.10) and, in addition, the $GaCl_4$ molecules, which accept electrons. The crystal is formed by double-stacked BETS molecules, and the structure contains the chains formed by these units. The electron delocalization and, correspondingly, the bulk conductivity occurs, because two BETS molecules act as donors, transferring electric charge to the acceptor $GaCl_4$ molecule. As a result, a half-filled orbital is formed.

The study performed by using scanning tunneling microscope (STM) shows that the sample and the chains contain molecular islands of different size (Clark et al., 2010). The STM technique allows determination of the presence of the superconducting

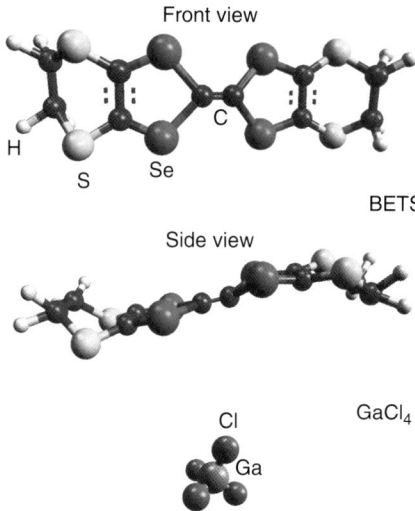

Fig. 14.10 Chemical structure of BETS and GaCl$_4$. (Reproduced from Clark et al., 2010, by permission from Macmillan Publishers, © 2010.)

energy gap and measurement of its value. It turns out that the gap can be observed even for two pairs of molecules. It means that the pairing, according to Clark et al. (2010), can be observed for the system, which contains only four(!) molecules. Such a small scale is interesting in itself, but is also promising for future applications in nano-electronics.

14.6 Pair correlation in aromatic molecules

Aromatic molecules (which are often called conjugated hydrocarbons) contain delocalized π-electrons. Figure 14.11 shows some structures.

π-electrons are moving in the field created by the so-called σ-core, formed by an ionic system. Each carbon atom supplies one such delocalized π-electron (for example, the hexabenzocoronene molecule contains thirty-two π-electrons). The delocalization makes π-electrons similar to conduction electrons in metals, though in this case we are dealing with a finite Fermi system.

π-electrons can form Cooper pairs, so that the molecules are in the superconducting state (Kresin, 1967; Kresin et al., 1975; see also Kresin and Wolf, 1990a). Of course, in this case one should not talk about conductivity and zero resistance: an isolated aromatic molecule is a small system. But the pair correlation similar to that in the usual superconductors exists and manifests itself in various phenomena.

The main manifestation is an anomalous diamagnetism which is a well-known feature of aromatic molecules. To describe this phenomenon one should write down

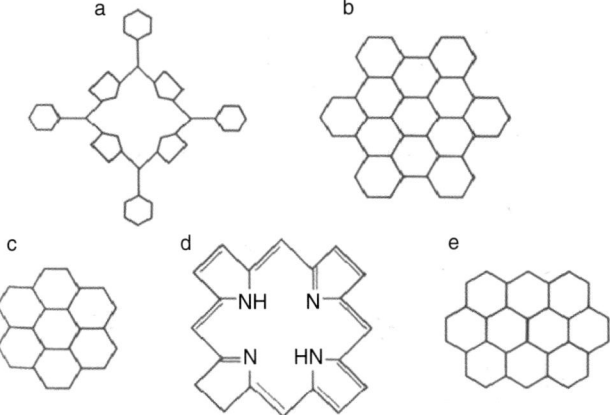

Fig. 14.11 Molecules with delocalized electrons: (a) tetraphenylporphin, (b) hexbenzcorenene, (c) corenene, (d) porphin, (e) ovalene.

the orbital part of the total Hamiltonian describing the impact of an external magnetic field:

$$\widehat{H} = \widehat{H}_p + \widehat{H}_d \tag{14.1}$$

where

$$\widehat{H}_p = \frac{ie}{2m} \int \psi^+ (\nabla \boldsymbol{A} + \boldsymbol{A}\nabla)\psi \mathbf{dr} \tag{14.1a}$$

$$\widehat{H}_d = -\frac{e^2}{2m} \int \psi^+ \psi \boldsymbol{A}^2 \mathbf{dr} \tag{14.1b}$$

$\boldsymbol{A}$ is a vector potential which can be written in the form: $\boldsymbol{A} = 1/2[\boldsymbol{Hr}]$. The expression for $<\widehat{H}_p>$ can be presented in the form:

$$<\widehat{H}_p> = \frac{ie}{2m} H_i \sum_{\lambda,\lambda'} \rho'_{\lambda,\lambda'} \widehat{M}_{i;\lambda,\lambda'} \tag{14.2}$$

where $\widehat{M}_i = -i\,[\boldsymbol{r}\nabla]_i$ (see eqn. (14.1a)), and ρ' is the change in the density matrix (Migdal, 1960, 1967) caused by the perturbation $\widehat{V}$ (in our case, $\widehat{V} = (ieh/4mc)\boldsymbol{HM}$):

$$\rho'_{\lambda,\lambda'} = \frac{(\xi_\lambda,\xi_{\lambda'} + \Delta^2 - \varepsilon_\lambda\varepsilon_{\lambda'})\,V_{\lambda\lambda'}}{2\varepsilon_\lambda\varepsilon_{\lambda'}(\varepsilon_\lambda + \varepsilon_{\lambda'})} \tag{14.3}$$

The total orbital magnetic susceptibility which can be evaluated with use of the relation $\chi = -\partial^2(\Delta E)/\partial H^2$, is a sum of two terms, so that $\chi = \chi_P + \chi_D$, where

$$\chi_P = \frac{e^2}{8m^2} \sum_{\lambda,\lambda'} \frac{\xi_\lambda, \xi_{\lambda'} + \Delta^2 - \varepsilon_\lambda \varepsilon_{\lambda'}}{2\varepsilon_\lambda \varepsilon_{\lambda'}(\varepsilon_\lambda + \varepsilon_{\lambda'})} \left|\widehat{M}_{z;\lambda\lambda'}\right|^2 \quad (14.4)$$

If the order parameter $\Delta = 0$, we obtain the expression for the so-called Van Vleck paramagnetism (1932):

$$\chi_P = \frac{e^2}{8m^2} \sum_{\lambda,\lambda'} \frac{1}{|\xi_\lambda| + |\xi_{\lambda'}|} \left|\widehat{M}_{z;\lambda\lambda'}\right|^2 ; \xi_\lambda \xi_{\lambda'} < 0 \quad (14.5)$$

We consider the case when the field is perpendicular to the plane of the molecule (z-axis). Expression (14.4) follows directly from eqns. (14.2) and (14.3). The expression for the diamagnetic term can be obtained easily from eqn. (14.1b). As is known, in usual metals the terms χ_P and χ_D almost cancel each other (one can observe only small residual Landau diamagnetism). As for superconductors, the presence of the gap parameter leads to the suppression of the paramagnetic term and to well-known anomalous diamagnetism (the Meissner effect). As for molecular systems, the study of their properties requires evaluation of both terms, χ_P and χ_D. However, the aromatic molecules of interest display strong diamagnetic response, and it means that the paramagnetic term is small, because of the impact of the pairing order parameter Δ.

It is worth noting that the paramagnetic term vanishes for the molecules with an axial symmetry. It follows directly from eqn. (14.5), because in this case the operator $\widehat{M}_z$ is diagonal. That is why, for example, the benzene molecule C_6H_6, which has almost axial symmetry, could be diamagnetic even in the absence of the pairing. However, for the molecules without an axial symmetry (such as ovalene; see Fig. 14.11) the paramagnetic term is suppressed only because of the pairing of the π-electrons.

The manifestation of pairing in aromatic molecules is similar to that for nucleons (protons and neutrons) in atomic nuclei—a well-established concept in nuclear physics (see, for example, Ring and Schuck, 1980; Broglia and Zelevinskii, 2013). This is not surprising, because in each case we are dealing with a finite Fermi system.

Both systems (nuclei and aromatic molecules) interact similarly with radiation. As we know, spectroscopy of the usual bulk superconductors is affected greatly by the opening of the energy gap (at $T = T_c$). As for the systems we are discussing in this section, their finite size leads to the discrete nature of the spectrum, regardless of the pair correlation. However, because of the pairing the energy spacing between the ground state and the first excited state (the 0–0' transition) greatly exceeds the energy of the 0'–0'' transition between the first and second excited states. In addition, one can observe the odd–even effect; that is, the lowest value of the frequency of the absorbed light is much smaller for the system with an odd number of electrons, because of the presence of one unpaired particle.

Note that the pair correlation of nucleons leads to a decrease in value of momenta of inertia relative to solid configuration. This phenomenon is similar to anomalous diamagnetism, because, as we know, the transition to the rotating coordinate system is equivalent to an appearance of the external magnetic field (see, for example, Landau and Lifshitz, 2000).

An interesting experimental study directly related to the pairing in aromatic molecules was carried our recently by Wehlitz et al. (2012); for a more detailed description, see Hartman et al. (2013). They employed the method of double ionization by a single photon (see, for example, Wehlitz, 2010; Hartman et al., 2012). The Cooper pairing is manifested in an appearance of humps in the double-to-single photoionization ratio. The effect has been observed for benzene, naphthalene, antracene, and coronene molecules (the structure of the coronene molecule can be seen in Fig. 14.11). It is interesting that for the pyrrole molecule with its five-member aromatic ring, the phenomenon has not been observed, so the manifestation of the effect is sensitive to the presence of unpaired electrons (odd–even effect).

The pairing of π-electrons impacts on the properties of biologically active systems. It is interesting that any such system contains conjugated networks and, therefore, mobile π-electrons. In connection with this, let us quote from *Quantum Biochemistry*, by Pullman and Pullman (1963):

> ... all the essential biochemical substances which are related to or perform the fundamental functions of the living matter are constituted of completely or, at least, partially conjugated systems ... The essential fluidity of life agrees with the fluidity of the electronic cloud in conjugated molecules. Such systems may thus be considered as both the cradle and the main backbone of life.

Superconducting transport can occur for the network formed by organic molecules. As we know, the oxidation process is a charge transfer of *two* electrons through the row of aromatic molecules. Such a transfer could be provided by Josephson tunneling between finite Fermi systems (see Section 15.4).

15

Pairing in nanoclusters: nano-based superconducting tunneling networks

This chapter contains a description of the superconducting state in metallic clusters. It turns out that the pairing can be strong if the cluster parameter satisfies certain but perfectly realistic conditions. Clusters constitute a new family of high-temperature superconductors. In principle, T_c can be increased to room temperatures.

The most recent and advanced study of nanoparticles was carried out by Tinkham and collaborators (see, for example, Tinkham et al., 1995, and the review by von Delft and Ralph, 2001)in which tunneling spectroscopy was employed. However, these nanoparticles were relatively large, of ~ 50–100 Å diameter; that is, they contain $\sim 10^4$–10^5 electrons.

Here we focus on much smaller nanoparticles—clusters containing $\sim 10^2$ delocalized electrons. The interest in such systems was prompted by the milestone discovery (Knight et al., 1984) of the shell structure of the spectrum in metallic nanoclusters.

15.1 Clusters: shell structure

Clusters are aggregates of atoms (or molecules) that have the structure A_k, where k is the number of atoms A (for example, Na_k, Ga_k); see Fig. 15.1. The number k of atoms can vary from 2 to any large value, and therefore the study of clusters allows tracking of the evolution from isolated atoms to bulk solids.

Metallic clusters contain delocalized electrons formed by the valence electrons of the atoms; the number N of delocalized electrons is the main parameter of the cluster. For simple monovalent metallic clusters, such as Na_k and Li_k, we have $N = k$; for more complicated clusters, $N = \nu k$, where ν is the number of valence electrons per atom; for example, $N = 135$ for Al_{45} clusters, because each Al atom contains three valence electrons. With the use of mass spectroscopy, the specific clusters can be selected.

The shell structure is a key feature of nanoclusters (see, for example, reviews by de Heer, 1993; V. V. Kresin and Knight, 1998; Frauendorf and Guet, 2001; Issendorff, 2010) As noted previously, Knight et al. (1984) discovered that the electronic states in many clusters form energy shells similar to those in atoms. The electronic states in atoms are classified by values of orbital momenta (s, p ... shells). A similar picture is valid for clusters, although, unlike atoms, clusters do not have point-like positive centers; the ions form a positive core (see Fig. 15.1).

Fig. 15.1 Cluster.

The shell structure in metallic clusters is also similar to those in atomic nuclei: proton and neutron states form energy shells. This is a well-known concept in nuclear physics (see, for example, Ring and Schuck, 1980).

Shell structure was discovered initially for simple monovalent alkali clusters, such as sodium (Na) and potassium (K) see, for example, Knight *et al.* (1984) and Wriggle *et al.* (2002). Later, this phenomenon was also observed for Al, Ga, In, Zn, and Cd clusters (Schriver *et al.*, 1990; Pellarin *et al.*, 1993; Baguenard, 1993; Katakuse *et al.*, 1992; Ruppel and Rademann, 1992; Lermé *et al.*, 1999). It is interesting that the presence of energy shells is not a universal feature. For example, for Nb clusters the spectrum does not display a shell structure.

The presence of shell structure is manifested in the appearance of so-called "magic" numbers with $N = N_m$; for example, $N_m = 20$, 40, 58, ..., 138, 168, ..., and so on. Such "magic" clusters possess completely filled electronic shells and, similarly to "inert" atoms, are most stable (Fig. 15.2).

It is essential that the "magic" clusters with a good accuracy have the spherical symmetry. Correspondingly, their electronic states are classified by the orbital momentum l and radial quantum number n. Note also that because of finite size, the electronic energy spectrum is discrete.

If the shell is not complete, the cluster undergoes a Jahn–Teller distortion and its shape becomes ellipsoidal. More specifically, as electrons are added to the lowest unoccupied shell (LUS), the shape deforms from spherical and becomes prolate. The energy levels are four-degenerate $(m, -m)$ and $(s_z, -s_z)$. If we continue to add electrons, then the shape initially continues to be prolate, but after crossing the half-shell threshold one can observe the transfer to the oblate configurations (Fig. 15.3.). Note that in reality the region near half-filling is a superposition of the prolate and oblate structure (see Appendix B).

The deformation splits the degenerate levels into sublevels (Fig. 15.4) labeled by the projection of the orbital momentum (m) An interesting and unique correlation between the number of electrons and the energy spectrum can therefore be observed. This is a remarkable feature of metallic clusters.

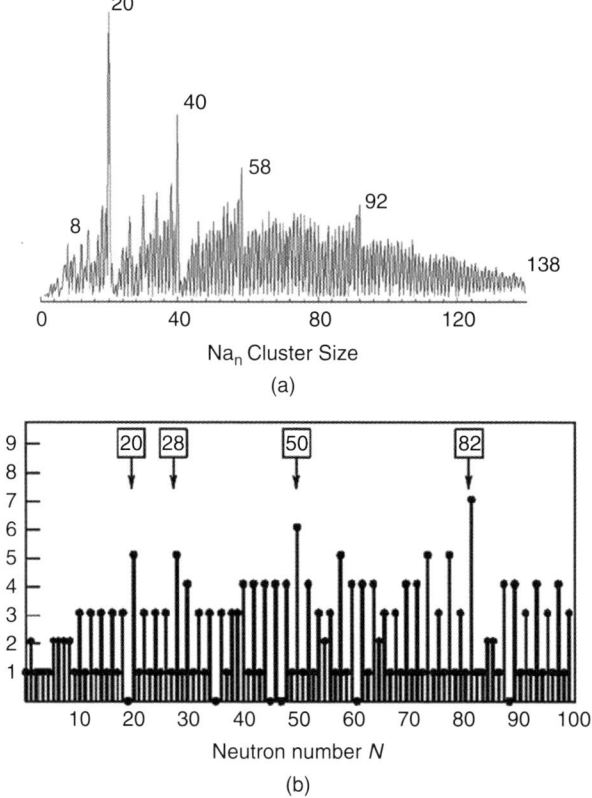

Fig. 15.2 The peaks in stability correspond to "magic" numbers: (a) in clusters; (b) in nuclei.

15.2 Pair correlation

The presence of shell structure leads to the possibility for electrons to display the phenomenon of pair correlation; in other words, the electrons form Cooper pairs. This possibility was discussed by Knight (1987) and especially by Friedel (1992). Below we follow the approach developed by Kresin and Ovchinnikov (2005, 2006). Afterward, various aspects of pairing in clusters were discussed by Groitory et al. (2011), by Garcia-Garcia et al. (2011), and by Lindenfeld et al. (2011).

15.2.1 Qualitative picture

As we know, the Cooper pairs in the bulk superconductors are formed by electrons with opposite momenta and spins. For the clusters of interest the momentum is not a quantum number, and the Cooper pairs are formed by electrons with opposite projections of angular momenta $(m, -m)$.

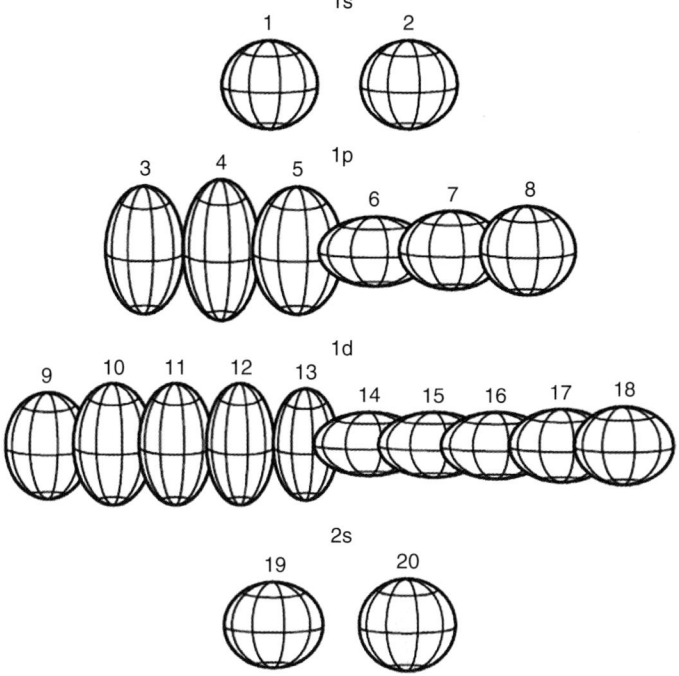

Fig. 15.3 Cluster shapes for various values of N.

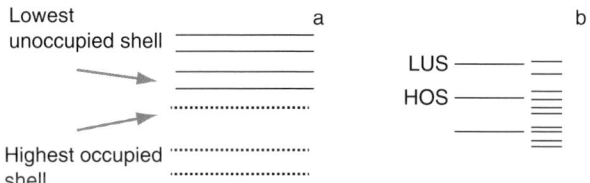

Fig. 15.4 (a) Energy spectra for a "magic" cluster, and (b) in the case of a nearly complete shell.

The pairing strongly affects the cluster's energy spectrum. We focus on "magic" clusters; the impact is also strong for clusters with slightly incomplete shells where the excitation energy in the absence of pairing can be rather small. As a whole, one should stress that the strength of pair correlation varies for different clusters. Correspondingly, the critical temperature, T_c, and the energy gap are strongly dependent upon the cluster's parameters, its shape, the strength of the coupling, and so on. Some specific cases will be described in what follows. It is important to stress that the superconducting state is not a universal feature of all metallic clusters (which is not surprising, as the same is true for bulk superconductors). However—and this is main conclusion—in some special but perfectly realistic situations one can obtain very large values of T_c.

Qualitatively, high values of T_c can be understood in the following way. Consider the highest occupied shell which is highly degenerate (large l; the degeneracy $g = 2(2l + 1)$); this means that it contains many electrons (Fig. 15.4). This feature can be viewed as a sharp peak in the density of states at the Fermi level. The situation is similar to that introduced and described by Labbe et al. (1967) for A-15 materials. The presence of a peak in the density of states (van Hove singularity) results in a noticeable increase in T_c; see Section 3.6.

Let us discuss one more point. As was noticed by Anderson (1959a), the superconducting state in small particles can manifest itself if the pairing energy gap ε exceeds or is comparable to the energy spacing caused by the finite size of the particle; that is, if $\varepsilon \gtrsim \delta E$ where $\delta E \approx E_F/N$ is the average energy level spacing, and ε is the energy gap.

Based on the numbers $E_F \approx 10^5$K and $\varepsilon \approx 2T_c \approx 10$ K, one can conclude that the number of electrons, N, should be relatively large: $N \lesssim 10^4$. As a result, if we are concerned with small nanoclusters, $N \approx 10^2 - 10^3$, it might seem that they do not display superconducting properties, because the average level spacing greatly exceeds the pairing energy gap. But such an estimation can be challenged. Indeed, for larger T_c (for example, $T_c \approx 10^2$ K for cuprates; for nanoclusters, T_c can be even higher) the energy gap is much larger than 10 K and this leads to a noticeable decrease in the limiting value of N. More importantly, the estimation by Anderson (1959a) is based on the assumption that the energy levels are approximately equidistant. This assumption can be justified for nanoparticles which do not display shell structure, such as for Pb nanoparticles (see, for example, Zolotavin and Guyot-Slonnest, 2010). However, the discovery of shell structure in many clusters totally invalidates this assumption: the pattern of electronic states is very different from an equally spaced level distribution (see, for example, Fig. 15.4). Instead, they contain highly degenerate energy levels, or groups of very close levels.

The superconductivity state can be caused not only by shell structure but also by the complex nature of the electronic spectrum. The clusters of the transition metal, Nb$_n$ ($n < 100$; shell structure is absent for N_b clusters) display ferroelectric properties which correlate with pairing (Moro et al., 2003; Yin et al., 2008). The presence of shell structure leads to the possibility of observing the high-T_c superconducting state, which is the main focus of this chapter.

In many aspects, the picture of pairing is similar to that in atomic nuclei (Bohr et al., 1958; Belyaev, 1959; Migdal, 1959, 1960, 1967; see the review by Ring and Schuck, 1980). In both cases (nuclei and clusters) we are dealing with finite Fermi systems and shells structures. The pairing states are labeled by similar quantum numbers $(m, -m)$. The manifestation of pairing also has similarities. However, the origin of the phenomenon is different: the pairing in clusters is caused by the electron–vibrational coupling, that is, the mechanism is similar to that in usual bulk superconductors.

15.2.2 Main equations: critical temperature

Let us write down a general equation describing the pairing in a metallic cluster. Such an equation for the pairing order parameter $\Delta(\omega_n)$ has the following form:

$$\Delta(\omega_n)Z = \eta \frac{T}{2V} \sum_{\omega_{n'}} \sum_s D(\omega_n - \omega_{n'}) F_s^+(\omega_{n'}) \tag{15.1}$$

Here, $\omega_n = (2n+1)\pi T$; $n = 0, \pm 1, \pm 2, \ldots$ and the thermodynamic Green's functions formalism is employed (see Chapter 3):

$$D(\omega_n - \omega_{n'}, \tilde{\Omega}) = \tilde{\Omega}^2[(\omega_n - \omega_{n'})^2 + \tilde{\Omega}^2]^{-1}; \quad F_s^+(\omega_{n'}) = \Delta(\omega_{n'})\left[\omega_{n'}^2 + \xi_s^2 + \Delta^2(\omega_{n'})\right]^{-1} \tag{15.1'}$$

are the vibrational propagator and the Gor'kov pairing function, respectively, and $\xi_s = E_s - \mu$ is the energy of the sth electronic state referred to the chemical potential μ, $\tilde{\Omega}$ is the characteristic vibrational frequency. V is the cluster volume, $\eta = <I>^2/M\tilde{\Omega}^2$ is the so-called Hopfield parameter (see, for example, Grimvall, 1981), $<I>$ is the electron–ion matrix element averaged over electronic states involved in the pairing, M is the ionic mass (we are using the harmonic approximation), and Z is the renormalization function (see Section 2.4).

The electron–vibrational interaction is considered as the major mechanism of pairing. The general equation (15.1) contains explicitly the vibrational propagator. We are using this equation because we want to go beyond the restriction $T_c \ll \tilde{\Omega}$ (weak coupling approximation).

Equation (15.1) looks similar to that in the theory of strong coupling superconductivity (see Section 3.3), but is different in two key aspects. First, it contains a summation over discrete energy levels E_s, whereas for a bulk superconductor one integrates over the continuous energy spectrum (over ξ). Another important difference is that, as opposed to a bulk superconductor, here we are dealing with a finite Fermi system, so that a number of electrons N is fixed. As a result, the position of the chemical potential differs from the Fermi level E_F and is determined by the values of N and T. Specifically, one can write

$$N = \sum_{\omega_n; s}\sum_s G_s(\omega_n) e^{i\omega_n \tau}|_{\tau \to 0} = \sum_s (u_s^2 \varphi_s^- + v_s^2 \varphi_s^+) \tag{15.2}$$

Here, $G_S(\omega_n)$ is the thermodynamic Green's function:

$$u_s^2, v_s^2 = 0.5(1 \mp \xi_s/\varepsilon_s); \quad \varphi_s^\mp = [1 + \exp(\mp \varepsilon_s/T)]^{-1}; \quad \varepsilon_s = \left(\xi_s^2 + \varepsilon_{0;s}^2\right)^{1/2} \tag{15.3}$$

$\varepsilon_{0;s}$ is the pairing gap parameter for the s th level; $\varepsilon_{0;s}$ is the root of the equation: $\varepsilon_{o;s} = \Delta(-i\varepsilon_s)$. Since $\xi_s = E_s - \mu$, eqn. (15.2) determines the position of the chemical potential for the given number of electrons N as well as the dependence $\mu(T)$. These dependences are important for the study of pairing in clusters.

Equations (15.1) and (15.2) are the main equations of the theory, allowing us to evaluate the critical temperature and the energy spectrum.

In a weak coupling case ($\eta/V \ll 1$ and, correspondingly, $\pi T_c \ll \tilde{\Omega}$), one should put in eqn. (15.1), $Z = 1$, $D = 1$, recovering the usual BCS scheme. Note also that the Coulomb term μ^* can be included in the usual way.

As noted previously, eqn. (15.1) contains a summation over all discrete electronic states. For "magic" clusters which have a spherical shape, one can replace summation over states by summation over the shells: $\sum_s = \sum_j g_j$, where g_j is the shell degeneracy: $g_j = 2(2l_j + 1)$, where l_j is the orbital momentum. If the shell is incomplete, the cluster undergoes a Jahn–Teller deformation, so that its shape becomes ellipsoidal, and the states "s" are classified by their projection of the orbital momentum $|m| \leq l$, so that each level contain up to four electrons $(for\ |m| \geq 1)$.

As noted previously, based on eqn. (15.1), one can evaluate the critical temperature: at $T = T_c$ one should put $\Delta = 0$ in the denominator of the expression (15.1'). Note also that since $\lambda_b = \eta v_b$ (see eqn. (3.23); λ_b and v_b are the bulk values of the coupling constant and density of states, and η is the Hopfield parameter; see eqn. (15.1)), one can write:

$$\Delta(\omega_n)Z = \tilde{\lambda}_b \pi T \sum_j \sum_{\omega_{n'}} g_j \frac{\tilde{\Omega}^2}{\tilde{\Omega}^2 + (\omega_n - \omega'_n)^2} \Delta(\omega_{n'}) \left[\omega_{n'}^2 + \Delta^2(\omega_{n'}) + \xi_j^2\right]^{-1} \Bigg|_{T=T_c}$$

$$\tilde{\lambda}_b = \lambda_b(2\pi\nu_b V)^{-1} = \left(\frac{2}{3\pi}\right)\lambda_b\left(\frac{E_F}{N}\right)$$

(15.4)

Indeed, $v_b = m^* p_F / 2\pi^2 = (3/4)n/E_F$, since $n \propto p_F^3$, and n is the carrier concentration. For clusters, the value of n is close to its bulk value (Ekardt, 1984). Note also that the characteristic vibrational frequency $\tilde{\Omega}$ is close to its value for the bulk materials. Indeed, the pairing is mediated mainly by the short-wavelength region of the vibrational spectrum, which is similar for clusters and bulk systems. In addition, the equality $\tilde{\Omega} = \Omega_D$ follows directly from the heat-capacity measurements performed by Hock et al. (2011).

As noted previously, eqn. (15.4), along with the equation for Z and eqn. (15.2), are the main equations that allow us to evaluate the value of the critical temperature. It is essential that these equations contain the experimentally measured parameters: N (number of delocalized electrons), E_F, $\tilde{\Omega}$, the degeneracy g_i, λ_b, and the energy spectrum (ξ_i). As a result, one can calculate the values of the critical temperature for various clusters.

It is convenient to write eqn. (15.4) in the dimensionless form:

$$\phi_n = \sum_{n'} K_{nn'} \phi_{n'}$$

$$K_{nn'} = g\tau_c \sum_s \left\{ \left[1 + (\tilde{\omega}_n - \tilde{\omega}'_n)^2\right]^{-1} - \delta_{nn'} Z \right\} \left(\tilde{\omega}_{n'}^2 + \tilde{\xi}^2\right)^{-1} \Bigg|_{T_c}$$

(15.5)

Here,

$$\phi_n = \Delta(\omega_n)\tilde{\Omega}^{-1},\ \tilde{\omega}_n = \omega_n \tilde{\Omega}^{-1},\ \tilde{\xi}_j = \xi_j \tilde{\Omega}^{-1},\ g = \lambda_b \left(4\pi \tilde{\Omega} \nu_b V\right)^{-1}$$

The calculation can be performed with the matrix method (see Section 3.4.4). One can demonstrate (Kresin and Ovchinnikov, 2006) that for perfectly realistic values of the parameters a high value of T_c can be obtained. For example, if we consider the "magic" cluster with the following typical parameter values: $\Delta E = E_H - E_L = 65$ meV, $\tilde{\Omega} = 25$ meV, $m^* = m_e$, $k_F = 1.5 \times 10^8 \text{cm}^{-1}$, $\lambda_b = 0.4$, the radius, $R = 7.5$ Å, and

$g_H + g_L = 48$ (for example, $l_H = 7$, $l_L = 4$), we will arrive at a value of $T_C \cong 10^2$ K (!) Note that a large degeneracy of LUS and HOS plays a key role.

The value of T_c is very sensitive to the cluster parameters. The most favorable case corresponds to:

1. A cluster with large values of the orbital momentum l for the highest occupied and lowest unoccupied shells, and hence with large degeneracies g_{HUS} and g_{LOS} (qualitatively it corresponds to a large effective density of states).
2. A relatively small energy spacing between these shells.

An increase in T_c can be achieved by changing the parameters (ΔE, $\tilde{\Omega}$, and so on) in the desired direction. For example, for $\Delta E \cong 0.2$ eV, $\lambda_b = 0.5$, $m \cong 0.5\,m_e$, $R \cong 5.5$ Å, $\tilde{\Omega} = 50\,meV$, $g_H + g_L = 60$, we obtain $T_c \cong 240$ K (!). In principle, T_c can be increased to room temperatures.

Based on eqns. (15.1) and (15.2) one can calculate the values of T_c for specific clusters. The results vary noticeably, depending on their parameters. For example, consider the cluster Al$_{56}$. It contains $N = 168$ electrons, since each Al atom contains three valence electrons. This is a "magic" cluster with the highest occupied shell (HOS) corresponding to $l = 7$. Indeed, according to Schriver et al. (1990), for $N = 168$ one can observe a sharp drop in the ionization potential. The next "magic" cluster contain $N = 198$ electrons and corresponds to hybridization of three shells with $l = 4$, $n = 2$; $l = 2$, $n = 2$; $l = 0$, $n = 4$; as a result, $g_H + g_L = 30$. According to Schriver et al. (1990), the spacing between HOS and LUS is $\Delta E \approx 0.1$ eV. With use of the parameters for Al$_{56}$ clusters ($R \approx 6.5$ Å, $\tilde{\Omega} = 350$ K, $\lambda_b = 0.4$, $m^* = 1.4\,m_e$, $k_F = 1.7 \cdot 10^8$ cm^{-1}), we obtain $T_c \cong 90$ K (!).

As for Ga$_{56}$ clusters, we need to use a different set of parameters ($\tilde{\Omega} = 270$ K, $\lambda_b = 0.4$, $m^* = 0.6\,m_e$, $k_F = 1.7 \cdot 10^8$ cm^{-1}), and we obtain $T_c \cong 145$ K (!). This greatly exceeds the value of T_c for the bulk sample ($T_{c;B} \cong 1.1$ K).

In a similar way, one can calculate T_c for clusters with slightly incomplete shells. Such clusters undergo a Jahn–Teller distortion, leading to the splitting of the degenerate levels. The splitting is described by the expression (see, for example, Migdal, 2000; Landau and Lifshitz, 1977):

$$\delta E_l^m = 2 E_l^{(0)} U r_l^m \\ r_l^m = 2\left[l(l+1) - 3m^2\right](2l+3)^{-1}(2l-1)^{-1} \qquad (15.6)$$

Here, U is the deformation potential, $E_l^{(0)} = E_H = E_{HOS}$. For example, for Zn$_{83}$ clusters ($N = 166$), one can obtain a value of $T_c \cong 105$ K.

15.2.3 Energy spectrum; fluctuations

Based on eqns. (15.1) and (15.2) one can also study the impact of pairing on the energy spectrum. Here one should remember that because of finite size the electronic spectrum is discrete even in the absence of pair correlation. That is why, contrary to the bulk case when the pairing leads to an appearance of the discrete feature of the spectrum (energy gap), for the nanoclusters, pairing is manifested in the change of the energy spectrum. Namely, one can observe an increase of the spacing between the ground and excited states, especially at low temperatures near $T = 0$ K. For "magic" clusters, the pairing is essential if it leads to a noticeable increase in the LUS–HOS spacing. The spectrum is determined by the equation:

$$\varepsilon_{os} = \Delta \left[i(\xi_s^2 + \varepsilon_{0;s}^2)^{1/2} \right] \quad (15.7)$$

ε_{os} is the gap parameter, and Δ and ξ_s are defined as in eqn. (15.1). For example, for the Cd$_{83}$ cluster one can obtain the following value of the excitation energy: $\Delta \varepsilon_{\min} = 34$ meV. This value greatly exceeds that for the same cluster in the absence of pairing: $\Delta \varepsilon_{\min} = 6.5$ meV.

A serious question can be raised about the role of fluctuations for the investigated transition of nanoclusters into the pairing state. Note that in this context it would be incorrect to consider the clusters as zero-dimensional systems. Indeed, the large values of T_c result in a relatively small coherence length, in fact comparable to the cluster size. The broadening of the transition $\delta T_c/T_c$ due to fluctuations can be evaluated rigorously (Kresin and Ovchinnikov, 2006). One can show that $(\delta T_c/T_c) \approx 5\%$. This width noticeably exceeds that for bulk superconductors, but is still relatively small.

15.3 How to observe the phenomenon?

Pairing leads to a strong temperature dependence of the excitation spectrum. Below T_c, and especially at low temperatures close to $T = 0$ K the excitation energy is strongly modified by pairing, and noticeably exceeds that in the region $T > T_c$. The shift appears to be especially dramatic for clusters with slightly unoccupied shells; for such clusters, the ratio $\Delta \varepsilon_{\min}/\Delta E_{\min}$ can be $\sim 6 - 7$.

A change of such magnitude in the excitation energy should be observable experimentally and would represent a strong manifestation of pair correlation. Nevertheless, we are talking about a rather difficult experiment. It requires the generation of beams of isolated metallic clusters at different temperatures in combination with mass selection. A measurement of the energy spectrum—in particular, a determination of the minimum excitation energy $\Delta E_{\min}$ (for example, by photoelectron spectroscopy; see, for example. Issendorff and Cheshnovsky, 2005)—would reveal a strong temperature dependence of the spectrum.

An interesting calorimetric study of clusters was described by Cao et al. (2008). They developed a special so-called multicollisional induced dissociative method that allows measuring the heat capacity of an isolated cluster. The study shows that for selected clusters—for example, for Al$_{45}^-$ ions—a peak in heat capacity can be observed. It is essential that the peak is not a universal feature and can be observed only for selected clusters. The peak was observed at $T \approx 200$ K(!); The result is highly reproducible. As we know, a jump in heat capacity is a signature of a phase transition. The Al clusters are not magnetic, and therefore a magnetic transition can be excluded. Special measurements of mobility also exclude any structural transition. It is natural to assume that we are dealing with a transition into a superconducting state. These heat-capacity data are the first indication for the transition.

One can expect that the spectroscopic method discussed previously will provide even stronger manifestation of the phenomenon.

15.4 Cluster-based tunneling network: macroscopic superconductivity

We have discussed various manifestations of pairing in isolated clusters. Let us now address the interesting question about the possibility of creating macroscopic supercurrents based on Josephson tunneling between nanoclusters.

Josephson tunneling between clusters needs to be analyzed with considerable care. Indeed, contrary to the usual bulk superconductors, the clusters are characterized by discrete energy spectra. Moreover, one should take into account the fact that the tunneling itself splits the energy levels and leads to the formation of symmetric and asymmetric terms. The direct calculation by Ovchinnikov and Kresin (2010, 2012) shows that charge transfer via Josephson tunneling is possible, and the effect of the discrete nature of the electronic spectrum is to increase the current relative to the bulk case. Note also that the discreteness of the spectrum results in an additional channel that is absent in the bulk case; namely, the possibility of resonant tunneling. This channel supports a current that is larger than in the non-resonant case. It turns out that the tunneling Hamiltonian formalism is not applicable in the case of resonant tunneling, and this case requires use of the general expression for the current operator.

The presence of a Coulomb blockade competes with the Josephson energy. One can show that for relatively short intercluster distances(~ 10 Å), Josephson energy greatly exceeds the impact of the Coulomb repulsion.

Let us also stress one point. As we know, the Josephson current results from the phase difference between superconducting electrodes (in our case, between clusters). However, the phase of a cluster does not have a specific value; this follows from the uncertainty relation between the phase and the number of particles. As a result, the phase is not defined for each of the clusters. Nevertheless, the phase difference that enters the expression for the Josephson current is defined. This fact was stressed by Cobert *et al.* (2004). As usual, the phase difference is determined by the value of the transmitted current.

Josephson junctions can be used to construct cluster-based networks (Fig. 15.5). It can be shown that the damping of the current (because of quantum fluctuations, for example) is quite small, due to strong renormalization of the capacitance (such renormalization was introduced by Larkin and Ovchinnikov (1983) and Eckern *et al.* (1984)).

The network would be made of clusters deposited on a surface. This implies that the cluster structure and energy spectrum are not destroyed by the surface–cluster interaction. The progress with the so-called "soft landing" technique based on the use of special organic substrates (see, for example, Meirwes-Broer, 2000; Duffee *et al.*, 2007) makes building such a network realistic.

Fig. 15.5 Cluster-based tunneling network.

Thus, charge transfer between clusters via Josephson coupling can give rise to strong macroscopic high-temperature superconducting currents.

15.5 Cluster crystals

A new type of 3D crystal (cluster metals) can be created by using metallic nanoclusters as building blocks. The lattice of such a crystal is formed by clusters instead of the usual point-like ions. Such solids are analogous to molecular crystals. The presence of pair correlation in an isolated cluster leads to the Josephson charge transfer between neighboring clusters, and as a result, such a cluster-based 3D crystal would display macroscopic superconductivity. The concept of such a superconducting crystal was introduced by Friedel (1992).

An interesting study of Ga-based cluster metals was described by Hagel *et al.* (2002) and by Bono *et al.* (2007). A crystal was formed by Ga_{84} clusters ($N = 252$; each Ga atom contains three valence electrons). The superconducting transition at $T_c \approx 7.2\,\text{K}$ has been observed; note that HOS-LUS spacing for such a cluster is rather large, and there is no reason to expect a very high value of T_c. The observed value of T_c is noticeably higher than $T_{c;b} \approx 1\text{K}$ for bulk Ga.

In principle, one can foresee a crystal synthesis out of different clusters (such as Ga_{56}) with rather strong internal pair correlation, and this will lead to macroscopic superconductivity with higher values of the critical temperature.

Appendices

Appendix A: Diabatic representation

According to the usual adiabatic theory (Born–Oppenheimer (BO) approximation; Section 2.2), the electronic Hamiltonian $\hat{H}_e = \hat{T}_r + V(r, R)$ is diagonal in the representation defined by eqn. (2.11). The analysis is based on the usual stationary Schrödinger equation (2.7). Equation (2.8) determines the electronic terms, $\varepsilon_m(R)$. In quantum chemistry the dependence $\varepsilon_m(R)$ is "visualized" as the potential-energy surface.

Assume that the energy terms have a structure as shown in Fig. A.1 (solid lines). The lowest term contains two regions. One of them (L) corresponds to the bound ionic state, and the second region (R) to the continuum spectrum.

Another example is the case when the potential has two close minima (double-well structure); Fig. A.2. The problem of propagation L→R can be solved with use of the usual method with matching at the boundaries, and so on. However, one can employ a different method: the so-called diabatic representation. Let us introduce a new potential, $\tilde{V}(r, R)$, so that the effect of the substitution $V \to \tilde{V}$ in the electronic equation (2.8) is to change the adiabatic terms to diabatic; that is, to the picture of crossing terms $\tilde{\varepsilon}_1$ and $\tilde{\varepsilon}_2$. In addition, the analysis in the diabatic representation is based not on the stationary but on the time-dependent Schrödinger equation. In other words, the L → R transfer can be described as a quantum transition between the terms 1 and 2. The total wavefunctions are sought in the form (see eqn. (2.39)):

$$\psi(r, R, t) = a(t)\psi_1(r, R) + b(t)\psi_2(r, R) \tag{A.1}$$

where

$$\begin{aligned} \psi_1(r, R) &= \psi_1(r, R)\phi_{1,v_1}(R) \\ \psi_2(r, R) &= \psi_2(r, R)\phi_{2,v_2}(R) \end{aligned} \tag{A.2}$$

It is essential that the functions $\phi_{1,v_1}(R)$ and $\phi_{2,v_2}(R)$ are not orthogonal, since they belong to different energy terms.

The approach is similar to that employed by Bardeen (1961) to describe the tunneling effect (see also, for example, Kane, 1969; Reittu, 1995). Instead of the usual method based on the stationary Schrödinger equation, Bardeen analyzed tunneling as a time-dependent phenomenon, and, more specifically, as a quantum transition between two (initial and final) states. The tunneling Hamiltonian formalism (Harrison, 1961; Cohen et al., 1962) appears to be a powerful tool in solid-state physics, and especially in the physics of superconductivity.

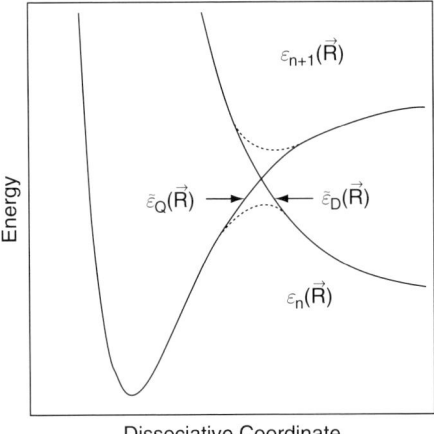

Fig. A.1 Adiabatic potential energy surfaces $\varepsilon_n(\mathbf{R})$ and $\varepsilon_{n+1}(\mathbf{R})$ from diagonalization of the electronic Hamiltonian, $\hat{H}_{el}$ (...). Diabatic potential energy surfaces $\tilde{\varepsilon}_Q(\mathbf{R})$ and $\tilde{\varepsilon}_D(\mathbf{R})$ which cross, from a non-diagonal representation of $\hat{H}_{el}(-)$.

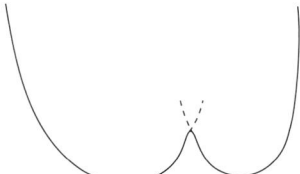

Fig. A.2 Diabatic representation; one-dimensional model. The solid line corresponds to the adiabatic potential energy term, and the dashed line corresponds to the crossing terms in the diabatic representation.

Assume that $a(0) = 1$ and $b(0) = 0$. Then the probability of the 1–2 transfer is described by eqn. (2.40), where $H_{el;1,2} = \int d\mathbf{R} \cdot L(\mathbf{R})\phi_2^*(\mathbf{R})\phi_1(\mathbf{R})$; $L(\mathbf{R}) = \int d\mathbf{r} \cdot \psi_2^*(\mathbf{r},\mathbf{R})\hat{H}_{el}\psi_1(\mathbf{r},\mathbf{R})$.

The function $L(\mathbf{R})$ depends slowly on $\mathbf{R}$. The main contribution to the matrix element comes from the region of the overlap of nuclear wavefunctions. As a result, one can write:

$$H_{el;1,2} \simeq L_0 F \tag{A.3}$$

where $L_0 = L(\mathbf{R}_0)$, $\mathbf{R}_0$ is the crossing point, and

$$F = \int \phi_2^*(\mathbf{R})\phi_1(\mathbf{R})d\mathbf{R} \tag{A.4}$$

is the Franck–Condon factor. As noted previously, it is essential that the nuclear wavefunctions $\tilde{\phi}_2$ and $\tilde{\phi}_1$ are not orthogonal.

The scale of the electronic wavefunction $\tilde{\psi}_i(\boldsymbol{r}, \boldsymbol{R})$ is much larger than that for the nuclear function $\tilde{\phi}_i(\boldsymbol{R})$, which is localized in the region $\delta R \sim a$, where a is the vibrational amplitude.

As noted previously, it is essential that, unlike the usual BO approximation, the operator $\widehat{H}_{el}$ has non-diagonal matrix elements. Such an element describes the transition (in our case, tunneling) between the terms.

The diabatic representation is a convenient tool in molecular spectroscopy and solid-state physics. It is employed to describe the polaronic state when each ion is characterized by two close minima (Section 2.62), the isotope effect (Chapters 9 and 10), and the quasi-resonant states of nanoclusters (Chapters 15).

Appendix B: Dynamic Jahn–Teller effect

As we know, a system containing degenerate electronic states is unstable. The degeneracy is removed via the Jahn–Teller (JT) effect, and as a result, one can observe the splitting of initially degenerate energy levels. One should distinguish static and dynamic JT effects.

The static JT effect (Fig. A.3) is caused by electron–lattice interaction. This interaction leads to the so-called JT static distortion, so that the ions move to different equilibrium positions. As a result, we are dealing with ionic configuration which has a lower symmetry, which leads to removal of the initial degeneracy. The scale of the distortion is determined by the interplay of the electron–lattice interaction and the elastic-energy terms. The shift is energetically favorable (static distortion energy) and as a result, distorted configuration corresponds to a new stable state. It is essential that the distortion energy exceed the zero-point energy of the corresponding vibrational mode. Indeed, only in this case can the system be permanently distorted (the static JT effect).

However, if this is not the case, then we are dealing with dynamic picture (the dynamic JT effect; see, for example, Salem, 1966), in which the system displays a dynamic interconversion between different distorted forms; that is, it oscillates between energy-equivalent distortions. There could be an intermediate case which can be treated as an interplay of the static and dynamic JT effects (see Fig. A.3).

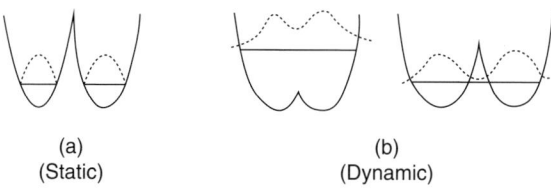

(a) (Static)   (b) (Dynamic)

Fig. A.3 (a) The static JT effect, by which the system is permanently distorted; (b) the dynamic JT effect, with rapid interconversion between two forms. (From Salem, 1966.)

The so-called "tunneling splitting" is a typical example of the dynamic JT effect (see Bersuker, 2006). The apical oxygen in the cuprates also displays such an effect, and the same is true for the oxygen ion in manganites (Section 12.5.3 and Fig. 12.8). For both these cases the dynamic JT effect is manifested in the peculiar isotopic dependence. We call it the "polaronic isotope effect" to stress the fact that the electronic and lattice degrees of freedom are not separable, contrary to the usual adiabatic scenario. This is a key ingredient of the dynamic polaron picture.

Shape deformation upon filling the energy shell in metallic nanoclusters is a manifestation of the Jahn–Teller effect (Section 15.1 and Fig. 15.3). At small filling of the unoccupied shell, the cluster acquires the prolate shape (static JT effect). However, near the half-shell filling the picture is more complicated and does not seem so straightforward as plotted on Fig. 15.3. This can be seen from the adsorption spectra (Borggreen, 1993). Indeed, as is known (see, for example, the review by Kresin V. V., 1992) the adsorption spectrum peaks at the cluster's plasmon frequency; and correspondingly, one can observe two peaks for the prolate configuration. For this configuration the intensity of the peak corresponding to the long axis should be twice as small relative to another peak, because the peak for the shorter axis is doubly degenerate. However, experimentally the intensities of both peaks near the half-shell filling are almost equal. This can be explained (Kresin and Friedel, 2011) by the dynamic JT effect. Indeed, the energy levels for the prolate and oblate configurations are becoming close near half-filling (see, for example, Ekardt and Penzar, 1991), and we are dealing with quantum oscillations between these two configurations. These oscillations represent the manifestation of the dynamic JT effect which is dominant in this region.

Note that the diabatic representation (Section 2.6.2 and Appendix A) is a very convenient tool for describing the dynamic JT effect.

References

Abrikosov, A. (1957), *JETP* 5, 114.
Abrikosov, A., Gor'kov, L. (1959), *JETP* 8, 1090.
——(1961), *JETP* 12, 1243.
Abrikosov, A., Gor'kov, L., Dzyaloshinski, I. (1975), *Methods of Quantum Field Theory in Statistical Physics*, Dover, NY.
Abrikosov, A., Gor'kov, L., Khalatnikov, I. (1959), *JETP* 8, 182.
Abrikosov, A., Khalatnikov, I. (1959), *Adv. Phys.* 8, 45.
Akamutsu, H., Inokuchi, H., Matsunaga, Y. (1954), *Nature* 173, 168.
Akimoto, T., Maruyama, Y., Maritomo, Y., Nakamura, A. (1998), *Phys. Rev. B* 55, 5594.
Alekseevskii, N., Vedeneev, S. (1967), *JETP Lett.* 6, 302.
Alexandrov, S., Ranninger, J. (1981), *Phys. Rev. B* 23, 1796.
Allen, P., Cohen, M. (1970), *Phys. Rev. B* 1, 1329.
Allen, P., Dynes, R. (1975), *Phys. Rev. B* 12, 995.
Allender, D., Bray, J., Bardeen, J. (1973), *Phys. Rev. B* 7, 1020.
Alloul, H., Ohno, T., Mendels, P. (1989), *Phys. Rev. Lett.* 67, 3140.
Alvarez, G., Mayr, M., Moreo, A., Dagotte, E. (2005), *Phys. Rev. B* 71, 014514.
Aminov, B., Hein, M., Miller, G., Piel, H., Ponomarev, Y., Wehler, D., Boockholt, M., Buschmann, L., Gunherodt, G. (1994), *Physica C* 235–40, 1863.
Anderson, P. (1959a), *J. Phys. Chem. Solids* 11, 59.
——(1959b), *Phys. Rev.* 115, 2.
——(1963), *Solid State Phys.* 14, 99.
——(1973), *Mater. Res. Bull.* 8, 153.
——(1987), *Frontiers and Borderlines in Many-Particle Physics (Proceedings of the Enrico Fermi International School of Physics)*, Schrieffer, J., Broglia, R. (eds.), North-Holland, Amsterdam.
——(1987), *Science* 235, 1196.
——(1988), *Proceedings of the Common Trends in Particle and Condensed Matter Physics, Physics Reports* 184, 2–4.
——(1991), *Phys. Rev. Lett.* 67, 2092.
Anderson, P., Baskaran, G., Zou, Z., Hsu, T. (1987), *Phys. Rev. Lett.* 58, 2790.
Anderson, P., Brinkman, W. (1973), *Phys. Rev. Lett.* 30, 1108.
Anderson, P., Hasegawa, H. (1955), *Phys. Rev.* 100, 675.
Andreev, A. (1964), *JETP* 19, 1998.
Ann, C., Bhattacharya, A., Di Ventra, M., Eckstein, J., Frisbie, C., Gershenson, M., Goldman, A., Inoue, I., Mannhart, J., Millis, A., Morpurgo, A., Natelson, D., Triscone, J.-M. (2006), *Rev. Mod. Phys.* 78, 1185.
Anselm, A. (1982), *Introduction to Semiconductor Theory*, Prentice Hall, NY.

Arovas, D., Schrieffer, R., Wilczek, F. (1984), *Phys. Rev. Lett.* 53, 722.
Ashcroft, N., Mermin, N. (1976), *Solid State Physics*, Holt, NY, B 64, 024502.
Auerbach, A., Levin, K. (1986), *Phys. Rev. Lett.* 57, 877.
Baguenard, B., Pellarin, M., Bordas, C., Lerme, J., Vialle, J., Broyer, M. (1993), *Chem. Phys. Lett.* 205, 13.
Babushkina, N., Inyushkin, A., Ozhogin, V., Taldenkov, A., Kobrin, I., Vorobeva, T., Molchanova, L., Damyanets, L., Uvarova, T., Kuzakov, A. (1991), *Physica C* 185–189, 901.
Bardeen, J. (1961), *Phys. Rev. Lett.* 6, 57.
Bardeen, J., Cooper, L., Schrieffer, J. (1957), *Phys. Rev.* 108, 1175.
Barisic, S., Barisic, I., Friedel, J. (1987), *Europhys. Lett.* 3, 1231.
Barnes, S. (1976), *J. Phys. F* 6, 1375.
Barone, A., Paterno, G. (1982), *Physics and Applications of the Josephson Junction Effect*, Wiley, NY.
Baskaran, G., Ruchenstien, A., Tosatti, E., Lu, Y. (eds.) (1991), *Progress in High-Temperature Superconductivity (Proceedings of the Adriatrico Research Conference and Miniworkshop- Strongly Correlated Electron Systems II)*, vol. 29. World Scientific, Singapore.
Bechgaard, K., Jacobsen, C., Mortenson, K., Pedersen, H., Thorup, N. (1979), *Solid State Commun.* 33, 1119.
Becker, T., Streng, C., Luo, Y., Moshnyaga, V., Damashke, B., Shannon, N., Samwer, K. (2002), *Phys. Rev. Lett.* 89, 237203.
Bedell, K., Coffey, D., Pines, D., Schrieffer, J. (eds.) (1990), *Physics of Highly Correlated Electron Systems*, Addison-Wesley, NY.
Bednorz, G., Mueller, K. (1986), *Z. Phys. B* 64, 189.
Bednorz, J., Mueller, K. (1987), *Angew Chemie* 100, 5.
Belova, L. (2000), *J. of Supercond. and Novel Magn.* 13, 305.
Belyaev, S. (1959), *Mat. Phys. Medd. Dan. Selsk.* 31, 131.
Bergemann, C., Tyler, A., Mackenzie, A., Cooper, J., Julian, S., Farrell, D. (1998), *Phys. Rev. B* 57, 14387.
Bergmann, G., Rainer, D. (1973), *Z. Physik* 263, 59.
Bergomi, L., Jollicoeur, T. (1994), *C.-R. de Academie des Sciences (Paris)* 318, II, 283.
Berk, N., Schrieffer, J. (1966), *Phys. Rev. Lett.* 17, 433.
Bermon, S. (1960), *Tech. Rept. 1*, University of Illinois, Urbana, National Science Foundation Grant NSF GP 1100.
Bersuker, I. (2006), *The Jahn–Teller Effect*, Cambridge University Press, Cambridge.
Bianconi, A. (1994a), *Solid State Commun.* 89, 933.
——(1994b), *Solid State Commun.* 91, 1.
Bill, A., Kresin, V., Wolf, S. (1998), *Phys. Rev. B* 57, 10814.
Bill, A., Morawitz, H., Kresin, V. (2003), *Phys. Rev. B* 68, 144519.
Billinge, S., Proffen, T., Petkov, V., Sarrao, J., Kycia, S. (2000), *Phys. Rev. B* 62, 1203.
Binnig, G., Baratoff, A., Hoenig, H., Bednorz, G. (1980), *Phys. Rev. Lett.* 45, 1352.
Binnig, G., Rohrer, H. (1982), *Helv. Phys. Acta* 55, 726.

Blanter, Y., Kaganov, M., Pantsulaja, A., Varlamov, A. (1994), *Phys. Rep.* 245, 161.
Blonder, G., Tinkham, M., Klapwijk, T. (1982), *Phys. Rev. B* 25, 4515.
Bogoluybov, N., Tolmachev, N., Shirkov, D. (1959), *A New Method in the Theory of Superconductivity*, Cons. Bureau, NY.
Bohr, A., Mottelson, B., Pines, B. (1958), *Phys. Rev.* 110, 936.
Bok, J., Bouvier, J. (2012), *J. of Supercond. and Novel Magn.* 25, 657.
Bono, D., Bakharev, O., Schepf, A., Hartig, J., Schockel, H., de Jongh, L. (2007), *Z. Anorg. Allg. Chem.* 633, 2173.
Booth, C., Bridges, F., Snyder, G., Geballe, T. (1996), *Phys. Rev. B* 54, 15606.
Booth, C., Bridges, F., Kwei, G., Lawrence, J., Cornelius, A., Neumeier, J. (1998), *Phys. Rev. Lett.* 80, 853.
Borggreen, J., Crowdhurry, P., Kebaili, N., Lundsberg-Nielsen, L., Lutzenkirchen, K., Nielsen, M., Pedersen, J., Rasmussen, H. (1993), *Phys. Rev. B* 48, 17507.
Born, M., Oppenheimer, R. (1927), *Ann. Phys.* 84, 457.
Born, M., Huang, K. (1954), *Dynamic Theory of Crystal Lattices*, Oxford University Press, NY.
Bornemann, H., Morris, D., Liv, H. (1991), *Physica C* 182, 132.
Bouquet, F., Wang, Y., Fisher, R., Hinks, D., Jorgensen, J., Junod, A., Phillips, N. (2001), *Europhys. Lett.* 56, 836.
Bozin, E., Kwei, G., Takagi, H., Billinge, S. (2000) *Phys. Rev. Lett.* 84, 5856.
Bozovic, I., Eckstein, J. (1998), *Solid State Commun.* 61–2, 9.
Bozovic, I., Eckstein, J., Virshup, G., Chaiken, A., Wall, M., Howell, R., Fluss, M. (1994), *J. Supercond.* 7, 187.
Bozovic, I., Logvenov, G., Verhoeven, M., Caputo, P., Goldobin, E., Beasley, M. (2004), *Phys. Rev. Lett.* 93, 157002.
Brenig, W., Kampf, A. (1991), *Phys. Rev. B* 43, 12914.
Broglia, R., Zelevinskii, V. (eds.) (2013), *50 Years of Nuclear BCS*, World Scientific, Singapore.
Brovman, Y., Kagan, Y. (1967), *JETP* 25, 365.
Bussman-Holder, A., Mueller, K. (2012), in *100 Years of Superconductivity*, H. Rogalla, P. Kes (eds.), Taylor & Francis, Boka Raton.
Buzdin, A., Bulaevskii, L., Panyukov, S. (1982), *JETP Lett.* 35, 178.
Bychkov, Y., Gor'kov, L., Dzyaloshinskii, I. (1966), *JETP* 23, 489.
Canfield, P., Gammel, P., Bishop, D. (1998), *Physics Today* 51, 40.
Cao, B., Neal, C., Starace, A., Ovchinnikov, Y., Kresin, V., Jarrell, M. (2008), *J. of Supercond. and Novel Magn.* 21, 163.
Cao, C., Hirschfeld, P., Cheng H. P. (2008), *Phys. Rev. B* 77 220506.
Carbotte, J. (1990), *Rev. Mod. Phys.* 62, 1027.
Carbotte, J., Greeson, M., Perez-Gonzales, A. (1991), *Phys. Rev. Lett.* 66, 1789.
Caretta, B., Lascialfari, A., Rigamonti, A., Rosso, A., Varlamov, A. (2000), *Phys. Rev. B* 61, 12420.
Carrington, A., Bonalde, I., Prozorov, R., Giannetta, R., Kini, A., Schlueter, J., Wang, H., Geiser, U., Williams, J. (1999), *Phys. Rev. Lett.* 83, 4172.
Chahara, K., Ohno, M., Kasai, M., Kosono, Y. (1993), *Appl. Phys. Lett.* 63, 1990.
Chakravarty, S., Nelon, D., Halperin, B. (1988), *Phys. Rev. Lett.* 60, 1057.

Chechersky, V., Nath, A., Isaak, I., Franck, K., Ghosh, K., Ju, H., Greene, R. (1999), *Phys. Rev. B* 59, 497.

Chevrier, J., Suck, J., Capponi, J., Perroux, M. (1988), *Phys. Rev. Lett.* 61, 554.

Chien, T., Wang, Z., Ong, N. (1991), *Phys. Rev. Lett.* 67, 2088.

Cho, K., Smith, B., Coniglio, W., Winter, L., Agosta, C., Schlueter, J. (2009), *Phys. Rev. B* 79, 220507.

Choi, H., Roundy, D., Sum, H., Cohen, M., Louie, S. (2002), *Nature* 418, 758.

Chu, C., Xue, Y., Wang, Y., Hellman, A., Lorenz, B., Meng, R., Cmaidalka, J., Dezaneti, L. (2000), *J. of Supercond. and Novel Magn.* 13, 679.

Clark, K., Hassanien, A., Khan, S., Braun, K., Tanaka, H., Hia, S. (2010), *Nature Nano* 41, 2061.

Clarke, J. (1969), *Proc. R. Soc., Ser. A* 308, 447.

Clougherty, D., Kohnson, K., Henry, M. (1989), *Physica C* 162–4, 1475.

Cobert, D., Schoolwock, U., von Delft, J. (2004), *Eur. Phys. J. B* 38, 501.

Coey, J., Viret, M., von Molnar, S. (1999), *Adv. Phys.* 48, 167.

Cohen, M. (1964), *Phys. Rev.* 134, A511.

——(1969), in *Superconductivity*, vol. 1, R. Parks (ed.), Marcel Dekker, NY.

Cohen, M., Falicov, L., Phillips, J. (1962), *Phys. Rev. Lett.* 8, 316.

Cohen, R., Pickett, W., Boyer, L., Krakauer, H. (1987), *Phys. Rev. Lett.* 60, 817.

Coleman, L., Cohen, M., Sandman, D., Yamagishi, F., Garito, A., Heeger, A. (1973), *Solid State Commun.* 12, 1125.

Coleman, P. (1984), *Phys. Rev. B* 29, 3035.

Cooper, L. (1956), *Phys. Rev.* 104, 1189.

——(1961), *Phys. Rev. Lett.* 6, 689.

Covaci, L., Marsiglio, F. (2006), *Phys. Rev. B* 73, 014503.

Crawford, M., Kunchur, M., Farneth, W., McCarron III, E., Poon, S. (1990), *Phys. Rev. B* 41, 282.

Croitoru, M., Buzdin, A. (2013), *J. Phys. Cond. Mat.* 25, 125702.

Croitoru, M., Shanenko, A., Kaun, C., Peeters, F. (2011), *Phys. Rev. B* 83, 214509.

Dagotto, E., Hotta, T., Moreo, A. (2001), *Phys. Rep.* 344, 1.

Dai, X., Fang, Z., Zhou, Y., Zhang, F. (2008), *Phys. Rev. Lett.* 101, 057008.

Damascelli, A. (2004), *Physica Scripta T* 209, 61–74.

Darhmaoui, H., Jung, J. (1998), *Phys. Rev. B* 57, 8009.

de Gennes, P. (1960), *Phys. Rev.* 118, 141.

——(1964), *Phys. Cond. Matter* 3, 79.

——(1966) *Superconductivity in Metals and Alloys*, Benjamin, NY.

de Heer, W. (1993), *Rev. Mod. Phys.* 65, 611.

Deutcher, G., Fenichel, H., Gershenson, M., Grunbaum, E., Ovadyahu, Z. (1973), *J. Low Temp. Phys.* 10, 231.

Deutscher, G. (1987), *Chance and Matter*, J. Souletie, J. Vannimenus, R. Stora (eds.), Elsevier, Amsterdam.

——(2005), *New Superconductors: From Granular to High T_c*, World Scientific, Singapore.

——(2005), *Rev. Mod. Phys.* 77, 109.

——(2012), *J. of Super. and Novel Magn.* 25, 581.

Deutscher, G., de Gennes, P. (1969), in *Superconductivity*, R. Parks (ed.), Marcel Dekker, NY.

DiCastro, C., Feiner, L., Grilli, M. (1991), *Phys. Rev. Lett.* 66, 3209.

Ding, H., Norman, M., Campusano, J., Randeria, M., Bellman, A., Yokoya, T., Takahashi, T., Mochiki, T., Kadowaki, K. (1996), *Phys. Rev. B* 54, 9678.

Ding, H., Yokota, T., Campuzano, J., Takahashi, T., Randeria, M., Norman, M., Mochiku, T., Kadowaki, K., Giapintzakis, J. (1996), *Nature* 382, 51.

Doniach, S., Engelsberg, S. (1966), *Phys. Rev. Lett.* 17, 750.

Duffee, S., Irawan, T., Bieletzki, M., Richter, T., Sieben, B., Yin, C., von Issendorff, B., Moseler, M., Hovel, H. (2007), *Eur. Phys. J. D* 45, 401.

Duprat, R., Bourbonnais, C. (2001), *Eur. Phys. J. B* 21, 219.

Dynes, R. (1972), *Solid State Commun.* 10, 615.

Dynes, R., Garno, J. (1981), *Phys. Rev. Lett.* 46, 137.

Dynes, R., Narayanamurti, V. (1973), *Solid State Commun.* 12, 341.

Dynes, R., Narayanamurti, V., Chin, M. (1971), *Phys. Rev. Lett.* 26, 181.

Dynes, R., Sharifi, P., Valles, J. (1992), in *Lattice Effects in High T_c Superconductors*, Bar-Yam, Y., Egami, T., Mustre-Leon, J., Bishop, A. (eds.), World Scientific, Singapore.

Dzaloshinskii, I., Polyakov, A., Wiegman, P. (1988), *Phys. Rev. Lett.* 127, 112.

Dzero, M., Gor'kov, L., Kresin, V. (1999), *Solid State Commun.* 112, 707.

——(2000), *Eur. Phys. B* 14, 459.

Dzyaloshinskii, I., Katz, E. (1968), *JETP* 55, 338.

Dzyaloshinskii, I., Larkin, A. (1972), *JETP* 34, 422.

Eagles, D. (1969), *Phys. Rev.* 186, 456.

Eckern, U., Schol, G., Ambegaokar, V. (1984), *Phys. Rev. B* 30, 6419.

Eilenberger, G. (1968), *Z. Phys.* 214, 195.

Einstein, A. (1922), *Naturwiss* 10, 105.

Eisenmenger, W., Dayem, A. (1967), *Phys. Rev. Lett.* 18, 125.

Eisenmenger, W. (1969), in *Tunneling Phenomena in Solids*, ed. E. Burstein and S. Lundqvist, Plenum, NY, p. 371.

Ekardt, W. (1984), *Phys. Rev. B* 29, 1558.

Ekardt, W., Penzar, Z. (1988), *Phys. Rev. B* 38, 4273.

——(1991), *Z. Phys. D* 19, 109.

Eliashberg, G. (1960), *JETP* 11, 696.

——(1961), *JETP* 12, 1000.

——(1963), *JETP* 16, 780.

Elsinger, H., Wosnitza, J., Wanka, S., Hagel, J., Schweitzer, D., Strunz, W. (2000), *Phys. Rev. Lett.* 84, 6098.

Emery, N., Herold, C., d'Astuto, M., Carcia, V., Bellin, C., Mareche, J., Lagrange, P., Loupias, G. (2005), *Phys. Rev. Lett.* 95, 078003.

Emery, V. (1987), *Phys. Rev. Lett.* 58, 3759.

Ewings, R., Perring, T., Bewley, R., Guidi, T., Pitcher, M., Parker, D., Clarke, S., Boothroyd, A. (2008), *Phys. Rev. B* 78, 220501.

Fath, M., Freisem, S., Menovsky, A., Tomioka, Y., Aarts, J., Mydosh, J. (1999), *Science* 285, 1540.

Feiner, L., de Leeuw, D. (1989), *Solid State Commun.* 70, 1165.
Feiner, L., Grilli, M., DiCastro, C. (1992), *Phys. Rev. B* 45, 1047.
Felner, I., Asaf, U., Levy, Y., Millo, O. (1997), *Phys. Rev. B* 55, 3374.
Felner, I., Tsindlekht, M., Gorodetsky, G., Rosenberg, F. (2000), *J. of Supercond. and Novel Magn.* 13, 659.
Felner, I., Wolf, O., Millo, O. (2013), *J. of Supercond. and Novel Magn.* 26, 511.
Fernandes, R., Haraldsen, J., Wolfle, P., Balatsky, A. (2013), *Phys. Rev. B* 87, 014510.
Ferraris, J., Cowan, D. O., Walatka, W., Perlstein, J. H. (1973), *J. Am. Chem. Soc.* 95, 948.
Ferrell, R. (1964), *Phys. Rev. Lett.* 13, 330.
Ferrell, R., Fulde, P. (1964), *Phys. Rev.* 135, A550.
Fert, A. (2007), Nobel Lecture.
Fetter, A. (1974), *Ann. Phys.* 88, 1.
Franck J., Isaac, I., Chen, W., Chrzanowski, J., Irwin, J. (1998), *Phys. Rev. B* 58, 5189.
Franck, J., Jung, J., Mohamed, A., Gydax, S., Sproule, G. (1991), *Phys. Rev. B* 44, 5318.
Franck, J. (1994), in *Physical Properties of High Temperature Superconductors*, D. Ginsberg (ed.), World Scientific, Singapore.
Fratini, M., Poccia, N., Ricci, A., Campi, G., Burghammer, M., Aeppli, G., Bianconi, A. (2010), *Nature* 466, 841.
Frauendorf, S., Guet, C. (2001), *Annu. Rev. Nucl. Part. Sci.* 51, 219.
Friedel, J. (1992), *J. Phys. (Paris)* 2, 959.
——(1999), *Philos. Mag.* 79, 1251.
Fröhlich, H. (1950), *Phys. Rev.* 79, 845.
——(1968), *J. Phys. C* 1, 544.
Fukumoto, N., Mori, S., Yamamoto, N., Moritomo, Y., Katsufuji, T., Chen, C., Cheong, S. (1999), *Phys. Rev. B* 60, 12963.
Galitski, V., Migdal, A. (1958), *JETP* 7, 96.
Gamble, F., McConnell, H. (1968), *Phys. Lett.* 26A, 162.
Ganzuly, B. (1973), *Z. Phys.* 265, 433.
Gao, Y., Xue, Y., Chen, F., Xiang, Q., Meng, R., Ramirez, D., Chu, C., Eggert, J., Mao, H. (1994), *Phys. Rev. B* 50, 4260.
Garcia-Garcia, A., Urbina, Y., Yuzbahyan, E., Richter, K., Altshuler, B. (2011), *Phys. Rev. B* 83, 014510.
Garland, J. (1963), *Phys. Rev. Lett.* 11, 111, 114.
Gedik, N., Yang, D., Logvenov, G., Bozovic, I., Zewait, A. (2007), *Science* 316, 425.
Geerk, J., Xi, X., Linker, G. (1988), *Z. Phys. B* 73, 329.
Geilikman, B. (1965), *JETP* 48, 1963.
——(1966), *Uspekhi* 8, 2032.
——(1971), *J. Low Temp. Phys.* 4, 189.
——(1973), *Uspekhi* 16, 141.
——(1975), *Uspekhi* 18, 190.
——(1976), *Solid State* 18, 54.
Geilikman, B., Kresin, V. (1966), *Solid State* 7, 8659.

——(1968a), *Semicond.* 2, 639.
——(1968b), *Solid State* 9, 2453.
——(1972), *Phys. Lett.* 40A, 123.
Geilikman, B., Kresin, V., Masharov, N. (1975), *J. Low Temp. Phys.* 18, 241.
Geilikman, B., Zaitsev, R., Kresin, V. (1967a), *Solid State* 9, 642.
——(1967b), *Fiz. Met and Metallov.* 21, 796.
——(1968), *Fiz. Met. and Metallov.* 25, 171.
Geshkenbein, V., Larkin, A., Barone, A. (1987), *Phys. Rev. B* 36, 235.
Giaever, I. (1960), *Phys. Rev. Lett.* 5, 464.
Giaever, I., Mergerle, K. (1961), *Phys. Rev.* 122, 1101.
Gilabert, A. (1977), *Ann. Phys.* 2, 203.
Gilabert, A., Romangan, J., Guyon, E. (1971), *Solid State Commun.* 9, 1295.
Gill, W., Greene, R., Street, G., Little, W. (1975), *Phys. Rev. Lett.* 35, 1732.
Ginsberg, D., Hebel, L. (1969), in *Superconductivity*, R. Parks (ed.), Marcel Dekker, NY.
Ginzburg, V. (1965), *JETP* 20, 1549.
Ginzburg, V., Kirzhnits, D. (eds.) (1982), *High Temperature Superconductivity*, Cons. Bureau, NY.
Ginzburg, V., Landau, L. (1950), *ZETP* 20, 1064.
Glover III, R., Tinkham, M. (1957), *Phys. Rev.* 108, 243.
Gomes, K., Pasupathy, A., Pushp, A., Ono, S., Ando, Y., Yazdani, A. (2007), *Nature* 447, 569.
Gor'kov, L. (1958), *JETP* 7, 505.
——(1959), *JETP* 9, 1364.
——(1960), *JETP* 10, 593.
——(2012), *Phys. Rev. B* 86, 060501.
Gor'kov, L., Kresin, V. (1998), *JETP Lett.* 67, 985.
——(1999), *J. of Super. and Novel Magn.* 12, 243.
——(2000), *J. of Supercond. and Novel Magn.* 13, 239.
——(2001), *Appl. Phys. Lett.* 78, 3657.
——(2002), *Physica C.* 367, 103.
——(2004), *Phys. Rep.* 400, 149.
Gor'kov, L., Melik-Barkhudarov, T. (1963), *JETP* 18, 1031.
Gor'kov, L., Sokol, A. (1987), *JETP Lett.* 46, 420.
Gordon, J., Marcenat, C., Franck, J., Isaac, I., Zhang, G., Lortz, R., Meingast, C., Bouquet, F., Fisher, R., Phillips, N. (2001), *Phys. Rev. B* 65, 024441.
Gorter, C., Casimir, H. (1934), *Phys. Zs.* 35, 963.
Gray, K., Hawley, M., Moog, E. (1987), in *Novel Superconductivity*, S. Wolf and V. Kresin (eds.), Plenum.
Greene, R., Street, G., Suter, L. (1975), *Phys. Rev. Lett.* 34, 577.
Grilli, M., Castellani, C., DiCastro, C. (1990), *Phys. Rev. B* 42, 6233.
Grilli, M., Kotliar, G., Millis, A. (1990), *Phys. Rev. B* 42, 329.
Grimvall, G. (1969), *Phys. Kondens. Materie* 9, 283.
——(1981), *The Electron–Phonon Interaction in Metals*, North-Holland, Amsterdam.

Groves, J., Arendt, P., Foltyn, S., DePaula, R., Wang, C., Hammond, R. (1999), *IEEE Trans. on Applied Super.* 9, 1964.

Gueron, S., Pothier, H., Birge, N., Esteve, D., Devoret, M. (1996), *Phys. Rev. Lett.* 77, 3025.

Gumeniuk, R., Schnelle, W., Rosner, H., Nicklas, M., Leithe-Jasper, A., Grin, Y. (2008), *Phys. Rev. Lett.* 100, 017002.

Guo, Y., Zhang, Y., Bao, X., Han, T., Tang, Z., Zhang, L., Zhu, W., Wang, E., Niu, Q., Qiu, Z., Jia, J., Zhao, Z., Xue, Q. (2004), *Science* 306, 1915.

Gurevich, V., Larkin, A., Firsov, Y. (1962), *Solid State* 4, 131.

Gutfreund, H., Little, W. (1979), *Highly Conducting One-Dimensional Solids*, Devresee, J., Evrard, R., den Doren, V. (eds.), Plenum, NY.

Haddon, R., Hebard, A., Rosseinsky, M., Murphy, D., Duclos, S., Lyons, K., Miller, B., Rosamilia, J., Fleming, R., Kortan, A., Glarum, S., Makhija, A., Muller, A., Elick, R., Zahurak, S., Tycko, R., Dabbagh, G., Thiel, F. (1991), *Nature* 350, 321.

Hagel, J., Kelemen, M., Fisher, G., Pilawa, B., Wosnitza, J., Dormann, E., Lohneysen, H., Schnepf, A., Schnockel, H., Neisel, U., Beck, J. (2002), *J. Low Temp. Phys.* 129, 133.

Haldane, F. (1981), *Phys. Lett.* 81A, 153.

Halperin, B. (1984), *Phys. Rev. Lett.* 52, 1583.

——(1991), *Phys. Rev. B* 49, 1111.

Hannay, N., Geballe, T., Mattias, B., Andres, K., Schmidt, F., MacNair, D. (1965), *Phys. Rev. Lett.* 14, 225.

Hardy, G., Hulm, J. (1953), *Phys. Rev.* 89, 884.

Hardy, W., Bonn, D., Morgan, D., Liang, R., Zhang, K. (1993), *Phys. Rev. Lett.* 70, 3999.

Harrison, W. (1961), *Phys. Rev.* 123, 85.

Hartman, T., Juranic, P., Collins, K., Reilly, B., Appathurai, N., Wehlitz, R. (2012), *Phys. Rev. Lett.* 108, 023001.

Hartman, T., Juranic, P., Collins, K., Reilly, B., Makoutz, E., Appaturai, N., Wehlitz, R. (2013), preprint.

Haskel, D., Stern, E., Hinks, D., Mitchell, D., Jorgenson, J. (1997), *Phys. Rev. B* 56, 521.

Hawley, M., Gray, K., Terris, B., Wang, H., Carlson, K., William, J. (1986), *Phys. Rev. Lett.* 57, 629.

Hebard, A. (1992), *Physics Today* 45, 26.

Hebard, A., Rosseinsky, M., Haddon, R., Murphy, D., Glarum, S., Palstra, T., Ramirez, A., Kortan, A. (1991), *Nature* 350, 600.

Heffner, R., Le, P., Hundley, M., Neumeier, J., Luke, G., Kojima, K., Nachumi, B., Uemura, Y., MacLaughlin, D., Cheong, S. (1996), *Phys. Rev. Lett.* 77, 1869.

Heid, R., Bohnen, K. (2005), *Phys. Rev. B* 72, 134527.

Hein, R., Gilson, J., Mazelsky, R., Miller, R., Hulm, J. (1964), *Phys. Rev. Lett.* 12, 320.

Helfand, E., Werthamer, N. (1964), *Phys. Rev. Lett.* 13, 686.

——(1966), *Phys. Rev.* 147, 288.

Hochheimer, H., Etters, R. (eds.) (1991), *Frontiers in High Pressure Research*, Plenum, NY.

Hock, K., Nickisch, H., Thomas, H. (1983), *Helv. Phys. Acta* 56, 237.

Hock, C., Schmidt, M., Issendorff, B. V. (2011), *Phys. Rev.* 1284, 113401.

Hohenberg, P. (1967), *Phys. Rev.* 158, 383.

Holcomb, M., Perry, C., Coleman, J., Little, W. (1996), *Phys. Rev. B* 53, 6734.

Holczer, K., Klein, O., Huang, S., Kaner, R., Fu, K., Whetten, R., Diederich, D. (1991), *Science* 252, 1154.

Holstein, T. (1959), *Ann. Phys.* 8, 343.

Homes, C. C., Timusk, T., Liang, R., Bonn, D. A., Hardy, W. N. (1993), *Phys. Rev. Lett.* 71, 1645–48.

Hosono, H. (2012), in *100 Years of Superconductivity*, H. Rogalla and P. Kes (eds.), CRC Press.

Hubbard, J. (1959), *Phys. Rev. Lett.* 3, 77.

——(1963), *Proc. Roy. Soc. (London) Ser. A* 276, 238; (1964), 281, 401.

Huebener, R., Luebbig, H. (2008), *A Focus of Discoveries*, World Scientific, Singapore.

Hulm, J., Ashkin, M., Deis, D., Jones, C. (1970), in *Progress in Low Temperature Physics*, Gorter (ed.), Plenum, NY.

Hybertsen, M., Schleuter, M., Christensen, N. (1989), *Phys. Rev. B* 39, 9028.

Ichimura, K., Takami, M., Nemura, K. (2008), *J. Phys. Soc. Jpn.* 77, 114107.

Iguchi, I., Yamaguchi, T., Sugimoto, A. (2001), *Nature* 412, 420.

Ihm, J., Cohen, M. L., Tuan, S. (1981), *Phys. Rev. B* 23, 3258.

Ishiguro, T., Yamagaji, K., Saito, G. (1998), *Organic Superconductivity*, Springer, Berlin.

Issendorff, B. (2011), *Handbook of Nanophysics*, K. Sattler, Ed. CRC Press, Boca Raton, 6–1.

Issendorff, B., Cheshnovsky, O. (2005), *Annu. Rev. Phys. Chem.* 56, 549.

Ito, H., Ishiguro, T., Komatsu, T., Matsukawa, N., Saito, G., Anzai, H. (1994), *J. Supercond.* 5, 667–9.

Izumi, M., Manako, T., Konishi, Y., Kawasaki, M., Tokura, Y. (2000), *Phys. Rev. B* 61, 18187.

Jahn, H., Teller, E. (1937), *Proc. Roy. Soc. A* 161, 220.

Jaime, M., Lin, P., Chun, S., Salamon, M., Dorsey, P., Rubinstein, M. (1999), *Phys. Rev. B* 60, 1028.

Jerome, D. (2012), *J. of Supercond. and Novel Magn.* 25, 633.

Jerome, D., Greuzet, F. (1987), in *Novel Superconductivity*, S. Wolf and V. Kresin (eds.), Plenum, NY.

Jerome, D., Mazaud, A., Ribault, M., Bechgaard, K. (1980), *J. Phys. Lett. (Paris)* 41, L95.

Ji, S., Zhang, T., Fu, Y., Chen, X., Ma, X., Li, J., Duan, W., Jia, J., Xue, Q. (2008), *Phys. Rev. Lett.* 100, 226801.

Jin, S., Tiefel, T., McCormack, M., Fastnacht, R., Ramesh, R., Chen, L. (1994), *Science* 264, 413.

Jollicoeur, T. (1998), in *Pair Correlations in Many-Fermion Systems*, V. Kresin (ed.), Plenum, NY.

Jonner, G., van Santen, J. (1950), *Physica* 16, 337.
Jorgensen, J., Veal, B., Paulikas, A., Nowicki, L., Crabtree, G., Claus, H., Kwok, W. (1990), *Phys. Rev. B* 41, 1863.
Josephson, B. (1962), *Phys. Lett.* 1, 251.
Jung, J., Boyce, B., Yan, H., Abdelhadi, M., Skinta, J., Lemberger, T. (2000), *J. Supercond.* 13, 753.
Jung, J., Kim, K., Noh, T., Choi, E., Yu, J. (1998), *Phys. Rev. B* 57, 11043.
Kadowaki, K., Campuzano, J. (2008), *Phys. Rev. Lett.* 101, 137002.
Kaiser, A., Zuckermann, M. (1970), *Phys. Rev. B* 1, 229.
Kalmeyer, V., Laughlin, R. (1987), *Phys. Rev. Lett.* 59, 2095.
Kamerlingh Onnes, H. (1911), *Leiden Commun.* 124C.
Kamihara, Y., Hiramatsu, H., Hirano, M., Kawamura, R., Yanagi, H., Kamiya, T., Hosono, H. (2008), *J. Am. Chem. Soc.* 130, 3296.
Kamihara, Y., Watanabe, T., Hirano, M., Hosono, H. (2006), *J. Am. Chem. Soc.* 128, 10012.
Kampf, A., Schrieffer, J. (1990), *Phys. Rev. B* 41, 11663.
——(1990), *Phys. Rev. B* 42, 6399.
Kanamori, J. (1961), *J. Appl. Phys. (Suppl.)* 31, 145.
Kane, E. (1969), in *Tunneling Phenomena in Solids*, E. Burshtein, Lundquist (eds.), Plenum, NY.
Kaneko, T., Yamauchi, H., Tanaka, S. (1991), *Physica C* 178, 377.
Kanigel, A., Chatterjee, U., Randeria, M., Norman, M., Koren, G., Kadowaki, K., Campusano, J. (2008), *Phys. Rev. Lett.* 101, 137002.
Kaplan, M., Vekhter, B. (1995), *Cooperative Phenomena in Jahn–Teller Crystals*, Plenum, NY.
Katakuse, I., Ichihara, T., Fujita, Y., Matsuo, T., Sakurai, T. (1986), *Int. J. of Mass Spectrometry and Ion Processes* 69, 109.
Kawano, H., Kajimoto, R., Yoshizawa, H., Tomioka, Y., Kuwahara, H., Tokura, Y. (1997), *Phys. Rev. Lett.* 78, 4253.
Keller, H. (1989), *IBM J. Res. Dev.* 33, 314.
——(2003), *Physica B*. 326, 283.
——(2005), in *Structure and Bonding*, K. Mueller and A. Bussmann-Holder (eds.), Springer, Heidelberg.
Khasanov, R., Eshchenko, D., Luetkens, H., Morensoni, E., Prokscha, T., Suter, A., Garifianov, N., Mali, M., Roos, J., Conder, K., Keller, H. (2004), *Phys. Rev. Lett.* 92, 057602.
Khasanov, R., Shengelaya, A., Conder, K., Morenzoni, E., Savic, I., Karpinski, J., Keller, H. (2006), *Phys. Rev. B* 74, 064504.
Khasanov, R., Shengelaya, A., Morenzoni, E., Angst, M., Conder, K., Savic, I., Lampakis, D., Liarokapis, E., Tatsi, A., Keller, H. (2003), *Phys. Rev. B* 68, 220506.
Kiefl, R., Brewer, J., Affleck, I., Carolan, J., Dosanjh, P., Hardy, W., Hsu, T., Kadano, R., Kempton, J., Kreitzman, S. L. Q., O'Reilly, A., Riseman, T., Schleger, P., Stamp, P., Zhou, H. (1990), *Phys. Rev. Lett.* 64, 2082.
Kihlstrom, K., Hovda, P., Kresin, V., Wolf, S. (1988), *Phys. Rev. B* 38, 4588.
Kivelson, S., Rokhsar, D., Sethna, J. (1987), *Phys. Rev. B* 35, 8865.

Klein, B., Cohen, R. (1992), *Phys. Rev. B* 45, 12405.
Kleiner, R., Mueller, P. (1994), *Phys. Rev. B* 49, 1327.
Kleiner, R., Steinmeyer, F., Kunkel, G., Muller, P. (1992), *Phys. Rev. Lett.* 68, 2394.
Knight, W., Clemenger, K., de Heer, W., Saunders, W., Chou, M., Cohen, M. (1984), *Phys. Rev. Lett.* 52, 2141.
Knight, W. (1987), in *Novel Superconductivity*, S. Wolf and V. Z. Kresin (eds.), Plenum, NY.
Kobayashi, H., Tomota, H., Udagawa, T., Naito, T., Kobayashi, A. (1995), *Synthet. Metals* 70, 867.
Kogan, V., Zhelezina, N. (2004), *Phys. Rev. B* 69, 132506.
Kondo, J., Asai, Y., Nagai, S. (1988), *J. Phys. Soc. Jpn.* 57, 4334.
Kosugi, T., Miyake, T., Ishibashi, S., Arita, R., Aoki, H. (2009), *J. Phys. Soc. Jpn.* 78, 113704.
Kotliar, G., Rickenstein, A. (1986), *Phys. Rev. Lett.* 57, 1362.
Kotliar, G., Lee, P., Read, N. (1988), *Physica C* 153–5, 528.
Kouznetsov, K., Sun, A., Chen, B., Katz, A., Bahcall, S., Clarke, J., Dynes, R., Gajewski, D., Han, S., Maple, M., Giapintzakis, J., Kim, J., Ginsberg, D. (1997), *Phys. Rev. Lett.* 79, 3050.
Krasnopolin, Y., Khaikin, M. (1973), *JETP* 37, 883.
Kresin, V. (1967), *Phys. Lett. A* 24, 749.
——(1972), *Solid State* 13, 2468.
——(1973), *J. Low Temp. Phys.* 11, 519.
——(1974), *Phys. Lett. A* 49, 117.
——(1982), *Phys. Rev. B* 25, 147.
——(1984), *Phys. Rev. B* 30, 450.
——(1986), *Phys. Rev. B* 34, 7587.
——(1987a), *Phys. Lett. A* 122, 434.
——(1987b), *Phys. Rev. B* 35, 8716.
——(1987c), *Solid State Commun.* 63, 725.
——(2010), *J. of Supercond. and Novel Magn.* 23, 179.
Kresin, V., Bill, A., Wolf, S., Ovchinnikov, Y. (1997), *Phys. Rev. B* 56, 107.
Kresin, V., Friedel, J. (2011), *Europhys. Lett.* 93, 13002.
Kresin, V., Gurfreund, H., Little, W. (1984), *Solid State Commun.* 51, 339.
Kresin, V., Lester, W., Jr (1984), *Chem. Phys.* 90, 935.
Kresin, V., Litovchenko, C., Panasenko, A. (1975), *J. Chem. Phys.* 63, 3613.
Kresin, V., Little, W. (eds.) (1990), *Organic Superconductivity*, Plenum, NY.
Kresin, V., Morawitz, H. (1990), *Phys. Lett.* 145, 368.
——(1991), *Phys. Rev. B* 1341, 2691.
Kresin, V., Morawitz, H., Wolf, S. (1993), *Mechanisms of Conventional and High T_c Superconductivity*, Oxford University Press, NY.
Kresin, V., Ovchinnikov, Y. (2006), *Phys. Rev. B* 74, 024514.
Kresin, V., Ovchinnikov, Y., Wolf, S. (2006), *Phys. Rep.* 431, 231.
——(2003), *Appl. Phys. Lett.* 83, 722.
——(2004), *Physica C* 408–10, 434.
Kresin, V., Parchomenko, A. (1975), *Solid State* 16, 2180.

Kresin, V., Tavger, B. (1966), *JETP* 23, 1124.
Kresin, V., Wolf, S. (1990a), *Fundamentals of Superconductivity*, Plenum, NY.
——(1990b), *Physica C* 169, 476.
——(1992), *Phys. Rev. B* 46, 6458.
——(1994), *Phys. Rev. B* 49, 3652.
——(1995), in *Anharmonic Properties of High T_c Cuprates*, D. Mihailovich, G. Ruani, E. Kaldis, K. Mueller (eds.), World Scientific, Singapore.
——(1995), *Phys. Rev. B* 51, 1229.
——(1998), *J. Appl. Phys.* 63, 7357.
——(2009), *Rev. Mod. Phys.* 81, 481.
——(2012), *J. of Supercond. and Novel Magn.* 25, 175.
Kresin, V., Wolf, S., Deutcher, G. (1992), *Physica C* 91, 9.
Kresin, V., Wolf, S., Ovchinnikov, Y. (1996), *Phys. Rev. B* 53, 118311.
Kresin, V., Zaitsev, G. (1978), *JETP* 47, 983.
Kresin, V. V., Knight, W. (1998), in *Pair Correlations in Many-Fermion Systems*, V. Z. Kresin (ed.), Plenum, NY.
Krusin-Elbaum, L., Greene, R., Holtzberg, F., Malozemoff, A., Yeshurun, Y. (1989), *Phys. Rev. Lett.* 62, 217.
Kubozono, Y., Mitamura, H., Lee, X., He, X., Yamanari, Y., Takahashi, Y., Suzuki, Y., Kaji, Y., Eguchi, R., Akaike, K., Kambe, T., Okamoto, H., Fujiwara, A., Kato, T., Kosugi, T., Aoki, H. (2011), *Phys. Chem. Chem. Phys.* 13, 16476.
Kudo, K., Miyoshi, Y., Sasaki, T., Kobayashi, N. (2005), *Physica C* 426, 251.
Kugel, K., Khomskii, D. (1973), *JETP* 37, 725.
——(1982), *Uspekhi* 25, 231.
Kusco, C., Zhai, Z., Hakim, N., Markiewicz, R., Sridhar, S., Colson, D., Viallet-Guillen, V., Nefyadov, Yu., Trunin, M., Kolesnikov, N., Maignan, A., Daignere, A., Erb, A. (2002), *Phys. Rev. B* 65, 132501.
Labbe, J., Barisic, S., Friedel, J. (1967), *Phys. Rev. Lett.* 19, 1039.
Labbe, J., Friedel, J. (1966), *J. Phys. (Paris)* 27, 153.
Landau, L. (1933), *Phys. Z. Sow.* 3, 664.
Landau, L., Lifshitz, E. (1977), *Quantum Mechanics*, Pergamon Press, Oxford.
——(2000), *Classical Theory of Field*, Elsevier, Oxford.
Landau, L., Lifshitz, E., Pitaevsky, L. (2004), *Electrodynamics of Continuous Media*, Elsevier, Amsterdam.
Landau, L., Pekar, S. (1946), *ZETP* 16, 341.
Lang, K., Madhavan, V., Hoffman, J., Hudson, E., Eisaki, Pan, S., H., Ushida, S., Davis, J. (2002), *Nature* 415, 412.
Lanzara, A., Bogdanov, P., Zhou, X., Kellar, S., Feng, D., Lu, E., Yoshida, T., Eisaki, H., Fujimori, A., Kishio, K., Shimoyama, J., Noda, T., Uchida, S., Hussain, Z., Shen, Z. (2001), *Nature* 412, 510.
Lanzara, A., Zhao, G., Saini, N., Bianconi, A., Conder, K., Keller, H., Mueller, K. (1999) *J. Phys. Cond. Mat.* 11, L541.
Larkin, A., Ovchinnikov, Y. (1964), *ZETF* 47, 1137.
——(1969), *JETP* 28, 1200.
——(1983), *Phys. Rev. B* 28, 6281.

——(1986), in *Non-Equilibrium Superconductivity*, Landenberg, D. and Larkin, A. (eds.), Elsevier, Amsterdam, p. 530.
Lascialfari, A., Mishonov, T., Rigamonti, A., Zucca, I., Berg, G., Loser, W., Drechsler, S. (2003), *Eur. Phys. J. B* 35, 326.
Lascialfari, A., Rigamonti, A., Romano, L., Tedesco, P., Varlamov, A., Embriaco, D. (2002), *Phys. Rev. B* 65, 1445232.
Laughlin, R. (1983), *Phys. Rev. Lett.* 50, 1395.
——(1986), *The Quantum Hall Effect*, Prange, R., Given, S. (eds.), Springer, NY.
——(1988), *Phys. Rev. Lett.* 60, 2677.
——(1988), *Science* 242, 525.
Lee, J., Fujita, K., McElroy, K., Slezak, J. A., Wang, M., Aiura, Y., Bando, H., Ishikado, M., Mazui, T., Zhu, J., Balatsky, A., Eisaki, H., Uchida, S., and Davis, J. (2006), *Nature* 442, 546.
Leggett, A. (1980), *J. Phys. (Paris)*, Colloq. 41, C7.
Lerme, J., Dugourd, P., Hudgins, R., Jarrold, M. (1999), *Chem. Phys. Lett.* 304, 19.
Li, G., Bridges, F., Booth, C. (1995), *Phys. Rev. B* 52, 6332.
Lifshitz, E., Pitaevsky, L. (1999), *Physical Kinetics*, Elsevier, Oxford.
——(2002), *Statistical Physics*, Part 2, Elsevier, Oxford, p. 227.
Lifshitz, I. (1960), *JETP* 11, 1130.
Lindenfeld, Z., Eisenberg, E., Lifshitz, R. (2011), *Phys. Rev. B* 84, 064532.
Little, W. (1964), *Phys. Rev.* 134, A1416.
——(1983), *J. de Phys. Coll.* C 3, 819.
Little, W., Collins, K., Holcomb, M. (1999), *J. of Super. and Novel Magn.* 12, 89.
Little, W., Holcomb, M., Ghiringhelli, G., Braicovich, L., Dallera, C., Piazzalunga, A., Tagliaferri, A., Brookes, N. (2007), *Physica C* 460, 40.
Liu, A., Mazin, I., Kortus, J. (2001), *Phys. Rev. Lett.* 87, 087005.
Liu, Q., Yang, H., Qin, X., Yu, Y., Yang, L., Li, F., Yu, R., Jin, C., Uchida, S. (2006), *Phys. Rev. B* 74, 100506.
Loeser, A., Shen, Z., Dessau, D., Marshall, D., Park, C., Fournier, P., Kapitulnik, A. (1996), *Science* 273, 325.
Lofland, S., Bhagat, S., Ghosh, K., Greene, R., Karabashev, S., Shulyatev, D., Arsenov, A., Mukovski, Y. (1997), *Phys. Rev. B* 56, 13705.
London, F., London, H. (1935), *Proc. Roy. Soc. A* 149, 71.
London, F. (1937), *J. Phys. (Paris)* 8, 397.
Loram, J., Mirza, K., Wade, J., Cooper, J., Liang, W. (1994), *Physica C*. 235–40, 134.
Louca, D., Egami, T. (1997a), *J. Appl. Phys.* 81, 5484.
——(1997b), *Physica B* 241, 842.
——(1999a), *Phys. Rev. B* 59, 6193.
——(1999b), *J. Supercond.* 12, 23.
Louie, S., Cohen, M. (1977), *Solid State Commun.* 10, 615.
Lubashevsky, Y., Garg, A., Sassa, Y., Shi, M., Kanigel, A. (2011), *Phys. Rev. Lett.* 106, 047002.
Luke, G., Fudamoto, Y., Kojima, K., Larkin, M., Merrin, J., Nachumi, B., Uemura, Y., Maeno, Y., Mao, Z., Mori, Y., Nakamura, H., Sigrist, M. (1998), *Nature* 394, 558.
Ma, F., Lu, Z. (2008), *Phys. Rev. B* 78, 033111.

Mackenzie, A., Haselwinner, R., Tyler, A., Lonzarich, G., Mori, Y., Nishizaki, S., Maeno, Y. (1998), *Phys. Rev. Lett.* 80, 161.

Mackenzie, A., Julian, S., Lonzarich, G., Carrington, A., Hughes, S., Liu, R., Sinclair, D. (1993), *Phys. Rev. Lett.* 71, 1238.

Maeno, Y., Hashimoto, H., Yoshida, K., Nishizaki, S., Fujita, T., Bednorz, J., Lichtenberg, F. (1994), *Nature* 372, 532.

Mahan, G. (1993). *Many-Particle Physics*, Plenum, NY.

Mahendiran, R., Ibarra, M., Maignan, A., Millange, F., Arulraj, A., Mahesh, R., Raveau, B., Rao, C. (1999), *Phys. Rev. Lett.* 82, 2191.

Maki, K. (1969), in *Superconductivity*, R. Parks (ed.), Marcel Dekker, NY.

Maki, K., Tsuneto, T. (1964), *Progr. Theor. Phys.* 31, 945.

Mannella, N., Rosenhahn, A., Booth, C., Marchesini, S., Mun, B., Yang, S., Ibrahim, K., Tomioka, Y., Fadley, C. (2004), *Phys. Rev. Lett.* 92, 166401.

Mannhart, J., Bednorz, J., Mueller, K., Schlom, D. (1991), *Z. Phys.* B83307.

Mannhart, J., Chaudhari, P. (2001), *Physics Today* 53, 48.

Maradudin, A. A., Montroll, E. W., Weiss, G. (1963), *Theory of Lattice Dynamics in the Harmonic Approximation*, Academic Press, NY.

Martinis, J., Hilton, G., Irwin, K., Wollman, D. (2000), *Nucl. Inst. and Methods in Phys. Res.* A444, 23.

Matheiss, L. (1987), *Phys. Rev. Lett.* 58, 1035.

Mattis, D. C., Bardeen, J. (1958), *Phys. Rev.* 111, 412.

Maxwell, E. (1950), *Phys. Rev.* 78, 477.

McMahan, A., Annett, J., Martin, R. (1990), *Phys. Rev. B* 42, 6268.

McMillan, W. (1968a), *Phys. Rev.* 167, 331.

——(1968b), *Phys. Rev.* 175, 537.

McMillan, W., Rowell, J. (1969), in *Superconductivity*, vol. 1, Parks R. (ed.), Marcel Dekker, NY.

——(1965), *Phys. Rev. Lett.* 14, 108.

Meirwes-Broer, K. (ed.) (2000), *Metal Clusters at Surfaces*, Springer, Berlin.

Meissner, H. (1960), *Phys. Rev.* 117, 672.

Meissner, W., Hom, R. (1932), *Z. Physik* 74, 715.

Meissner, W., Ochsenfeld, R. (1933), *Naturwissenschafen* 21, 787.

Merchant, L., Ostrick, J., Barber, B., Jr., Dynes, R. (2001), *Phys. Rev. B* 63, 134508.

Mesot, J. and Furrer, A (1997), *J. Supercond.* 10, 623.

Michel, C., Raveau, B. (1984), *Rev. Chem. Miner.* 21, 407.

Migdal, A. (1958), *JETP* 7, 996.

——(1959), *Nucl. Phys.* 13, 655.

——(1960), *JETP* 10, 176.

——(1967), *Theory of Finite Fermi Systems and Application to Atomic Nuclei*, Interscience, NY.

——(2000), *Qualitative Methods in Quantum Theory*, Pergamon Press, NY.

Mihailovich, D., Kabanov, V., Mueller, K. (2002), *Europhys. Lett.* 57, 254.

Miller, R., Mein, R., Gibson, J., Hulm, J., Jaones, C., Mazersky, R. (1965), *Proc. of the Ninth Intern. Conf. on Low Temperature Physics*, J. Daunt, D. Eduards, F. Milford, M. Jagut (eds.), Plenum, NY.

Millis, A., Lee, P. (1987), *Phys. Rev. B* 35, 3394.
Millis, A., Shraiman, B., Mueller, R (1996a), *Phys. Rev. Lett* 77, 175.
——(1996b), *Phys. Rev. B* 54, 5389.
Millis, A., Littlewood, P., Shraiman, B. (1995), *Phys. Rev. Lett.* 74, 5144.
Mitsuhashi, R., Suzuki, Y., Yamanari, Y., Mitamura, H., Kambe, T., Ikeda, N., Okamoto, H., Fujiwara, A., Yamaji, M., Kawasaki, N., Maniwa, Y., Kubozono, Y. (2010), *Nature* 464, 76.
Miyake, K., Schmitt-Rink, S., Varma, C. (1986), *Phys. Rev. B* 34, 3544.
Morawitz, H., Bozovic, I., Kresin, V., Rietveld, G., van der Marel, D. (1993), *Z. Phys. B: Cond. Matter* 90, 277.
Morel, P., Anderson, P. (1962), *Phys. Rev.* 125, 1263.
Moreo, A., Yunoki, S., Dagotto, E. (1999), *Science* 283, 2034.
Moritomo, Y., Akimoto, T., Fujishiro, H., Nakamura, A. (2001), *Phys. Rev. B* 64, 064404.
Moro, R., Xu, X., Yin, S., de Heer, W. (2003), *Science* 300, 1265.
Morse, R., Bohm, H. (1957), *Phys. Rev.* 108, 1094.
Moskalenlo, V. (1959), *Fiz. Met. Metalloved* 8, 503.
Mota, A., Visani, P., Pollini, A. (1989), *J. Low Temp. Phys.* 76, 465.
Mott, N. (1968), *Rev. Mod. Phys.* 40, 677.
Mueller, J., Olsen, J. (1988), *Proceedings of Materials and Mechanisms of High Temperature Superconductivity I*, North-Holland, Amsterdam.
Mueller, K. (1990), *Z. Phys. B* 80, 193.
——(1995), *Nature* 377, 133.
——(2007), *J. Phys. Cond. Matter* 19, 251002.
——(2012), *J. of Supercond. and Novel Magn.* 25, 2101.
Mueller, K., Burkard, H. (1979), *Phys. Rev. B* 19, 3593.
Mueller, K., Shengelaya, A. (2013), *J. of Supercond. and Novel Magn.* 26, 486.
Muller, P. (1994), *Physica C* 35–40, 289.
Murakami, H., Ohbuchi, S., Aoki, R. (1994), *J. Phys. Soc. Jpn.* 63, 2653.
Nagaev, E. (1996), *Uspekhi* 39, 781.
Nagamatsu, J., Nakagawa, N., Muranaka, T., Zenitani, Y., Akimitsu, J. (2001), *Nature* 410, 63.
Nicol, J., Shapiro, S., Smith, P. (1960), *Phys. Rev. Lett.* 5, 461.
Niedermayer, C., Bernard, C., Blasius, T., Golnik, A., Moodenbaugh, A., Budnick J. (1998), *Phys. Rev. Lett.* 80, 3843.
Nowack, A., Poppe, U., Weger, M., Schweitzer, D., Schwenk, H. (1987), *Z. Physik B* 68, 41.
Nowack, A., Weger, M., Schweitzer, D., Keller, H. (1986), *Solid State Commun.* 60, 199.
Nozieres, P. (1995), in *Bose–Einstein Condensation*, A. Griffin, D. Snoke, S. Stringari (eds.), Cambridge University Press, Cambridge.
Nozieres, P., Schmitt-Rink, S. (1985), *J. Low Temp. Phys.* 59, 195.
Nuecker, N., Romberg, H., Xi, X., Fink, J., Hahn, D., Zetterer, T., Otto, H., Renk, K. (1989), *Phys. Rev. B* 39, 6619.
O'Malley, T. (1967), *Phys. Rev.* 152, 98.

Okimoto, Y., Tokura, Y. (2000), *J. of Supercond. and Novel Magn.* 13, 271.

Okuda, T., Asamitsu, A., Tomioka, Y., Kimura, T., Taguchi, Y., Tokura, Y. (1998), *Phys. Rev. Lett.* 81, 3203.

Olsen, C., Liu, R., Lynch, D., List, R., Arko, A., Veal, B., Chang, Y., Jiang, P., Paulikas, A. (1990), *Phys. Rev. B* 42, 381.

Onari, S., Kontani, M. (2010), *Phys. Rev. Lett.* 104, 157001.

Osofsky, M., Soulen Jr., R., Wolf, S., Broto, J., Rakoto, H., Ousset, J., Coffe, G., Ashkenazy, S., Pari, P., Bozovic, I., Eckstein, J., Virshup, G. (1993), *Phys. Rev. Lett.* 71, 2315.

Ovchinnikov, Y., Kresin, V. (1996), *Phys. Rev. B* 54, 1251.

——(2000), *Europhys. J. B* 14, 203.

——(2002), *Phys. Rev. B* 65, 214507.

——(2005), *Eur. Phys. J. B* 45, 5.

——(2010), *Phys. Rev. B* 81, 214505.

——(2012), *Phys. Rev. B* 85, 064518.

Ovchinnikov, Y., Wolf, S., Kresin, V. (1999), *Phys. Rev. B* 60, 4329.

——(2001), *Phys. Rev. B* 63, 064524.

Owen, C., Scalapino, D. (1971), *Physica* 55, 691.

Palstra, T., Zhou, O., Iwasa, Y., Sulewski, P., Fleming, R., Zegarski, B. (1995), *Solid State Commun.* 93, 327.

Parker, C., Pushp, A., Pasupathy, A., Gomes, K., Wen, J., Xu, Z., Ono, S., Gu, G., Yazdani, A. (2010), *Phys. Rev. Lett.* 104, 117001.

Parkin, S. (1995), *Annu. Rev. Mater. Sci.* 25, 357.

Parmenter, H. (1968), *Phys. Rev.* 166, 392.

Pellarin, M., Baguenard, B., Bordas, C., Broyer, M., Lerme, J., Vialle, J. (1993), *Phys. Rev. B* 48, 17645.

Pellarin, M., Baguenard, B., Broyer, M., Lermé, J., Vialle, J.-L. (1992), *J. Chem. Phys.* 98, 944.

Peña, V., Gredig, T., Santamaria, J., Schuller, I. (2006), *Phys. Rev. Lett.* 97, 177005.

Pickett, W., Singh, S. (1996), *Phys. Rev. B* 53, 1146.

Pickett, W. (1989), *Rev. Mod. Phys.* 61, 433.

Pineiro, A., Pardo, V., Baldomir, D., Rodriguez, A., Cortes-Gil, R., Gomes, A., Aries, J. (2012), *J. Phys. Cond. Mat.* 24, 275503.

Plevert, L. (2011), *Pierre-Gilles de Gennes: A Life in Science*, World Scientific, Singapore.

Pokkia, N., Fratini, M., Ricci, A., Campi, G., Barba, L., Vittertini-Orgeas, A., Bianconi, G., Aeppli, G., Bianconi, A. (2011), *Nature Mater.* 10, 733.

Pokrovsky, V. (1961), *JETP* 13, 447.

Ponomarev, Y., Tsokur, E., Sudakova, M., Tchesnokov, S., Shabalin, S., Lorenz, M., Hein, M., Muller, G., Piel, H., Aminov, B. (1999), *Solid State Commun.* 111, 513.

Pullman, B., Pullman, A. (1963), *Quantum Biochemistry*, Interscience, NY.

Qadri, S., Toth, L., Osofsky, M., Lawrence, S., Gubser, D., Wolf, S. (1987), *Phys. Rev. B.* 35, 7235.

Ralph, D., Black, C., Tinkham, M. (1995), *Phys. Rev. Lett.* 74, 3241.

Ramirez, A., Kortan, A., Rosseinsky, M., Duclos, S., Mujsce, A., Haddon, R., Murphy, D., Makhija, A., Zahurak, S., Lyons, K. (1992), *Phys. Rev. Lett.* 68, 1058.

Randeria, M., Campuzano, J. (1998), in *Proceedings of the International School of Physics "Enrico Fermi" on Conventional and High Temperature Superconductors*, G. Iadonisi, J. R. Schrieffer, and M. L. Chiafalo (eds.), IOS Press, Amsterdam.

Rashba, E. (1982), in *Excitons*, Rashba, E. and Sturge, M. (Eds.), North-Holland, Amsterdam.

Raveau, B., Michel, C., Hervieu, M., Groult, D. (1991), *Crystal Chemistry of High-T_c Superconducting Copper Oxides*, Springer, Berlin.

Razzoli, E., Drachuk, G., Keren, A., Radovic, M., Plumb, N., Chang, J., Huang, Y., Ding, H., Mesot, J., Shi, M. (2013), *Phys. Rev. Lett.* 110, 047004.

Read, N. (1985), *J. Phys. C* 18, 2651.

Read, N., Newns, D. (1983), *J. Phys. C* 16, 3273.

Reeves, M., Ditmars, D., Wolf, S., Vanderah, T., Kresin, V. (1993), *Phys. Rev. B* 47, 6065.

Reger, J., Young, A. (1988), *Phys. Rev. B* 37, 5978.

Reittu, H. (1995), *Am. J. Phys.* 63, 940.

Renker, B., Compf, F., Ewert, D., Adelmann, P., Schmidt, H., Gering, E., Hinks, H. (1989), *Z. Phys. B: Condens. Matter* 77, 65.

Renker, B., Compf, F., Gering, E., Nucker, N., Ewert, D., Reichardt, W., Rietschel, H. (1987), *Z. Phys. B* 67, 15.

Renner, Ch., Revaz, B., Genoud, J., Kadowaki, K., Fisher, O. (1998), *Phys. Rev. Lett.* 80, 149.

Reynolds, C., Serin, B., Wright, W., Nesbitt, L. (1950), *Phys. Rev.* 78, 487.

Rice, T. (1965), *Phys. Rev.* 140, A889.

Ring, P., Schuck, P. (1980), *The Nuclear Many-Body Problem*, Springer, NY.

Romberg, H., Nücker, N., Alexander, M., Fink, J., Hahn, D., Zetterer, T., Otto, H., Renk, K. (1990), *Phys. Rev. B* 41, 2609.

Rosseinsky, M., Ramirez, A., Glarum, S., Murphy, D., Haddon, R., Hebard, A., Palstra, T., Kortan, A., Zahurak, S., Makhija, A. (1991), *Phys. Rev. Lett.* 66, 2830.

Ruppel, M., Rademann, K. (1992), *Chem. Phys. Lett.* 197, 280.

Ruvalds, J. (1981), *Adv. Phys.* 30, 677.

Sabo, J. (1969), *Phys. Rev.* 131, 1325.

Salem, L. (1966), *The Molecular Orbital Theory of Conjugated Systems*, W. Benjamin, NY.

Santiago, R., de Menezes, O. (1989), *Solid State Commun.* 70, 835.

Scalapino, D. (1969), in *Superconductivity*, Parks R. (ed.), Marcel Dekker, NY, p. 449.

——(1990), *Physics of Highly Correlated Electron Systems*, Bedell, K., Correy, D., Pines, D., Schrieffer, J. (eds.), Addison-Wesley, NY.

Scalapino, D., Loh, E., Hirsch, J. (1986), *Phys. Rev. B* 34, 8190.

——(1987), *Phys. Rev. B* 35, 6694.

Scalapino, D., Schrieffer, J., Wilkins, J. (1966), *Phys. Rev.* 148, 263.

Scalettar, R., Loh, E., Gubernatis, J., Moreo, A., White, S., Scalapino, D., Sugar, R., Dagotto, E. (1989), *Phys. Rev. Lett.* 62, 1407.

Schafroth, M. (1955), *Phys. Rev.* 100, 463.

Schenck, A. (1985), *Muon Spin Rotation Spectroscopy: Principles and Applications in Solid State Physics*, Adam Hilger, Bristol.

Schenkel, T., Lo, C., Weis, C., Schuh, A., Persaud, A., Bokor, J. (2009). *Nuclear Instruments and Methods in Physics Research B* 267, 2563.

Scher, H., Zallen, R. (1970), *J. Chem. Phys.* 53, 3759.

Schiffer, P., Ramirez, A., Bao, W., Cheong, S. (1995), *Phys. Rev. Lett.* 75, 3336.

Schilling, A., Cantoni, M., Guo, J., Ott, H. (1993), *Nature* 363, 56.

Schilling, J., Klotz, S. (1992), in *Physical Properties of High Temperature Superconductors*, vol. 3, D. Ginzberg (ed.) World Scientific, Singapore.

Schirber, J., Northrup, C. (1974), *Phys. Rev. B* 10, 3818.

Schluter, M., Lannoo, M., Needles, M., Baraff, G., Tomanek, D. (1992), *Phys. Rev. Lett.* 68, 526.

Schmiedeshoff, G., Detwiler, J., Beyermann, W., Lacerda, A., Canfield, P., Smith, J. (2001), *Phys. Rev. B* 63, 134519.

Schneider, R., Geerk, J., Reitschel, H. (1987), *Europhys. Lett.* 4, 845.

Schooley, J., Hosler, W., Cohen, M. (1964), *Phys. Rev. Lett.* 12, 474.

Schrieffer, J., Wen, X., Zhang, S. (1989), *Phys. Rev. B* 39, 11663.

Schriver, K., Persson, J., Honea, E., Whetten, R. (1990), *Phys. Rev. Lett.* 64, 2539.

Shanenko, A., Croitoru, M., Peeters, F. (2006), *Europhys. Lett.* 76, 498.

Sharma, R., Xiong, G., Kwon, C., Ramesh, R., Greene, R., Venkatesan, T. (1996), *Phys. Rev. B* 54, 10014.

Shaw, W., Swihart, J. (1968), *Phys. Rev. Lett.* 20, 1000.

Shelton, R., Harrison, W., Philipps, N., (eds.) (1989), *Proceedings of Materials and Mechanisms of High Temperature Superconductivity II*, North-Holland, Amsterdam.

Shen, Z., Dessau, D. (1995), *Phys. Rep.* 253, 1.

Shen, Z., Dessau, D., Wells, B., King, D., Spicer, W., Arko, A., Marhshall, D., Lombardo, L., Kapitulnik, A., Dickinson, P., Doniach, S., DiCarlo, J., Loeser, T., Park, C. (1993), *Phys. Rev. Lett.* 70, 1553.

Shiina, Y., Shimada, D., Mottate, A., Ohyagi, Y., Tsuda, N. (1995), *J. Phys. Soc. Jpn.* 64, 2577.

Shim, H., Chaudhari, P., Logvenov, G., Bozovic, I. (2008), *Phys. Rev. Lett.* 101, 247004.

Shimada, D., Tsuda, N., Paltzer, U., de Wette, F. (1998), *Physica C* 298, 195.

Shirane, G., Birgeneau, R., Endoh, Y., Gehring, P., Kastner, M., Kitazawa, K., Kojima, H., Tanaka, I., Thurston, T., Yamada, K. (1989), *Phys. Rev. Lett.* 63, 330.

Shklovskii, B., Efros, A. (1984), *Electronic Properties of Doped Semiconductor*, Springer, Berlin.

Sigai, S., Shamoto, S., Sato, M., Takagi, H., Uchida, S. (1990), *Solid State Comm.* 76, 371.

Shraiman, B., Siggia, E. (1988), *Phys. Rev. Lett.* 61, 467.

——(1989), *Phys. Rev. Lett.* 62, 1564.

Shukhareva, I. (1963), *JETP* 16, 828.

Sigmund, E., Mueller, K. (eds.) (1994), *Phase Separation in Cuprate Superconductors*, Springer, Berlin.

Simon, R., Chaikin, P. (1981), *Phys. Rev. B* 23, 4463.

Simon, R., Dalrymple, B., Van Vechten, D., Fuller, W., Wolf, S. (1987), *Phys. Rev. B* 36, 1962.

Singleton, J., Mielne, C. (2002), *Contemp. Phys.* 43, 63.

Skalski, S., Betbeder, O., Weiss, P. (1964), *Phys. Rev. A* 136, 1500.

Sleight, A., Gillson, J., Bierstedt, P. (1975), *Solid State Commun.* 17, 27.

Sluchanko, N., Glushkov, V., Demishev, S., Samarin, N., Savchenko, A., Singleton, J., Hayes, W., Brazhkin, V., Gippius, A., Shulgin, A. (1995), *Phys, Rev. B* 51, 1112.

Sommerfeld, A., Bethe, H. (1933), *Electronen Theorie der Metalle*, Verlag, Berlin.

Sonier, I., Ilton, M., Pacradouni, V., Kaiser, C., Sabok-Sayr, S., Ando, Y., Komiya, S., Hardy, W., Bonn, D., Liang, R., Atkinson, W. (2008), *Phys. Rev. Lett.* 101, 117001.

Sonier, J., Brewer, J., Kiefl, R., Miller, R., Morris, G., Stronach, G., Gardner, J., Dunsiger, S., Bonn, D., Hardy, W., Liang, R., Heffner, R. (2001), *Science* 292, 1692.

Sparn, G., Thompson, J., Whetten, R., Huang, S., Kaner, R., Diederich, F., Grüner, G., Holczer, K. (1992), *Phys. Rev. Lett.* 68, 1228.

Speilman, S., Fesler, K., Eom, B., Geballe, T., Gejer, M., Kapitulnik, A. (1990), *Phys. Rev. Lett.* 65, 123.

Stratonovich, R. (1957), *Doklady* 115, 1097.

Strongin, M. R. Kammerer, O., Paskin, A. (1965), *Phys. Rev. Lett.* 14, 949.

Stuffer, D., Aharony, A. (1992), *Introduction to Percolation Theory*, Taylor & Francis, London.

Sugai, S., Shamoto, S., Sato, M., Ido, T., Takagi, H., Uchida, S. (1990), *Solid State Commun.* 76, 371.

Suh, B., Borsa, F., Torgenson, D., Cho, B., Canfield, P., Johnston, D., Rhee, J., Harmon, B. (1996), *Phys. Rev. B* 54, 15341.

Suhl, H., Mattias, B., Walker, L. (1958), *Phys. Rev. Lett.* 3, 552.

Sun, X., Ono, S., Abe, Y., Komiya, S., Segawa, K., Ando, Y. (2006), *Phys. Rev. Lett.* 96, 017008.

Suzuki, M., Watanabe, T., Matsuda, A. (1999), *Phys. Rev. Lett.* 82, 5361.

Tachiki, M., Muto, Y., Syono, Y. (eds.) (1991), *Proceedings of Materials and Mechanisms of High Temperature Superconductivity III*, North-Holland, Amsterdam.

Takigawa, M., Hammel, P., Heffner, R., Fisk, Z., Smith, J., Schwarz, R. (1989), *Phys. Rev. B* 39, 300.

Tanigaki, K., Ebbesen, T., Saito, S., Mizuki, J., Tsai, J., Kubo, Y., Kuroshima, S. (1991), *Nature* 352, 222.

Tao, H., Lu, F., Wolf, E. (1997), *Physica C* 282, 1507.

Temprano, D., Mesot, J., Janssen, S., Conder, K., Furrer, A., Mutka, H., Mueller, K. (2000), *Phys. Rev. Lett.* 84, 1990.

Tersoff, J., Hamann, D. (1985), *Phys. Rev. B* 31, 85.

Tewari, S., Gumber, P. (1990), *Phys. Rev.* 41, 2619.

Tinkham, M. (1996), *Introduction to Superconductivity*, McGraw-Hill, NY.

Tinkham, M., Hergenrother, J., Lu, J. (1995), *Phys. Rev. B* 51, 12649.

Tokura Y. (2003), *Physics Today*, 56, 50.

Tokura, Y., Tomioka, Y (1999), *J. Magn. Magn. Mater.* 200, 1.

Tomic, S., Jérome, D., Mailly, D., Ribault, M., Bechgaard, K. (1983), *J. Phys.* 44, C3:1075.

Torrance, J., Metzger, R. (1989), *Phys. Rev. Lett.* 63, 1515.
Tou, H., Maniwa, Y., Koiwasaki, T., Yamanaka, S. (2001), *Phys. Rev. Lett.* 86, 5775.
Tou, H., Maniwa, Y., Yamanaka, S. (2003), *Phys. Rev. B* 67, 100509.
Toyli, D., Weis, C., Fuchs, G., Schenkel, T., Awschalon, D. (2010). *Nano Lett.* 10, 3168.
Toyozawa, Y., Shinozuka, Y. (1983), *J. Phys. Soc. Jpn.* 52, 1446.
Tranquada, J., Axe, J., Ichikawa, N., Moodenbaugh, A., Nakamura, Y., Uchida, S. (1997), *Phys. Rev. Lett.* 78, 338.
Tranquada, J., Heald, S., Moodenbaugh, A. (1987), *Phys. Rev. B* 36, 5263.
Tranquada, J., Heald, S., Moodenbaugh, A., Suenaga, M. (1987), *Phys. Rev. B* 35, 7187.
Tranquada, J., Sternlieb, B., Axe, J., Nakamura, Y., Uchida, S. (1995), *Nature* 375, 561.
Tsuda, N., Shimada, D., Miyakawa, N. (2007), in *New Research on Superconductivity*, B. Martins (ed.), Nova Science, NY, p. 105.
Tsuda, S., Yokoya, T., Takano, Y., Kito, H., Matsushita, A., Yin, F., Itoh, J., Harima, H., Shin, S. (2003), *Phys. Rev. Lett.* 91, 127001.
Tsuei, C., Kirtley, J. (2000), *Rev. Mod. Phys.* 72, 969.
Tsuei, C., Kirtley, J., Chi, C., Yu-Jahnes, L., Gupta, A., Shaw, T., Sun J., Ketchen, M. (1994), *Phys. Rev. Lett.* 73, 593.
Tyablikov, S., Tolmachev, V. (1958), *JETP* 7, 867.
Uchiyama, H., Shen, K., Lee, S., Damacelli, A., Lu, D., Feng, D., Shen, Z.-X., Tajima, S. (2002), *Phys. Rev. Lett.* 88 157002.
Uehara, M., Mori, S., Chen, S., Cheong, S. (1999), *Nature* 399, 560.
Uemura, Y., Luke, G., Sternlieb, B., Brewer, J., Carolan, F., Hardy, W., Kadono, R., Kempton, J., Kiefl, R., Kreitzman, S., Mulhern, P., Riseman, T., Williams, D., Yang, B., Uchida, S., Takagi, H., Gopalakrishnan, J., Sleight, A., Subramanian, M., Chien, C., Cieplak, M., Gang Xiao, Lee, V., Statt, B., Stronach, C., Kossler, W., Yu, X. (1989), *Phys. Rev. Lett.* 62, 2317.
Urushibara, A., Moritomo, Y., Arima, T., Asamitsu, A., Kido, G., Tokura, Y. (1995), *Phys. Rev. B* 51, 14103.
Usadel, K. (1970), *Phys. Rev. Lett.* 25, 507.
Van Harlingen, D. (2012), in *100 Years of Superconductivity*, H. Rogalla, P. Kes (eds.), Taylor & Francis, Boka Raton.
Van Hove, L. (1953), *Phys. Rev.* 89, 1189.
Van Vleck, J. H. (1932), *The Theory of Electric and Magnetic Susceptibilities*, Oxford University Press, Oxford.
Varelogiannis, G. (2002), *Phys. Rev. Lett.* 88, 117005.
Varma, C., Zaanen, J., Raghavachari, K. (1991), *Science* 254, 989.
Vinetskii, V. (1961), *JETP* 13, 1023.
Viret, D., Ranno, L., Coey, J. (1997), *Phys. Rev. B* 55, 8067.
Vishik, I., Hashimoto, M., He, R., Lee, W., Schmitt, F., Lu, D., Moore, R., Zhang, C., Meevasana, W., Sasagawa, T., Uchida, S., Fujita, K., Ishida, S., Ishikato, M., Yoshida, Y., Eisaki, H., Hussain, Z., Devereaux, T., Shen, Z. (2012), *Proc. Nat. Acad. Sci.* 109, 18332.

Visscher, P., Falicov, L. (1971), *Phys. Rev. B* 3, 2541.
von Delft, D., Ralph, D. C. (2001), *Phys. Rep.* 345, 61.
von Helmholt, R., Wecker, J., Holzareff, B., Schultz, L., Samuer, K., Fert, A. (2008), *Uspekhi* 178, 1336.
von Issendorf, B. (2010), in *Handbook of Nanophysics: Clusters and Fullerenes*, K. Sadder (ed.), Taylor & Francis, Oxfordshire.
von Issendorff, B., Cheshnovsky, O. (2005), *Annu. Rev. Phys. Chem.* 56, 549.
Wade, J., Loram, J., Mirza, K., Cooper, J., Tallon, J. (1994), *J. Supercond.* 7, 261.
Waldram, J. (1996), *Superconductivity of Metals and Cuprates*, Inst. of Phys. Pull., Bristol.
Walstedt, R., Warren, W. (1990), *Science* 248, 1082.
Wang, S., Evanson, W., Schrieffer, J. (1969), *Phys. Rev.* 23, 92.
Wang, Y., Li, L., Naughton, M., Gu, G., Uchida, S., Ong, N. (2005), *Phys. Rev. Lett.* 95, 247002.
Warren, W., Walstedt, R., Brennert, G., Cava, R., Tycko, R., Bell, F., Dabbagh, G. (1989), *Phys. Rev. Lett.* 62, 1193.
Weber, W. (1988), *Z. Phys. B* 70, 323.
Wehlitz, R. (2010), in *Advances in Atomic, Molecular, and Optical Physics*, E. Arimando, P. Berman, C. Lin (eds.), Academic Press, NY.
Wehlitz, R., Juranic, P., Collins, K., Reilly, B., Makoutz, E., Hartman, T., Appathurai, N., Whitfield, S. (2012), *Phys. Rev. Lett.* 109, 193001.
Weht, R., Filippelli, A., Pickett, W. (1999), *Europhys. Lett.* 48, 320.
Wen, X., Wilczek, F., Zee, A. (1989), *Phys. Rev. B* 39, 11413.
Weyeneth, S., Mueller, K. (2011), *J. of Supercond. and Novel Magn.* 24, 1235.
Wilczek, F. (1982), *Phys. Rev. Lett.* 49, 957.
Wolf, E. (2012), *Principles of Electron Tunneling Spectroscopy*, Oxford University Press, NY.
Wolf, S., Cukauskas, E., Rachford, F., Nisenoff, M. (1979), *IEEE Trans. on Magn.* 15, 595.
Wolf, S., Kresin, V. (2012), *J. of Supercond. and Novel Magn.* 25, 165.
Wolf, S., Lowrey, W. (1977), *Phys. Rev. Lett.* 39, 1038.
Wriggle, G., Astruc Hoffman, M., von Issendorff, B. (2002), *Phys. Rev. A* 65, 063201.
Wu, M., Ashburn, J., Torng, P., Hor, P., Meng, R., Gao, L., Huang, Z., Wang, Y., Chu, C. (1987), *Phys. Rev. Lett.* 58, 908.
Yamanaka, S., Hotehama, K., Kawaji, H. (1998), *Nature* 392, 580.
Yan, H., Jung, J., Darhmaoui, H., Ren, Z., Wang, J., Kwok, W. (2000), *Phys. Rev. B* 61, 11711.
Yanson, I. (1974), *JETP* 39, 506–513.
Yin, S., Xu, X., Liang, A., Bowlan, J., Moro, R., de Heer, W. (2008), *J. of Supercond. and Novel Magn.* 21, 265.
Yu, G., Lee, C., Heeger, A., Herron, N., McCarron, E., Cong, L., Spalding, G., Nordman, C., Goldman, A. (1992), *Phys. Rev. B* 45, 4964.
Yu, J., Freeman, A., Xu, J. (1987), *Phys. Rev. Lett.* 58, 1035.
Zaanen, J., Oles, A., Feiner, L. (1991), *Dynamics of Magnetic Fluctuations in High T_c Superconductors*, vol. 246 of NATO Advanced Study Institute, Series B: Physics, Reiter, G., Horsch, P., Psaltakis, G. (eds.), Plenum Press, NY.

Zaanen, J., Sawatzky, G., Allen, J. (1985), *Phys. Rev. Lett.* 55, 418.
Zaanen, J. (2000), *Nature* 404, 714.
Zavaritskii, N., Itskevich, E., Voronovskii, A. (1971), *JETP* 33, 762.
Zech, D., Conder, K., Keller, H., Kaldis, E., Mueller, K. (1996), *Physica B* 219, 136.
Zech, D., Keller, H., Conder, K., Kaldis, E., Liarokapis, E., Poulakis, N., Mueller, K. (1994), *Nature* 371, 681.
Zeljkovic, I., Xu, Z., Wen, J., Gu, G., Markiewicz, R., Hoffman, J. (2012), *Science* 337, 320.
Zener, C. (1951a), *Phys. Rev* 81, 440.
——(1951b), *Phys. Rev.* 82, 403.
Zhang, F., Rice, T. (1988), *Phys. Rev. B* 57, 3759.
Zhang, J., Chen, Y., Jiao, L., Grameniuk, R., Nicklas, M., Chen, Y., Yang, L., Fu, B., Schnelle, W., Rosner, H., Leithe-Jaspers, A., Grin, Y., Steglich, F., Yuan, H. (2013), *Phys. Rev. B* 87, 064502.
Zhao, G., Conder, K., Keller, H., Mueller, K. (1996), *Nature* 381, 676.
Zhao, G., Hunt, M., Keller, H., Mueller, K. (1997), *Nature* 385, 236.
Ziman, J. (1960), *Electrons and Phonons*, Clarendon Press, Oxford, NY.
Zolotavin, P., Guyot-Slonnest, P. (2010), *ACS Nano* 4, 5599.
Zubarev, N. (1960), *Uspekhi* 3, 320.

Index

A
A15 compounds, 39, 201–202
Acoustic plasmons, 52, 55, 58, 144, 205
a.c. transport, pseudogap state, 153, 162–163
Adiabatic approximation, 5–9
   Born–Oppenheimer, 5, 7–8
   "crude" ("clamped"), 8–9
   Hamiltonian, 5–6
$Al_{1-x}Si_x$, 29
Amorphous carbon films, 211
Andreev scattering, 117
Angle-resolved photoemission (ARPES), 100
Anharmonicity, 123–124
Anomalous diamagnetism, 148–150, 214–216
Antracene, 217
Anyon models, 86–87
Apical oxygen, 62, 138, 157, 170, 232
Aromatic molecules, 214–217
ARPES, 100

B
$Ba_{1-x}K_xFe_2As_2$, 197
$BaPb_{1-x}Bi_xO_3$, 42
BCS (Bardeen–Cooper–Schrieffer) theory, 3, 21–22, 26, 36, 39–40, 92
Bechgaard salts, 207–209, 210
BEDT-TTF, 209–210
Benzene, 217
Beta-tungsten structure, 201
$(BETS)_2GaCl_4$, 213–214
BiPbCaSrCuO, 128
Bipolarons, 39–40
Bismuth, 135
$Bi_2Sr_2CaCu_2O_8$, 141
$Bi_2Sr_2CaCu_2O_{8+\delta}$, 137, 143, 157
$Bi_2Sr_2CuO_6$, 135
Born–Oppenheimer approximation, 5, 7–8
Borocarbides, 167
Bose–Einstein condensation, 39, 40
Buckminster fullerene, 212

C
C_{60} cluster, 212, 213
Chalcogenide superconductors, 197–199
Charge-density wave (CDW), 156, 158
"Clamped" approximation, 8–9

Clusters
   C_{60}, 212, 213
   crystals, 228
   see also Nanoclusters
Coherence length, 65–66, 105, 114, 141
Colossal magnetoresistance effect, 172–173, 177–178
Conjugated hydrocarbons, 214–217
Cooper pairs, 39–40
   multigap superconductors, 103
   nanoclusters, 220–226
   π-electrons, 214–217
Cooper theorem, 20
Coronene, 215, 217
Coulomb interaction, 29–31, 32, 34
Coulomb pseudopotential, 30, 91–94
   isotope effect, 122–123
Coupling constant, 12, 13, 14, 15, 21, 24, 28, 29, 38, 40, 41, 49
   non-diagonal (interband), 103–104
   tunneling spectroscopy, 94–95
Critical field, 135–136, 167
Critical temperature (T_c), 26
   carrier concentration, 41
   Coulomb interaction, 29–31, 32, 34
   gapless state, 121
   general case, 33–35
   granular films, 203
   increase in maximum over time, 2
   intermediate coupling, 28–29
   intrinsic, 158, 159, 168
   magnetic impurities, 120, 125
   multigap superconductivity, 104–105
   nanoclusters, 221, 222–225
   non-magnetic scattering, 210
   phonon softening, 29, 35
   polaronic effect, 42–46
   proximity effect, 115–119, 120
   tunneling spectroscopy, 90
   upper limit, 29, 35, 47–48
   very strong coupling, 31–32
   weak coupling, 26–28
"Cross"-diagrammatic techniques, 120
"Crude" approximation, 8–9
Cs_3C_{60}, 212
Cs_2RbC_{60}, 212
Cuprate superconductors, 131–146
   bulk preparation, 133

Cuprate superconductors (cont.)
  compared to Fe-based pnictides, 197–199
  compared to manganites, 172, 174, 180, 195–196
  critical field, 135–136
  electron–lattice interaction, 140–144
  electron mechanisms, 144
  energy scales, 157
  Fermi arcs, 100
  film preparation, 133
  historical background, 3–4, 131
  isotope effect, 138–140
  localized vs itinerant aspects, 60–62
  mechanism of high T_c, 128, 140–146
  phase diagram, 134
  photoemission experiments, 143
  plasmons, 144
  properties, 134–137
  scanning tunneling microscopy, 143
  structure, 132
  symmetry of the order parameter, 136–137
  thermodynamic properties, 143–144
  tunneling spectroscopy, 141–143
  two-gap spectrum, 136
  ultrafast electron crystallography, 143
  Van Hove singularity, 39
Curie temperature, 156
Cyclotron resonance, 11

D
de Haas–van Alphen effect, 11
Delocalized (π) electrons, 207, 212, 214–217
"Demons", 58
Density of states, 117
Diabatic representation, 17–18, 229–231
Diamagnetic islands, 148, 155, 156, 158
Diamagnetic transition, 158
Diamagnetism, 148–150, 160–162, 214–216
"Dirty" semiconductors, 171
Dopant ordering, 169–170
Double-exchange, 172, 176–177
Double-well structure, 17, 42–44, 138
d-wave pairing, 59
d-wave symmetry, 136–137, 199, 204
Dynamic Jahn–Teller effect, 231–232
Dynamic polarons, 17–19

E
e_{2g} electron, 174, 176–177, 183
Effective average phonon frequency, 25
Effective Hamiltonian, 69–70, 78
Effective mass, 11, 14
Einstein, A., 3
Electron-energy-loss spectroscopy, 55–56
Electronic sound, 55, 57
Electronic terms, 16–17
  crossing, 17–18
Electron–phonon coupling, 9–11; see also Coupling constant
Electron–phonon interaction, 5–6, 11–15
Electron–phonon vertex, 15, 16
Electrons
  delocalized (π), 207, 212, 214–217
  e_{2g}, 174, 176–177, 183
Eliashberg equations, 91–94
Energy gap, 22
  critical temperature, 36–38
  gapless state, 120
  induced, 117, 119
  induced two-gap superconductivity, 112–113
  infrared spectroscopy, 97–98
  inhomogeneity, 155, 158
  nanoclusters, 221, 222
  overlapping bands, 103
  pseudogap state, 150–152
  small-scale organic superconductivity, 214
  spin-density wave, 64
  strong coupling, 36, 37, 38
  tunneling spectroscopy, 90–91
  two-gap state, 41, 105–107, 108
  ultrasonic attenuation, 99–100
  weak coupling, 27
Energy scales, 157–159
Energy shells, 219
Energy spectrum
  multigap superconductors, 105–107
  nanoclusters, 225–226
ET salts, 209–210

F
Fe-based pnictides, 197–199
Fermi arcs, 100
Fermi liquid-based theories, 59, 62–85
  slave boson, 82–85
  spin-bag model, 62–66
  spiral distortions, 77–82
  t-J model, 66–70
  two-dimensional Hubbard model, 70–77
FFLO state, 210
Filled-skutterudite compounds, 111
Films
  amorphous carbon, 211
  cuprates, 133
  granular materials, 203
  manganites, 173
Fractional statistics, 86–87
Franck–Condon factor, 18–19, 45
Frohlich Hamiltonian, 15
Fullerides, 207, 212–213

G
Ga-based clusters, 228
$GaCl_4$, 213
Gapless superconductivity, 120–121
$Gd_{1.4}Ce_{0.6}RuSr_2Cu_2O_{10-\delta}$, 204
GeTe, 41
"Giant" Josephson effect, 152–153, 163–166
"Giant" magnetoresistance effect, 176, 177–178

Ginzburg–Landau theory, 2, 110
g-ology, 49
Granular superconductors, 167–168, 202–203
Graphite, intercalated, 211
Green's function, 12, 22, 23, 24, 32, 104, 115, 160, 194, 223
  one-particle, 71–72, 75
  two-particle, 52, 72–73
Grimvall function, 14

H
Half-filled band case, 63
Half-metallic state, 177, 193
Hall effect, 86
Hamiltonian
  BCS theory, 21
  e_{2g} electron, 183
  effective, 69–70, 78
  effective hopping, 69
  electron–phonon interaction, 5–6, 9–11
  Frohlich, 15
  Hartree–Fock, 63
  Heisenberg, 61, 69
  Holstein, 17, 40
  Hubbard, 61, 190–191
  non-linear sigma model, 79
  pairing interaction, 104
  S–N–S Josephson effect, 194
  t-J model, 66, 67, 78
  tunneling, 229
Hartree–Fock Hamiltonian, 63
Heat capacity, 12, 13–14, 38, 144
  magnesium diboride, 200
  manganites, 186
  nanoclusters, 226
  organic superconductors, 210
  two-gap structures, 108, 111, 112
Heisenberg Hamiltonian, 61, 69
Hexbenzocoronene, 215
HgBaCaCuO, 3–4
Highest-energy scale, 158
HoBa$_2$Cu$_4$O$_8$, 152, 166
Holons, 86
Holstein Hamiltonian, 17, 40
Hopfield parameter, 223
Hubbard model, 60–61, 67, 86, 190–191
  two-dimensional, 70–77
Hubbard–Stratonovich field, 70–71
Hund's interaction, 174, 176–177, 183

I
Impurities
  electronic mechanism, 50
  gapless states, 120–121
  isotope effect, 125
  multigap superconductors, 105, 107
  pair-breaking, 119–121, 155–156
  scanning tunneling microscopy and spectroscopy, 96–97

Induced energy gap, 117, 119
Induced superconductivity, two-band structures, 112–113; *see also* Proximity effect
Infinite clusters, 155, 156, 159, 168, 178, 179–180
Infrared spectroscopy, 97–99
Inhomogenous state, 154–157, 211
Interband coupling constants, 103–104
Intercalated materials, 205, 210–212
Intermediate coupling, 28–29
Intervalley carrier transitions, 40–41
Intraband transition, 104
Intrinsic critical temperature, 158, 159, 168
Iron-based compounds, 197–199
Isotope effect, 3, 122–130, 170
  anharmonicity, 123–124
  Coulomb pseudopotential, 122–123
  cuprates, 138–140
  discovery, 3, 122
  magnetic impurities, 125
  manganites, 187–189
  multi-component lattice, 123
  penetration depth, 128–130, 140
  polaronic, 126–128, 138–139
  proximity systems, 124–125
  pseudogap state, 152, 166–167

J
Jahn–Teller effect, 16, 17, 62, 183, 184, 231
  dynamic, 231–232
Josephson effect
  historical background, 3
  nanoclusters, 227–228
  pseudogap state, 152–153, 163–166
  S–N–S junctions, 118, 163, 164, 193–195

K
K$_3$C$_{60}$, 212, 213
K$_8$C, 211
Kamerlingh Onnes, H., 1
K-doped picene, 211–212
Knight shift, 151
Kramers–Kronig transform, 98
Kubo–Greenwood expression, 189

L
La(Ba,Sr)CuO, 3
La$_{0.67}$Ca$_{0.33}$MnO$_3$, 179
La$_{0.8}$Ca$_{0.2}$MnO$_3$, 187
La$_{1-x}$Ca$_x$MnO$_3$, 180, 181, 182
La$_2$CuO$_4$, 61
La$_2$CuO$_{4+\delta}$, 143
LaFeAsO, 199
LaFePO, 197
LaMnO$_3$, 173, 174
Large Hadron Collider, 202
LaSrCuO, 163, 168
La$_{1.84}$Sr$_{0.16}$CuO$_4$, 143
La$_{2-x}$Sr$_x$CuO$_4$, 128, 137, 148, 152
La$_{0.7}$Sr$_{0.3}$MnO$_3$, 156

$La_{1-x}Sr_xMnO_3$, 174, 180, 182, 193
Lattice instability, 15
Layered conductors, 52, 53–57
Layered electron gas, 53
$La_{1-x}Zn_xMnO_3$, 181
Lifshitz 2.5 singularity, 186
$Li_x(THF)_yHfNCl$, 205
Little model, 47–50
Local pairs, 39–40
Lock-in technique, 89
London equations, 2, 207
Luttinger liquid, 86

M
McMillan equation, 28–29
McMillan tunneling model, 115, 118–119, 124
Madelung energy, 61, 62
"Magic" clusters, 219, 223–224, 225, 226
Magnesium diboride (MgB_2), 105, 109, 199–200
Magnetic impurities, 96
  critical temperature, 120, 125
  gapless state, 120–121
  isotope effect, 125
  pair-breaking, 119–121
Magnetic scattering, 130
Magnetoresistance
  colossal effect, 172–173, 177–178
  "giant" effect, 176, 177–178
Magnons, 186
Manganites, 172–196
  band approach, 190
  chemical composition, 172
  colossal magnetoresistance effect, 172–173, 177–178
  compared to cuprates, 172, 174, 180, 195–196
  conductivity/magnetism correlation, 175, 177
  doping, 174, 176
  double-exchange, 172, 176–177
  dynamic Jahn–Teller effect, 232
  e_{2g} electron, 174, 176–177, 183
  ferromagnetic metallic state, 184–190
  half-metallic state, 177, 193
  heat capacity, 186
  historical background, 172
  Hubbard model, 190–191
  insulating phase, 190–192
  insulating regions, 181
  isotope substitution, 187–189
  low doping, 178–179, 191–192
  magnetic order, 176
  magnons, 186
  main interactions, 183–184
  metallic A-phase, 182, 193–195
  neutron pulsed experiments, 181
  optical properties, 189–190
  parent compound, 190–191
  percolation phenomena, 156, 178–183
  phase separation, 178–179
  polarons, 191–192
  structure, 173–175
  t-core, 174, 177
  two-band structure, 183, 184–186
Martensitic transition, 202
Matrix method, 31, 224
Matsubara frequencies, 72, 76
Mattis–Bardeen theory, 97
Meissner effect, 1
Metallic A-phase, 182, 193–195
Metallic clusters, *see* Nanoclusters
MgB_2, 105, 109, 199–200
Microwave properties, pseudogap state, 153
Migdal theorem, 15–16
MnO_6, 173
Monte Carlo techniques, 70–77
Mössbauer spectroscopy, 182
Multicollisional induced dissociative method, 226
Multigap superconductors, 103–113
  critical temperature, 104–105
  energy spectrum, 105–107
  experimental data, 111–112
  Ginzburg–Landau equations, 110
  heat capacity, 108, 111, 112
  impurities, 105, 107
  induced, 112–113
  penetration depth, 108–109
  properties, 108–112
  surface resistance, 108–109
  symmetry of the order parameter, 113
  tunneling spectroscopy, 111
Multi-valley semiconductors, 51–52
Muon-spin precession, 87
Muon spin relaxation/resonance, 100–102, 148, 150

N
Nanoclusters, 218–228
  calorimetric study, 226
  critical temperature, 221, 222–225
  crystals, 228
  dynamic Jahn–Teller effect, 232
  energy gap, 221, 222
  energy shells, 219
  energy spectrum, 225–226
  heat capacity, 226
  incomplete shells, 219, 221, 225
  Josephson tunneling, 227–228
  macroscopic superconductivity, 227–228
  "magic" clusters, 219, 223–224, 225, 226
  observables, 226
  pair correlation, 220–226
  shape deformation, 219, 232
  shell structure, 218–219
Naphthalene, 217
Narrow bands, 78, 85
Nb, 28
Nb_3Ge, 95, 202

NbN, 203
Nb$_3$Sn, 39, 202
NbTe–Cu(Ag) systems, 114
Nd$_{1-x}$Sr$_x$MnO$_3$, 182
Nearest-neighbor hopping term, 61
Neutron pulsed experiments, 181
Nitrides, intercalated, 205
Non-adiabatic effect, 7, 8, 11, 20–21
Non-diagonal coupling constants, 103–104
Non-Fermi liquid models, 59, 60
   anyon models, 86–87
   resonant valence bonds, 85–86
Non-linear sigma model Hamiltonian, 79
Nuclear magnetic resonance, 151

O

Octahedral tilt mode, 62
Odd–even effect, 216, 217
One-electron self-energy, 76
One-particle Green's function, 71–72, 75
Orbital degeneracy, 62
Ordering of dopants, 169–170
Organic superconductors, 206–217
   aromatic molecules, 214–217
   ET salts, 209–210
   fullerides, 207, 212–213
   heat capacity, 210
   historical background, 3–4, 47, 206–207
   inhomogeneity, 211
   intercalated materials, 210–212
   mechanism of superconductivity, 210
   small-scale, 213–214
   structure and properties, 207–210
   (TMTSF)$_2$X systems, 207–209, 210
Ovalene, 215, 216
Overlapping bands, 51–52, 103, 105
Oxide molecular beam deposition, 133
Oxide superconductors, 16, 41–42

P

Pair-breaking, 119–121
   borocarbides, 167
   inhomogeneity, 155–156
Pair correlation, 220–226
Pair field propagator, 72–73
Pairing self-energy, 22–23
Partition function, 70–71
Pauli principle, 79
Pb, 96–97, 203
Pb+Ag system, 167–168
Pb(Zr$_x$Ti$_{(1-x)}$)O$_3$, 170
Pd-H, 123
Penetration depth, 50
   isotope dependence, 128–130, 140
   two-gap superconductors, 108–109
Percolation theory, 155, 159
   manganites, 156, 178–183
   "site" problem, 179–180
Percolation threshold, 179–180

Percolative transition, 155, 156, 180
Periodic stripe structure, 158
Phase separation, 156, 178–179
Phonon frequency, 15, 25, 28, 29
Phonon mechanisms, main equations, 22–26
Phonon softening, 29, 35
Photoemission techniques, 100
   cuprates 143
   pseudogap state, 152
Picene, 211–212
π-electrons, 207, 212, 214–217
Plasmons, 52–58
   acoustic, 52, 55, 58, 144, 205
   cuprates, 144
   "demons", 58
   dispersion law, 55
   electronic sound, 55, 57
   intercalated nitrides, 205
   pairing, 57–58
Pnictides, 197–199
Polarization, 23
Polarons
   concept, 16–17
   critical temperature, 42–46
   dynamic, 17–19
   isotope effect, 126–128, 138–139
   manganites, 191–192
Porphin, 215
Proximity effect, 114–121
   coherence length, 114
   critical temperature, 115–119, 120
   isotope effect, 124–125
   "Orsay school" approach, 114
   pair-breaking, 119–121
   Pb+Ag system, 168
   S–N system, 114–118, 120, 121, 124, 129–130
   S_α–S_β system, 118, 121
   tunneling model, 115, 118–119, 124
   two-gap model and, 119
PrPt$_4$Ge$_{12}$, 111
Pseudogap state, 134, 147–153
   a.c transport, 153, 162–163
   borocarbides, 167
   diamagnetism, 148–150, 160–162
   energy gap, 150–152
   Fe-based pnictides, 199
   "giant" Josephson effect, 152–153, 163–166
   granular superconductors (Pb+Ag system), 167–168
   inhomogeneity, 154–157
   isotope effect, 152, 166–167
   photoemission spectroscopy, 152
   theoretical analysis, 159–167
   transport properties, 153, 162–166
   tunneling spectroscopy, 151
p-wave superconductivity, 102
Pyrrole, 217

Q
Quasi-linearization method, 36–38
"Quasi-one-dimensional" system, 49
Quasiparticle tunneling, 3

R
Random-phase approximation, 53–54, 64
Rb_3C_{60}, 212, 213
Renormalization effect, 11–15
Reservoir columns, 169
Resistive transition, 159
Resonant valence bonds, 85–86
Room-temperature superconductivity, 169–170
$RuSr_2GdCu_2O_8$, 205
Ruthenates, 203–204
Ruthenium cuprates, 204–205

S
"Sandwich" excitonic mechanism, 50
Scanning SQUID microscopy, 148
Scanning tunneling microscopy and spectroscopy, 96–97, 105
  cuprates, 143
  inhomogeneity, 157
  manganites, 181
  pseudogap state, 151
  small-scale organic superconductivity, 213–214
Schwinger spin boson, 78
s–d evolution, 137
Self-energy parts, 12, 104, 115
  one-electron, 76
  pairing, 22–23
Self-supported superconductivity, 205
Semiconductors, 40–42
Shell structure, 218–219
Single vertex, 15
"Site" problem, 179–180
Slave boson, 82–85
$SmFeAsO_{1-x}F_x$, 197
$Sm(Nd)FeAsO_{1-x}$, 197
"Soft-landing" technique, 227
Sommerfeld constant, 12, 13, 95, 152, 181, 205
Spin-bag model, 62–66
Spin-density wave (SDW), 50, 63, 64, 199, 210
Spin–flip scattering, 136
Spinons, 86
Spinor, 78
Spin-relaxation rate, 151
Spiral distortions, 77–82
$Sr_{(2-x)}Ba_xCuO_4$, 131
$Sr_{(1-x)}Ba_xTiO_3$, 170
$SrMnO_3$, 182
$SrRuO_3$, 203
Sr_2RuO_4, 203–204
$Sr_3Ru_2O_7$, 203
$SrTiO_3$, 41, 42, 170
  Nb-doped, 41, 105
$Sr_2VFeAsO_3$, 197

Strong coupling
  critical temperature, 31–32
  main equations, 22–25
  properties of superconductors, 36–39
Superconducting "islands", 134, 154–155, 158, 159, 160, 162–163, 164, 165, 166, 168, 169
Superconductivity
  as a "giant" non-adiabatic phenomenon, 20–21
  historical background, 1–4
  at room temperature, 169–170
  self-supported, 205
Surface resistance, 108–109
s-wave pairing, 59, 109, 210
s-wave symmetry, 65, 204
Symmetry of the order parameter, 102
  cuprates, 136–137
  multigap superconductors, 113

T
T_c, see Critical temperature
TCNQ, 207
t-core, 174, 177
Tetracyanoguinodimethan, 51
Tetrahydrofuran, 205
Tetraphenylporphin, 215
Thermal-difference-reflectance spectroscopy, 98
Thermodynamic pairing order parameter, 24
Three-dimensional systems, electronic mechanism, 50–52
Tight-binding approximation, 43, 184, 194
t-J model, 66–70
$Tl_2Ba_2CaO_{+\delta}$, 150
$Tl_2Ba_2CuO_6$, 135
$(TMTSF)_2PF_6$, 204
$(TMTSF)_2X$ systems, 207–209, 210
Torque magnetometry, 150
Transition-metal atom chain, 49
TTF, 207
Tunneling model (McMillan), 115, 118–119, 124
Tunneling spectroscopy, 88–95
  critical temperature, 90
  cross-stripe junction, 88
  cuprates, 141–143
  electron–phonon coupling constant, 94–95
  energy gap, 90–91
  inversion of gap equation, 91–94
  pseudogap state, 151
  two-gap superconductors, 111
Tunneling splitting, 232
Tunnel junctions, 3
Two-fluid model, 2, 120
Two-gap superconductors, 103
  cuprates, 136
  energy spectrum, 105–107
  experimental data, 111–112
  first observation, 41, 105
  gapless, 121
  Ginzburg–Landau equations, 110

heat capacity, 108, 111, 112
induced, 112–113
penetration depth, 108–109
properties, 108–112
proximity effect, 119
surface resistance, 108–109
tunneling spectroscopy, 111
Two-particle Green's function, 52, 72–73
Type I/II superconductors, 2

U
Ultrafast electron crystallography, 143
Ultrasonic attenuation, 99–100
Underdoped region, 134, 140, 156

V
Van Hove singularity, 39
Van Vleck paramagnetism, 216
Very strong coupling, 31–32
Vibrational states, 51
V_3Si, 39, 95, 201

W
Wannier function, 68
Weak coupling
critical temperature, 26–28
main equations, 25–26

X
X-ray adsorption fine-structure spectroscopy, 182

Y
YBCO, 158, 168, 169
YBaCuO, 3, 102
$YBa_2Cu_3O_7$, 131, 132, 133, 134
$YBa_2Cu_3O_{(7-\delta)}$, 137
$YBa_2Cu_3O_{(7-y)}$, 61
YNi_2B_2C, 167
$Y_{1-x}Pr_xBa_2Cu_3O_7$, 139

Z
Zhang–Rice singlet, 62